# Establishment of State-of-the-Art Methods/Techniques and Identification of Novel Biomarkers in Brain Tumor Diagnosis

# Establishment of State-of-the-Art Methods/Techniques and Identification of Novel Biomarkers in Brain Tumor Diagnosis

Guest Editor

**Dimitrios Kanakis**

MDPI

Basel • Beijing • Wuhan • Barcelona • Belgrade • Novi Sad • Cluj • Manchester

*Guest Editor*
Dimitrios Kanakis
Department of Basic and
Clinical Sciences
University of Nicosia
Medical School
Nicosia
Cyprus

*Editorial Office*
MDPI AG
Grosspeteranlage 5
4052 Basel, Switzerland

This is a reprint of the Special Issue, published open access by the journal *Biomedicines* (ISSN 2227-9059), freely accessible at: https://www.mdpi.com/journal/biomedicines/special_issues/ZPOI929G90.

For citation purposes, cite each article independently as indicated on the article page online and as indicated below:

| Lastname, A.A.; Lastname, B.B. Article Title. *Journal Name* **Year**, *Volume Number*, Page Range. |
|---|

**ISBN 978-3-7258-6141-5 (Hbk)**
**ISBN 978-3-7258-6142-2 (PDF)**
**https://doi.org/10.3390/books978-3-7258-6142-2**

# Contents

# About the Editor

**Dimitrios Kanakis**

Dimitrios Kanakis is a Professor of Pathology and Neuropathology and the Head of Pathology at the University of Nicosia Medical School, Cyprus, and an Honorary Senior Lecturer at CITY St George's, University of London. He earned an MD from the University of Novi Sad and a PhD, magna cum laude, from Otto von Guericke University Magdeburg. He holds a specialty certification in neuropathology from the Medical Association of North Rhine, Düsseldorf (September 2011), and in general pathology from the Region of Thessaly, Larissa (January 2013). In 2016 at the University of Nicosia, he established the MD pathology course and refined the MBBS curriculum. He teaches large group lectures, small group tutorials, and laboratory practical sessions and contributes to assessment design, curriculum quality assurance, and the mentoring of junior faculty and students. His earlier roles include being a Lecturer in Pathology at the Democritus University of Thrace (December 2013 to September 2016) and a Scientific Associate at the Aristotle University of Thessaloniki (March to December 2013). These appointments followed residency training in pathology and neuropathology across Germany and Greece (September 2001 to November 2012), where he developed broad diagnostic expertise and contributed to collaborative research. His scholarship spans general pathology and neuropathology. He edited *Pituitary Adenoma: Pathophysiology, Diagnosis and Treatment Options* (NOVA, 2016) and co-edited *Frontiers in Neurology Research Topic: Recent Advances in Diagnosis and Treatment of Brain Tumors*, which was published as an e-book in 2025. He has authored four book chapters, including three for the *Hellenic Society of Pathology Manuals* and one for Springer's *Biomarkers in Cancer*, 2015. He has published peer-reviewed papers and serves as a reviewer and editorial board member for several journals. He has also been a Guest Editor for six Special Issues, in the *Biomedicines* and the *International Journal of Molecular Sciences*.

# Preface

Recent advances in molecular biology and genetics have profoundly influenced medicine, particularly oncology. These developments have reshaped the diagnostic approach to tumors of the central nervous system (CNS). Because research in this area advances rapidly, diagnostic frameworks continue to evolve.

Accordingly, the updated World Health Organization (WHO) classification of CNS tumors endorses an integrated diagnosis that combines histopathological criteria with tumor-specific molecular features.

This Special Issue was motivated by the need to highlight advances across molecular biology, pathology, neuroimaging, and clinical assessment, including the discovery and validation of biomarkers, the development of new diagnostic techniques, and the implementation of state-of-the-art methods for tumor detection and classification.

Our goal was to provide a venue for comparing methods on analytical validity and clinical utility, for clearly describing datasets and protocols, and for offering practical guidance on implementation. By focusing on techniques that improve accuracy while reducing complexity, we aim to narrow the gap between discovery and clinical impact.

Following a thorough review process, 14 articles were accepted, spanning original research and comprehensive reviews. Topics include molecular profiling strategies; immunohistochemical and imaging biomarkers with radiogenomic correlations; intraoperative diagnostic tools; liquid biopsy approaches; and analytical frameworks that integrate multiple data types.

This Special Issue is intended for neuropathologists, neuroradiologists, neurosurgeons, neuro-oncologists, molecular biologists, and clinical laboratory scientists. It is equally relevant to computational and data scientists working on diagnostic algorithms.

We thank the authors and reviewers who contributed to this collection. We hope the articles that follow will inform current practice and stimulate future research in integrated brain tumor diagnostics.

**Dimitrios Kanakis**
*Guest Editor*

biomedicines

MDPI

*Article*

# Identification of *MDK* as a Hypoxia- and Epithelial–Mesenchymal Transition-Related Gene Biomarker of Glioblastoma Based on a Novel Risk Model and In Vitro Experiments

**Minqi Xia [1,†], Shiao Tong [2,†] and Ling Gao [1,*]**

1 Department of Endocrinology & Metabolism, Renmin Hospital of Wuhan University, Wuhan 430060, China
2 Department of Neurosurgery, Renmin Hospital of Wuhan University, Wuhan 430060, China
* Correspondence: ling.gao@whu.edu.cn
† These authors contributed equally to this work.

**Abstract:** Background: Tumor cells are commonly exposed to a hypoxic environment, which can easily induce the epithelial–mesenchymal transition (EMT) of tumor cells, further affecting tumor proliferation, invasion, metastasis, and drug resistance. However, the predictive role of hypoxia and EMT-related genes in glioblastoma (GBM) has not been investigated. Methods: Intersection genes were identified by weighted correlation network analysis (WGCNA) and differential expression analyses, and a risk model was further constructed by LASSO and Cox analyses. Clinical, immune infiltration, tumor mutation, drug treatment, and enrichment profiles were analyzed based on the risk model. The expression level of the *MDK* gene was tested using RT-PCR, immunohistochemistry, and immunofluorescence. CCK8 and EdU were employed to determine the GBM cells' capacity for proliferation while the migration and invasion ability were detected by a wound healing assay and transwell assay, respectively. Results: Based on the GBM data of the TCGA and GTEx databases, 58 intersection genes were identified, and a risk model was constructed. The model was verified in the CGGA cohort, and its accuracy was confirmed by the ROC curve (AUC = 0.807). After combining clinical subgroups, univariate and multivariate Cox regression analyses showed that risk score and age were independent risk factors for GBM patients. Furthermore, our subsequent analysis of immune infiltration, tumor mutation, and drug treatment showed that risk score and high- and low-risk groups were associated with multiple immune cells, mutated genes, and drugs. Enrichment analysis indicated that the differences between high- and low-risk groups were manifested in tumor-related pathways, including the PI3K-AKT and JAK-STAT pathways. Finally, in vivo experiments proved that the hypoxia environment promoted the expression of *MDK*, and *MDK* knockdown reduced the proliferation, migration, and EMT of GBM cells induced by hypoxia. Conclusions: Our novel prognostic correlation model provided more potential treatment strategies for GBM patients.

**Keywords:** glioblastoma; risk model; EMT; hypoxia; immune; drug treatment

## 1. Introduction

Glioblastoma (GBM) is the most common primary malignant brain tumor in adults, with 3.2 cases per 100,000 people in the United States [1]. The total incidence of GBM has been growing year after year and will continue to rise as the population ages [2]. GBM, as an advanced glioma, has high invasiveness, a poor clinical prognosis, frequent recurrence, and a high mortality rate [3]. Patients with GBM, which accounts for 60% of gliomas, have a 5-year survival rate of only 4.7% [1]. Current treatment for GBM consists of maximal removal of the tumor followed by radiation and chemotherapy [4]. Despite recent progress in treatment strategies, GBM is prone to recurrence and has a near 100% fatality rate [5]. Therefore, a new diagnostic and therapeutic technique for GBM is urgently needed.

Hypoxia is an almost universal sign of malignant tumor growth, and tumor cells have a strong capacity to adapt in the face of the harsh environment of hypoxia. Studies have shown that the epithelial–mesenchymal transition (EMT) [6] of tumor cells can be induced by activating the hypoxia-inducible factors (HIFs) signaling pathway [7,8] in the face of hypoxia stress and further enhance the migration and invasive capacity of tumor cells and their adaptation to their surroundings. For example, melanoma, a highly advanced skin tumor, survives in the hypoxic environment and performs EMT by activating HIF-1$\alpha$ and blocking important immune system signaling pathways [9]. The accumulation of AKT in mitochondria can induce the phosphorylation of PDK1 at specific sites, reduce tumor cell death, maintain tumor cell proliferation, and induce EMT in the hypoxic environment [10]. The tumor environment is complex and variable, and the underlying mechanism of hypoxia-induced EMT in GBM patients has not been fully resolved. Targeted studies on GBM patients based on the characteristics of hypoxia-induced EMT are more comprehensive and feasible compared with separate studies on hypoxia or EMT in tumors.

Bioinformatics investigations of genome-wide RNA expression have revealed biological pathways that govern cell function and disease development at the molecular level [11,12], allowing for more accurate and personalized molecularly targeted therapeutics [13]. In this study, we used weighted gene co-expression network (WGCNA) and differential expression genes (DEGs) analyses to process the expression data of GBM in The Cancer Genome Atlas (TCGA) and Genotype-Tissue Expression (GTEx) databases. Then, based on least absolute shrinkage and selection operator (LASSO) and Cox analyses, we established a risk model consisting of the *MDK* and *STC1* genes, and we subjected this model to the analyses of clinical, immune, mutation, drug treatment, and functional enrichment. Finally, we selected *MDK* for further in vitro experiments. Our results suggest that the development of a differential expression risk score based on two genes (*STC1* and *MDK*) has potential value in predicting the prognosis and guiding the treatment of GBM patients.

## 2. Materials and Methods

### 2.1. Data Gathering

In this study, three public cohorts were gathered. Among them, the TCGA database (https://portal.gdc.cancer.gov/ (accessed on 10 October 2022), mRNA expression data and clinical details of GBM patients) [14] and GTEx database (https://www.gtexportal.org/home/index.html (accessed on 10 October 2022), mRNA information for normal brain tissues) [15] were used to establish a risk model for prognosis evaluation of GBM patients. The Chinese Glioma Genome Atlas (CGGA) database (http://www.cgga.org.cn (accessed on 10 October 2022), mRNA expression data and clinical details of GBM patients) [16,17] was used to further validate the risk model. Clinical information for all GBM patients is shown in Supplementary Table S1. In order to minimize any potential batch effects within or between the three cohorts, the "normalizeBetweenArrays" of the "limma" packages was used [18]. Moreover, we downloaded 863 hypoxia-related genes, 1253 EMT-related genes, and the c2.cp.kegg.v7.5.1.symbols dataset from the Molecular Signatures Database (MsigDB) (http://www.gsea-msigdb.org/gsea/downloads.jsp (accessed on 10 October 2022)) [19] for subsequent expression analysis and functional enrichment analysis. In addition, 300 chemokine genes and 149 immune checkpoint genes were obtained from the National Center for Biotechnology Information (NCBI) database (https://www.ncbi.nlm.nih.gov/ (accessed on 10 October 2022)) [20] for immune-related analysis.

### 2.2. Weighted Gene Co-Expression Network Analysis (WGCNA)

The weighted gene co-expression network analysis (WGCNA) [21] was executed employing the R package "WGCNA". Expression data from the TCGA and GTEx databases were subjected to clustering, utilizing an optimal soft power set to 14, leading to the construction of a dendrogram. Within the clustering results, genes exhibiting comparable characteristics were grouped into the same module.

### 2.3. Detection of Differentially Expressed Genes (DEGs)

The "limma" packages were employed for the identification of differentially expressed genes (DEGs) in the comparative analysis between GBM and normal brain samples. Subsequently, the 'ggplot2' package was utilized to visualize the DEGs through a volcano diagram. The criteria for defining DEGs were set at $|\text{logFC}| > 1$ and $p < 0.001$.

### 2.4. Risk Model Construction and Clinical Correlation Analysis

Firstly, by using the "Venn" package, intersection genes from the DEGs and WGCNA modules were obtained. The Venn diagram was drawn using the "Venn" package. Subsequently, prognosis-related genes were gained using univariate Cox analysis ($p < 0.05$) based on survival status, survival time, and expression levels of intersection genes in GBM patients, followed by LASSO regression analysis using the "glmnet" package to avoid overfitting. Furthermore, based on regression analysis and multivariate Cox analysis, the optimal risk model and coef values of genes in the model were obtained. We calculated a risk score for each patient according to the formula below:

$$\text{Risk Score} = \sum_{i=1}^{n} \text{coef}(i)^{*}\text{x}(i) \quad (1)$$

where n is the number of prognosis-related genes, coef (i) is the regression coefficient, and x(i) is the gene expression level. The median risk score (1.0155–1.0161) of the TCGA cohort divided all GBM samples into high- and low-risk groups. Survival differences between high- and low-risk groups were calculated by the "survival" and "survminer" packages. Cox regression and Wilcoxon rank sum tests were used to evaluate the differences in survival probability and risk scores in clinical subgroups. The "survivalROC" package was used to calculate and visualize the AUC value to evaluate the accuracy of the risk model. Univariate and multivariate Cox analyses with clinical information were performed using the "survival" package. The "rms" package was utilized to develop a nomogram based on multivariate Cox regression coefficients.

### 2.5. Immune Infiltration Analysis

Firstly, the association between risk score and multiple immune cells was assessed using seven algorithms (XCELL, TIMER, QUANTISEQ, MCPCOUNTER, EPIC, CIBERSORT-abs, and CIBERSORT). Then, we used the "estimate", "ggpubr", and "limma" packages to compare the differences in the tumor microenvironment between the high- and low-risk groups, and we utilized Pearson correlation analyses to assess the association between risk score and DNA/RNAss. Following this, immune cell and functional differences between the high- and low-risk groups were further compared using single-sample gene set enrichment analysis (ssGSEA). Finally, the expression of the immune checkpoint and chemokine genes in the two groups was investigated by the "limma" package.

### 2.6. Mutation Analysis

Information on somatic mutations in GBM was downloaded from the TCGA database for 168 patients. Tumor mutation burden (TMB) was defined as the numbers of insertions/deletions and substitution mutations per million bases. Mutation in the high- and low-risk groups was analyzed by the "maftools" package. According to the median TMB score, all samples were divided into a high-TMB group and low-TMB group. Associations between risk score and TMB were shown in scatter plots, and differences in survival probability between the high- and low-TMB groups were analyzed by the Kaplan–Meier survival analysis.

### 2.7. Analysis of Drug Therapy

To ascertain the relationship between a drug and gene expression levels, the Genomics of Drug Sensitivity in Cancer (GDSC) (https://www.cancerrxgene.org/ (accessed on 10 October 2022)) [22] database and the Cancer Therapeutics Response Portal (CTRP) (https://portals.broadinstitute.org/ctrp/ (accessed on 10 October 2022)) [23] database were used. Furthermore, drug sensitivity and CCLE expression data from the PRISM database (https://depmap.org/repurposing (accessed on 10 October 2022)) [24] were collected to assess the relationship between drug sensitivity and risk score, as well as the differences in drug sensitivity between high- and low-risk groups.

### 2.8. Enrichment Analysis

Gene Ontology (GO), Kyoto Encyclopedia of Genes and Genomes (KEGG), and Gene Set Enrichment Analysis (GSEA) analyses used the "clusterProfile" package, and a $p$-value < 0.05 with FDR < 0.05 was regarded as statistically significant for the GO and KEGG enrichment analyses. In the GSEA enrichment analysis, the minimum to maximum number of genes was set at 15 to 500. After cyclic sampling 1000 times, the functional enrichment results of the high- and low-risk groups were obtained, and FDR < 0.25 was considered significant.

### 2.9. Immunohistochemistry (IHC)

The collected brain tissue paraffin sections were deparaffinized by xylene and dehydrated in ethanol. After blocking with 5% BSA, these sections reacted with anti-*MDK* (1:100, ZENBIO, R22526) overnight at 4 °C and reacted with the secondary antibody for 1.5 h. Then, these sections were visualized with a DAB kit (Servicebio, Wuhan, China) and captured using an Olympus fluorescent microscope.

### 2.10. Cell Culture and Transfection

GBM cell lines U251 and U87 were obtained from the Shanghai Institute of Biochemistry and Cell Biology (Shanghai, China). The cells were grown in High-Glucose DMEM containing 15% Fetal Bovine Serum (FBS, Gibco, CA, USA), 100 U/ML Penicillin, and 100 g/ML Streptomycin (Biosharp, CA, USA) at a suitable environment of 37 °C and 5% $CO_2$. To specifically knock down the expression of *MDK*, the Si-NC and Si-*MDK* plasmids (sense: GGGAAGGGAAAGGACUAGA

TT; anti-sense: UCUAGUCCUUUCCCUUCCCTT) were purchased from RiboBio (Guangzhou, China) and were transfected to cells with a Lipofectamine3000 Kit (Invitrogen, Waltham, MA, USA) according to the guidance of the manufacturer.

### 2.11. RNA Extraction and Real-Time PCR

First, we collected all the RNA of GBM cells, and then these RNAs were combined as cDNA using the RevertAidTM cDNA Synthesis Kit (Thermo, Waltham, MA, USA). Finally, the SYBR® Green Master Mix was utilized for RT-PCR to detect the total GAPDH and *MDK* levels according to the reagents' guidance. All primers in this study were as follows: the forward primer of GAPDH, CCTTCCGTGTCCCCACT; the reverse primer of GAPDH, GCCTGCTTCACCACCTTC; the forward primer of *MDK*, CACCCCTAAGTGCCCAAA; the reverse primer of *MDK*, TGGGGAAGAACAAAAGCG.

### 2.12. Western Blotting (WB)

RIPA buffer (Servicebio, Wuhan, China) was utilized for cell lysis to obtain total cellular proteins, and 10 μg of these proteins were separated through SDS-PAGE electrophoresis according to different protein molecules and transferred to PVDF membranes. Next, these membranes were sealed with a quick blocking solution (Servicebio, Wuhan, China) and incubated with the corresponding primary antibodies and HRP-conjugated secondary antibodies in turn. Finally, the enhanced ECL Reagent (Biosharp, CA, USA) was utilized to expose and visualize the blots. The primary antibodies in this study included anti-*MDK* (1:1000, R22526, ZENBIO, Suzhou, China), anti-E-cadherin (1:1000, Proteintech, 20874-1-AP), anti-Vimentin (1:1000, 10366-1-AP, Proteintech, Waltham, MA, USA), anti-snail (1:1000, 13099-1-AP, Proteintech, Waltham, MA, USA), and anti-tubulin (1:5000, M20005, Absmart, Shenzhen, China).

### 2.13. Immunofluorescence Staining

The treated cells were immobilized for 15 min by paraformaldehyde and permeabilized for 10 min by 0.2% Triton. Next, these cells were blocked for 45 min by 5% BSA (Sigma, Shanghai, China) and incubated overnight with anti-*MDK* (1:100, R22526, ZENBIO) at 4 °C. These cells were exposed to Alexa Fluor 488-conjugated secondary antibody (1:100, GB25301, Servicebio) for 30 min and were then captured and analyzed with an Olympus fluorescent microscope.

### 2.14. Cell Viability and EdU Assay

A CCK8 kit (Biosharp, CA, USA) was utilized to detect the cell viability. The transfected cells were planted in 96-well plates with 5000 cells per well. After a period of culture, these cells were incubated with CCK8 reagent for 1 h. Finally, the absorbance (450 nm) was measured and analyzed using a microplate reader (Molecular Devices, CA, USA). For the EdU assay, after incubation with the EdU solution (Beyotime, Shanghai, China) in a dark environment at 37 °C for 1 h, the cells were immobilized with paraformaldehyde for 20 min, and DAPI (Sigma, Shanghai, China) was utilized to visualize the nucleus. Subsequently, EDU-positive cells were captured and counted using an Olympus fluorescent microscope.

### 2.15. Wound Healing and Transwell Assays

For wound healing, in short, the transfected cells were planted in 6-well plates and cultivated to the density of 80%. A pipette tip (100 μL) was used to create scratches and the cells were incubated with serum-free medium for 24 h. Finally, the wound healing photographs were captured using an inverted microscope (Olympus, Tokyo, Japan) and the movement distance of cells was calculated using Image J. For the transwell assay, the transfected cells were planted in the upper transwell chamber supplemented with the serum-free medium, and the serum medium was added into the lower transwell chamber. After incubation at 37 °C for 24 h, these cells were immobilized and stained. Subsequently, an inverted microscope (Olympus, Japan) was utilized to take images and count the cell numbers.

### 2.16. Statistical Analysis

R studio (version 4.2.0) and GraphPad Prism (9.0.0) software were used to analyze all the data. To analyze the differences between the two groups, Student's *t*-test was used. For comparisons involving three or more groups, one-way analysis of variance (ANOVA) was employed. A statistically significant difference was defined as $p < 0.05$.

## 3. Results

### 3.1. The Flowchart of the Study

The research flow chart is shown in Figure 1. GBM expression data from the TCGA and GTEx cohorts were used to obtain 330 hypoxia-related DEGs, 607 EMT-related DEGs, and 2314 genes from the MEdarkgrey module, respectively. The intersection genes were further used for subsequent clinical, immune infiltration, mutation, drug treatment, and enrichment analysis. Furthermore, we selected the prognostic related gene *MDK* for further in vitro validation, including the expression of *MDK* in the environment of hypoxia, the proliferation and migration situation of GBM cells after *MDK* knockdown, and the EMT of GBM cells after *MDK* knockdown under a hypoxia condition.

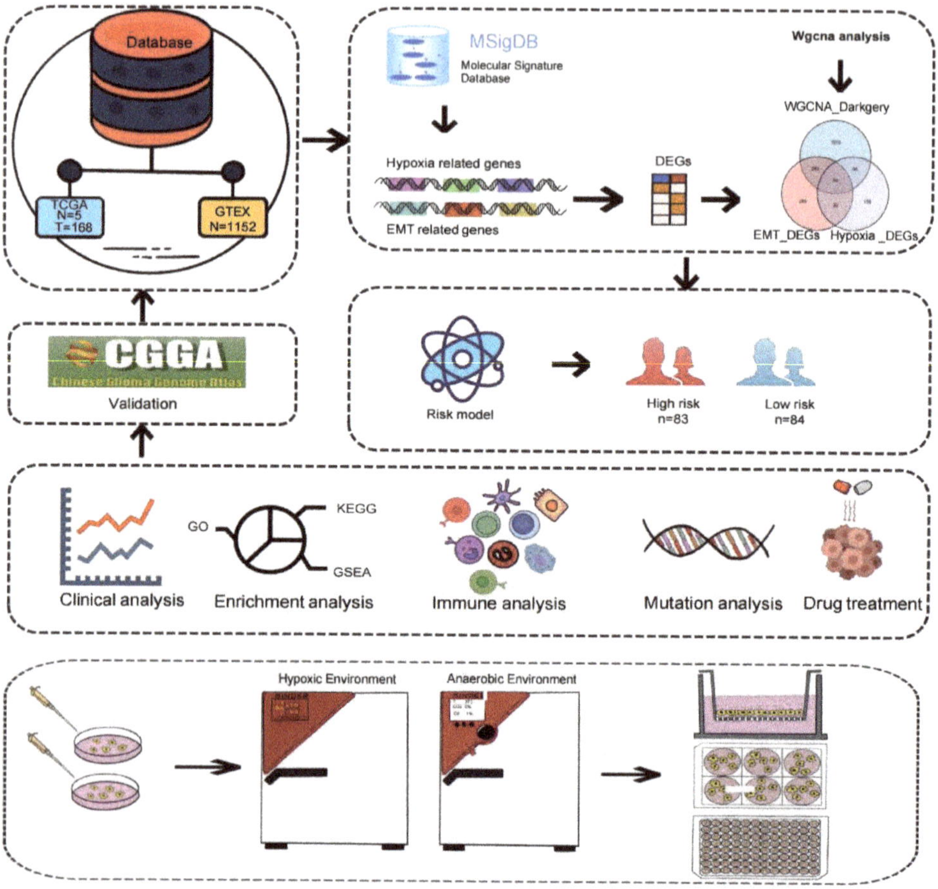


**Figure 1.** Flow chart of this study. Normal and tumor data of GBM patients in the TCGA and GTEx databases were used for WGCNA and differential analysis. The 59 genes obtained were further analyzed by Cox and LASSO regression analyses to construct a risk model. Based on the risk model, clinical correlation analysis, enrichment analysis, immune analysis, mutation analysis, and drug treatment analysis were continued. Finally, the *MDK* gene associated with prognosis was used for in vitro experiment verification.

### 3.2. Identification of DEGs in Collected Datasets

We combined the TCGA and GTEx cohorts to analyze the DEGs between GBM and healthy samples. We obtained 607 EMT-related DEGs (Figure 2A) and 330 hypoxia-related DEGs (Figure 2B) by setting $|\log FC| > 1$ and $p < 0.001$ as the criteria. The TCGA and GTEx combined cohort was then used to create a co-expression network using the "WGCNA" package. Based on two evaluation criteria (scale independence and average connectivity), a soft threshold of 14 was set (Figure 2C,D) and genes were divided into 11 modules, with the MEdarkgrey module having the highest correlation with gene characteristics, comprising 2314 genes (Figure 2E,F). The 59 intersection genes in the three sets of filtered genes were then identified using Venn diagrams (Figure 2G).

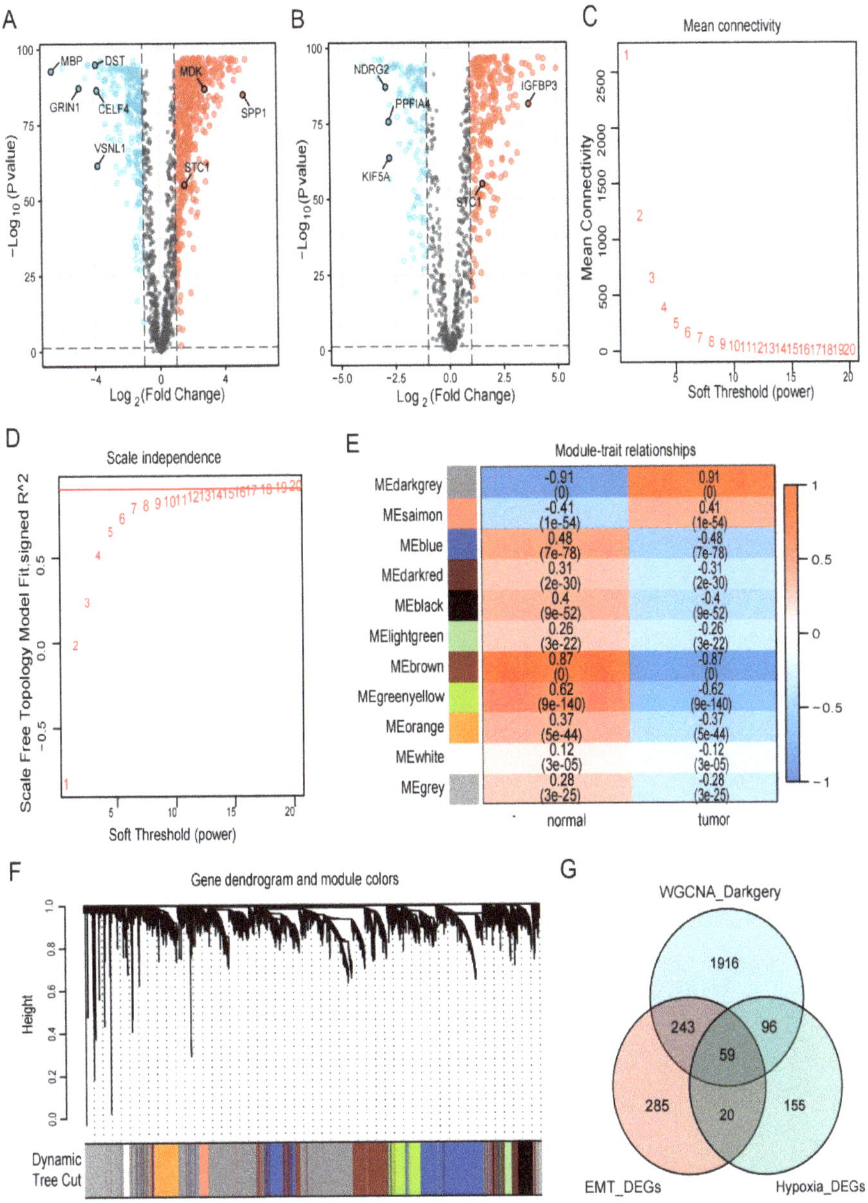


**Figure 2.** Identification of DEGs in the collected datasets. (**A**,**B**) Based on the TCGA and GTEx cohorts, EMT- and hypoxia-related DEGs were screened, respectively. Red represents upregulated genes with logFC > 1, black represents genes with no significant difference, and blue represents downregulated genes with logFC < −1. The $p$ for both red and blue distributed genes was less than 0.001. (**C**,**D**) The selection of soft-thresholding power and construction of a scale-free network. (**E**) Module–trait relationships, where the ME darkgrey module has the highest correlation (cor = 0.91 and $p$ = 0). (**F**) Cluster dendrogram in which similar genes are grouped into the same module. (**G**) Venn diagram screening 59 DEGs in 3 GBM cohorts.

### 3.3. Construction of Risk Models and Clinical Correlation Analysis

We used intersection genes to perform univariate Cox analysis, LASSO regression analysis, and multivariate Cox analysis in order to create the most efficient risk model for analyzing clinical traits and prognosis prediction. Screening by the criteria of $p < 0.05$, we obtained 22 genes that were connected to prognosis (Figure 3A). Further LASSO regression analysis and multivariate Cox regression analysis identified two genes (*STC1* and *MDK*) associated with prognosis (Figure 3B,C). Based on the level of the two genes and matching coefficients, the risk score was calculated: risk score = (0.2138 × *SEC1*) + (0.2702 × *MDK*). The TCGA-GBM cohort's median risk score (1.0155 to 1.0161) was used to classify patients into two risk subgroups: high- and low-risk groups. The risk score, survival status, and two gene expression levels of GBM patients in the TCGA cohort are shown in Figure 3D. In addition, the Kaplan–Meier survival curve shows that all GBM patients in the high-risk group had significantly shorter survival probability than those in the low-risk group (Figure 3E). The results of the CGGA cohort's risk score, survival status, two genes, and Kaplan–Meier survival curve were in line with those from the TCGA cohort (Figure 3F,G).

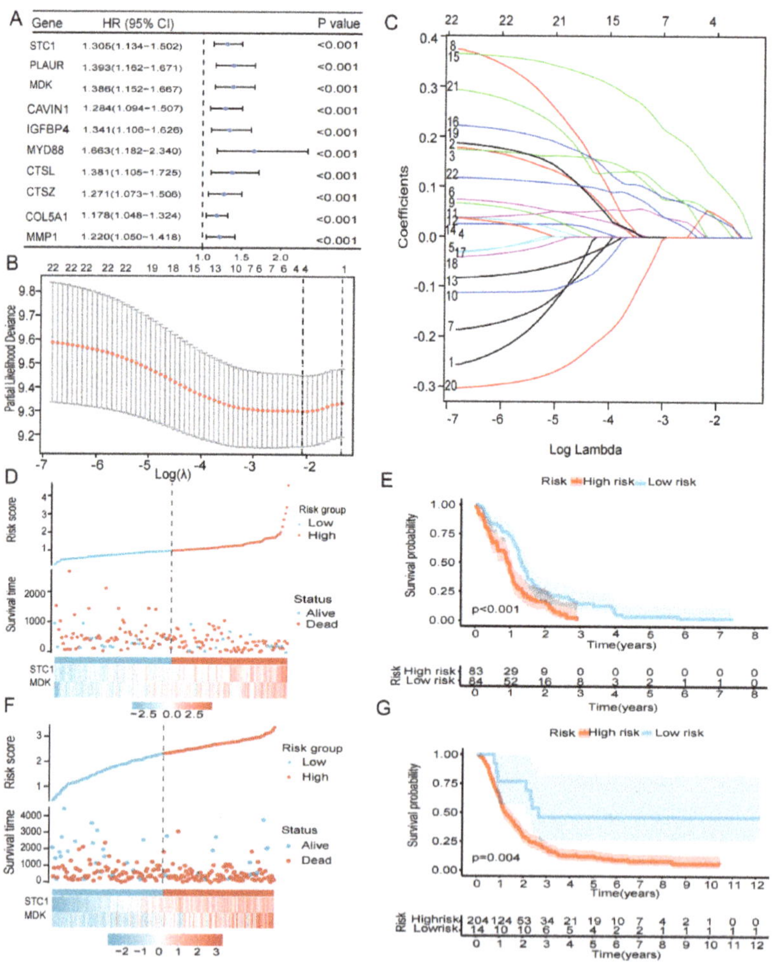


**Figure 3.** Construction and verification of a 17-gene risk signature. (**A**) Univariate Cox analysis of 59 DEGs. (**B**) LASSO coefficient profiles of the 59 DEGs. (**B**,**C**) Cross-validation for optimal parameter

selection in the LASSO regression model. (**D**) The risk score, survival status, and risk genes in the TCGA cohort. (**E**) Kaplan–Meier survival analysis between the low- and high-risk groups in TCGA cohort, $p < 0.05$ was considered significant. (**F**) The risk score, survival status, and risk genes in the CGGA cohort. (**G**) Kaplan–Meier survival analysis between the low- and high-risk groups in CGGA validation cohort, $p < 0.05$ was considered significant.

To investigate the relationship between the clinical subgroup and the probability of survival, the difference in the risk score between each clinical subgroup, we extracted the clinical data of age, gender, IDH status, 1p19q status, and MGMTp status, excluded samples with missing information, and used Cox regression analysis on the remaining samples, yielding the correlation between clinical subgroups and the probability of survival (Figure 4A–E) and the correlation between clinical subgroups and risk score (Figure 4F–J). Our results showed that patients aged $\geq$ 40 years had a significantly lower probability of survival compared to the patients aged < 40 years ($p < 0.001$). Patients with the IDH wild type have a significantly higher probability of survival than those with the IDH mutation ($p < 0.05$). In contrast, there appeared to be no significant differences in survival probability for gender (male and female), 1p19q status (codel and non-codel), and MGMTp status (unmethylated and methylated). In addition, we analyzed the differences in the risk score between subgroups. Age, IDH status, 1p19q status, and MGMTp status were different in the high- and low-risk groups. Risk scores were higher in the age $\geq$ 40 years, IDH wild type, 1p19q non-codel, and MGMTp non-methylation groups.

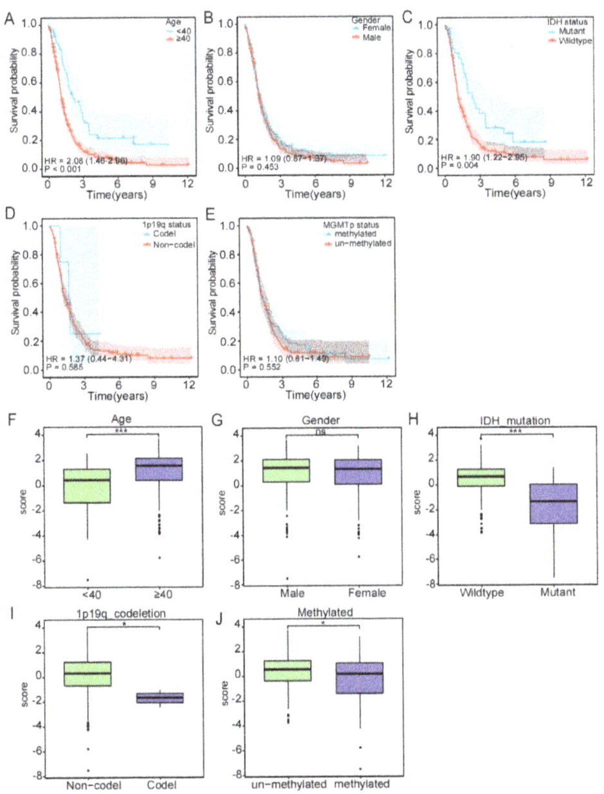


**Figure 4.** Subgroup clinical traits analysis. (**A–E**) Association of each clinical trait (age, gender, stage, TNM stage) with survival probability. (**F–J**) Association of each clinical trait (age, gender, stage, TNM stage) with risk score (ns is considered not significant, * $p < 0.05$, *** $p < 0.001$, $p < 0.05$ was considered significant).

The ROC curve then confirmed that the risk score-based model had an AUC value of 0.807, indicating that it was reliable (Figure 5A). As shown in Figure 5B, a nomogram combining the risk score with other important clinical traits was constructed to predict patient survival at 1, 3, and 5 years in the TCGA cohort (C-index = 0.6410, 95% CI: 0.5911–0.6909, $p = 3.1124 \times 10^{-8}$). Furthermore, in both the univariate and multivariate Cox analyses, the $p$ value of the risk score was less than 0.05, with HR > 1, indicating that the risk score was an independent risk factor for the prognosis of GBM patients (Figure 5C,D). In the CGGA validation cohort, the AUC value of the risk score-based model was 0.653 (Figure 5E). The C-index of the constructed nomogram was 0.6093, 95% CI: 0.5598–0.6588, $p = 1.4823 \times 10^{-5}$ (Figure 5F). Univariate and multivariate Cox analyses equally demonstrated that risk score was an independent prognostic factor for GBM patients (Figure 5G,H).

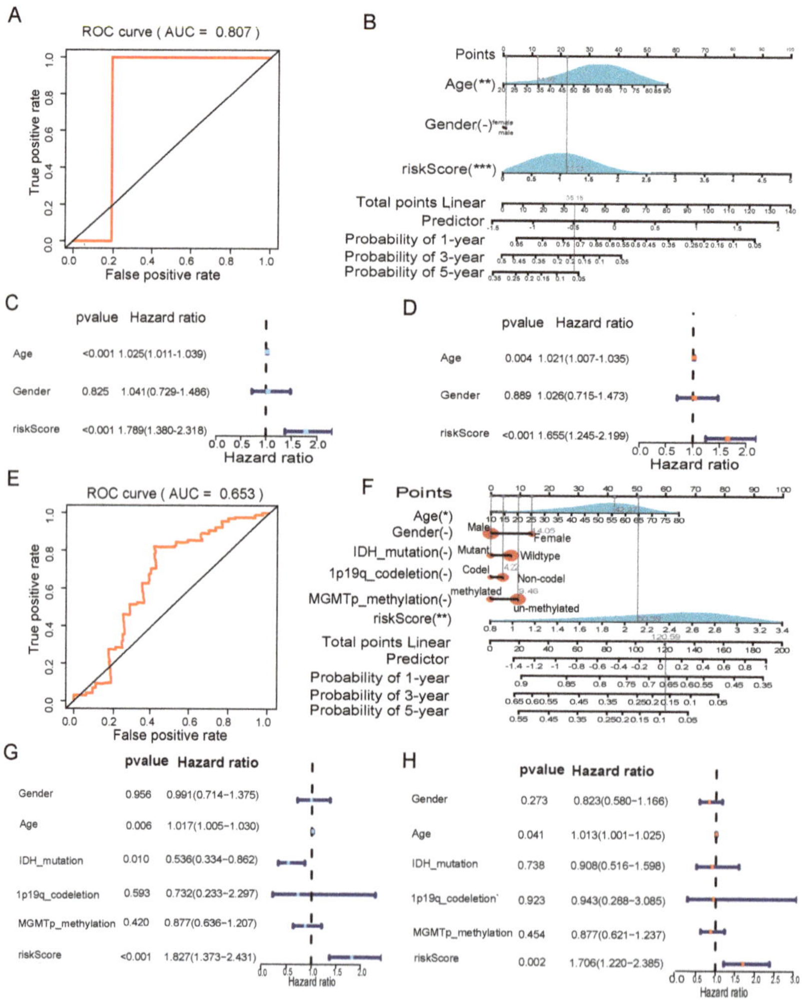


**Figure 5.** The risk score could be an independent factor for predicting the overall survival of GBM patients. (**A**) ROC curve for the risk model in the TCGA cohort. (**B**) Nomogram was based on clinical

traits and risk score in CGGA validation cohort. (**C**) Univariate Cox analysis in the TCGA cohort. (**D**) Multivariate Cox analysis in the TCGA cohort. (**E**) ROC curve for the risk score in the CGGA validation cohort. (**F**) Nomogram was based on clinical traits and risk score in CGGA validation cohort. (**G**) Univariate Cox analysis in the CGGA validation cohort. (**H**) Multivariate Cox analysis in the CGGA validation cohort. (" – " is considered not significant, * $p < 0.05$, ** $p < 0.01$, *** $p < 0.001$, $p < 0.05$ was considered significant).

### 3.4. Overview of Immune Cell Infiltration

The relationship between the risk score and immune cells was examined. Firstly, we used seven algorithms to calculate the correlation between the risk score and multiple immune cells/functions. The results revealed that cancer-associated fibroblasts had a significant positive correlation with risk score (Figure 6A). As seen in Figure 6B, the Stromal score and Estimate Score for the high-risk group were significantly higher than in the low-risk group ($p < 0.05$). Additionally, the relationship between the risk score and DNAss/RNAss is depicted in Figure 6C,D, and the findings reveal a significant negative relationship ($R = -0.57$, $p < 0.001$) between the risk score and RNAss. Finally, the expression of 16 immune cell types in the high- and low-risk groups were also examined using the ssGSEA algorithm. The box plot demonstrates that the proportion of immune cells were significantly higher in the high-risk group compared to the low-risk group (Figure 6E).

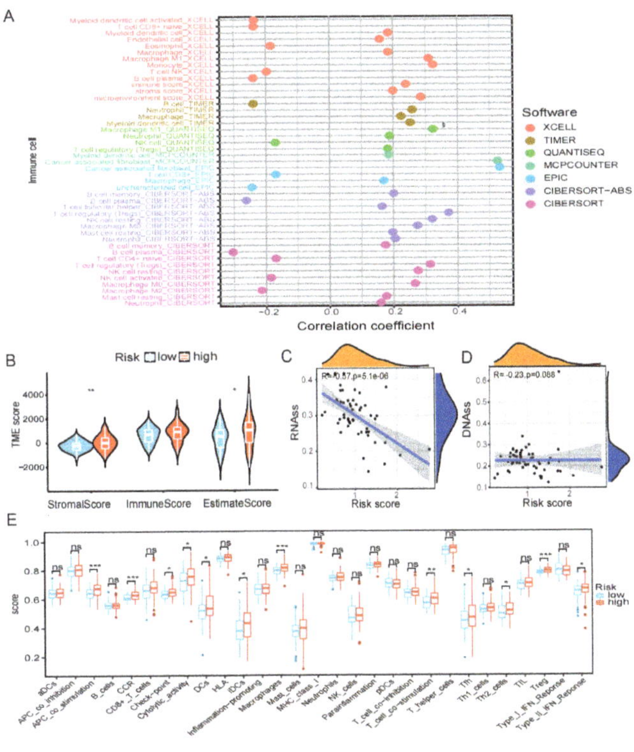


**Figure 6.** Immune infiltration analysis. (**A**) Seven algorithms (XCELL, TIMER, QUANTISEQ, MCP-COUNTER, EPIC, CIBERSORT-ABS, CIBERSORT) were used to determine the relationship between risk score and immune cells. (**B**) TME score for high- and low-risk groups. (**C**) Relationship between risk score and RNAss. (**D**) Relationship between risk score and DNAss. (**E**) Differences in immune cells and function between high- and low-risk groups (ns is considered not significant, * $p < 0.05$, ** $p < 0.01$, *** $p < 0.001$, $p < 0.05$ was considered significant).

Immune checkpoints and chemokines are noted to be major factors of tumor immunotherapy because they regulate immune cell function. We looked into the variations in chemokines and immune system checkpoints between the high- and low-risk groups. The immune checkpoint-related genes, such as LIF, VEGFA, and SPP1, were found to be significantly elevated in the high-risk group (Figure 7A,B). And the chemokines' related genes, for instance, CXCL8, CXCL2, THBS1, and POSTN, were elevated in the high-risk group when PI16 and ADCYAP1R1 were found to be significantly elevated in the low-risk group (Figure 7C,D).

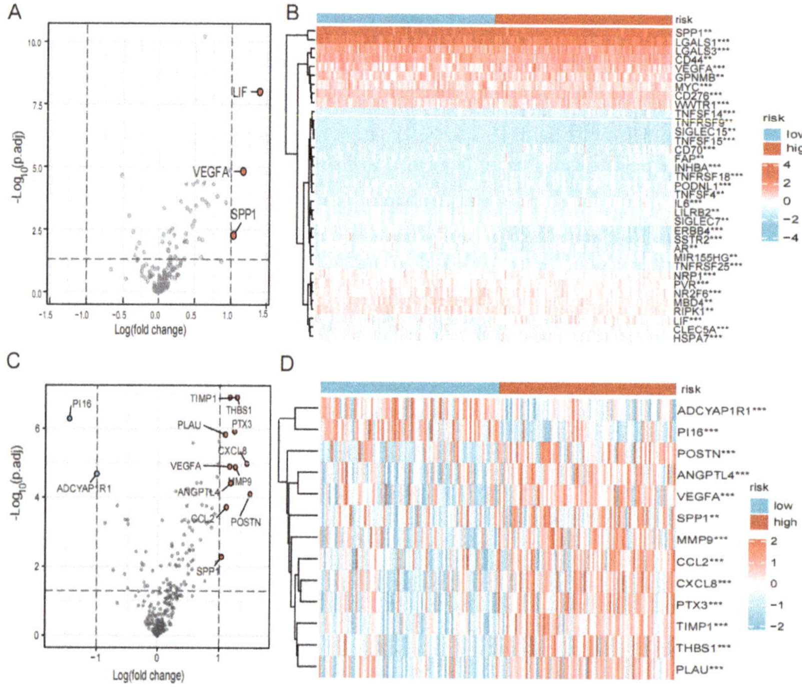

**Figure 7.** Expression of immune checkpoints, chemokines in high- and low-risk groups. (**A**) Volcano plot of the distribution of immune checkpoint-associated genes in high- and low-risk groups. The right side of the vertical axis is $p < 0.05$, logFC >1, and the left side is $p < 0.05$, logFC $< -1$. (**B**) Heatmap of the expression of immune checkpoint-related genes in high- and low-risk groups. (**C**) Volcano plot of the distribution of chemokines in high- and low-risk groups. The right side of the vertical axis is $p < 0.05$, logFC > 1, and the left side is $p < 0.05$, logFC $< -1$. (**D**) Heatmap of the expression of chemokines in high- and low-risk groups. (** $p < 0.01$, *** $p < 0.001$, $p < 0.05$ was considered significant).

### 3.5. Mutation Situation

We started with identifying the 15 most frequently mutated genes in the high- and low-risk groups (Figure 8A,B). Following that, TMB levels in the high- and low-risk groups were compared, and the relationship between risk score and TBM was investigated (Figure 8C,D). The findings revealed no discernible difference in TMB between the two groups and no significant relationship between the risk score and TMB score. According to the median TMB score, patients were split into two groups: high and low TMB score. Patients with high TMB had a higher probability of survival than those with low TMB, as shown in the Kaplan–Meier survival curve (Figure 8E). The worst prognosis was seen in the group of a low TMB score combined a high risk score (Figure 8F).

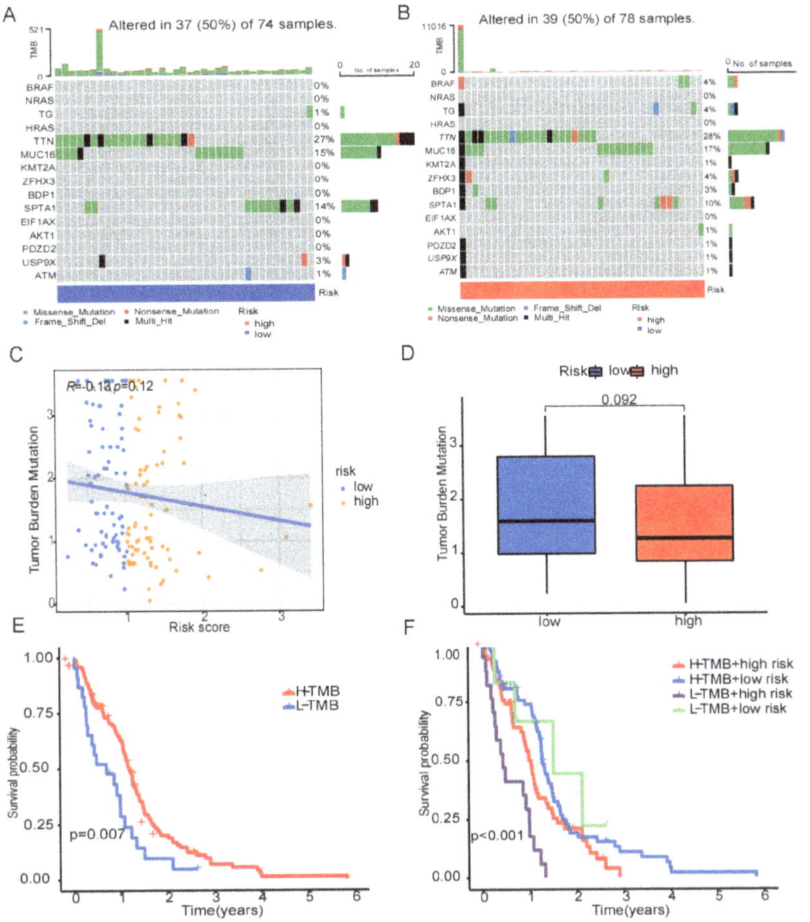


**Figure 8.** Tumor mutation status. (**A**) The 15 mutation genes in the high-risk group. (**B**) The 15 mutation genes in the low-risk group. (**C**) Relationship between risk scores and TMB. (**D**) Differences in TMB in high- and low-risk groups. (**E**) Differences in survival probability between the high- and low-TMB groups. (**F**) Differences in survival probability between high TMB in the low-risk group, high TMB in the high-risk group, low TMB in the low-risk group, and low TMB in the high-risk group. ($p < 0.05$ was considered significant).

### 3.6. Drug Treatment

Using the GDSC and CTRP databases, we summarized the associations between gene expression and drug sensitivity in different cancers, with 30 drugs having the strongest correlations as listed (Figure 9A,B). We also analyzed the relationship between risk score and responsiveness to drugs (Figure 9C). The results showed that 20 drugs, LY2090314, TMC647055, TMC647055, methscopolamine, 3-AQC, niraparib, chloroxine, mCPP, L-732,138, BRD4770, AZD2858, nifekalant, simeprevir, hydrocortisone-hemisuccinate, CEP-32496, atiprimod, homochlorcyclizine, ciclesonid, AMG-232, CGM097, and idasanutlin, were positively associated with the risk score. Compared with the low-risk group, idasanutlin, CGM097, AMG-232, and CEP-32496 were significantly upregulated while LY2090314 and 3-AQC were significantly downregulated in the high-risk group (Figure 9D–I).

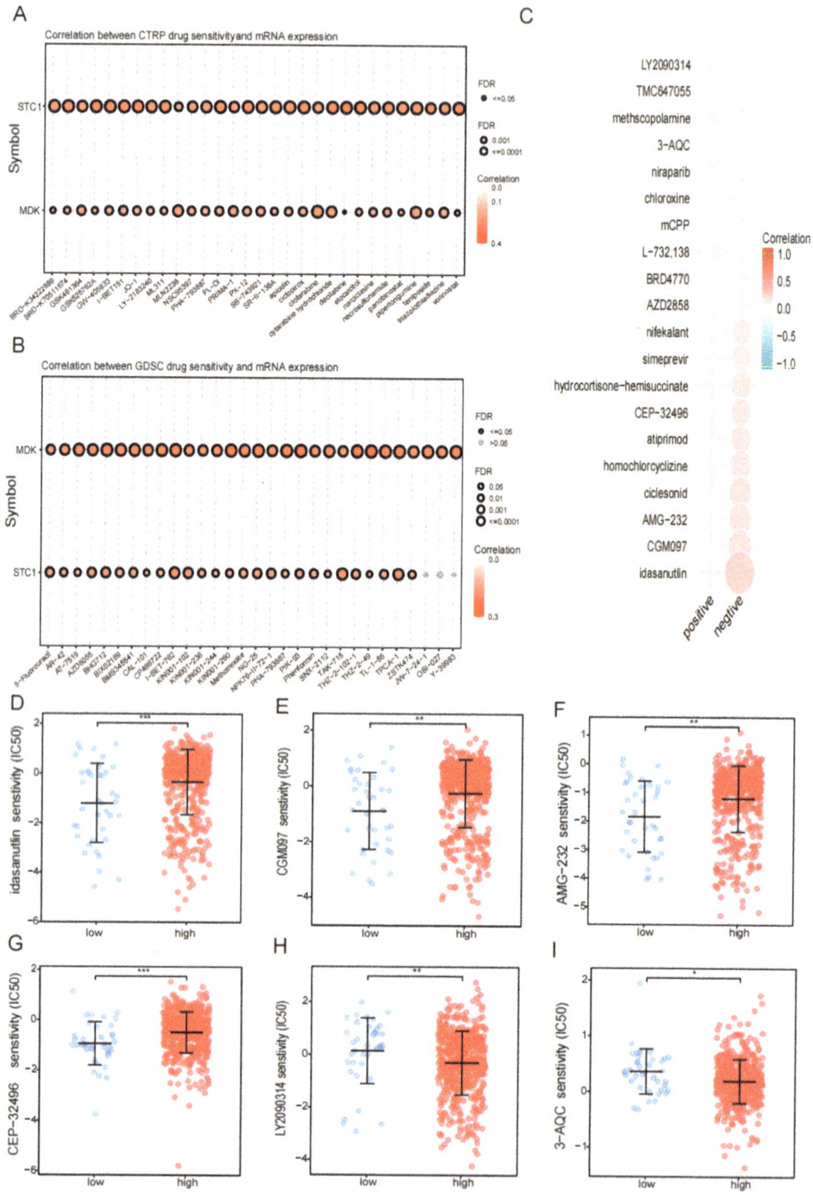

**Figure 9.** Drug treatment. (**A**) Relationship between risk score and drug sensitivity in the CTRP database. (**B**) The relationship between risk score and drug sensitivity in the GDSC database. (**C**) Relationship between risk scores and drug sensitivity in the PRISM database. (**D–I**) Relationship between risk scores and drug sensitivity in drugs (idasanutlin, CGM097, AMG-232, CEP-32496, LY2090314, 3-AQC). (* $p < 0.05$, ** $p < 0.01$, *** $p < 0.001$, $p < 0.05$ was considered significant.)

### 3.7. Results of Enrichment Analysis

We sought to elucidate the potential biological functions and pathways associated with the two prognostic genes (*STC1* and *MDK*). According to the expression of the high- and low-risk groups, we identified eight DEGs (|logFC| > 1 and $p < 0.05$). Then, KEGG enrichment analysis showed that the differences between high- and low-risk groups were mainly focused on cancer signaling pathways such as the PI3K-AKT and JAK-STAT pathways (Figure 10A). GO enrichment analysis showed that the differences between two groups were mainly related to the extracellular matrix (ECM) and cytokines (Figure 10B). Furthermore, GSEA enrichment analysis showed that the high-risk group was mainly enriched in cancer signaling pathways and the extracellular matrix while the low-risk group was mainly enriched in amino acid metabolism (Figure 10C).

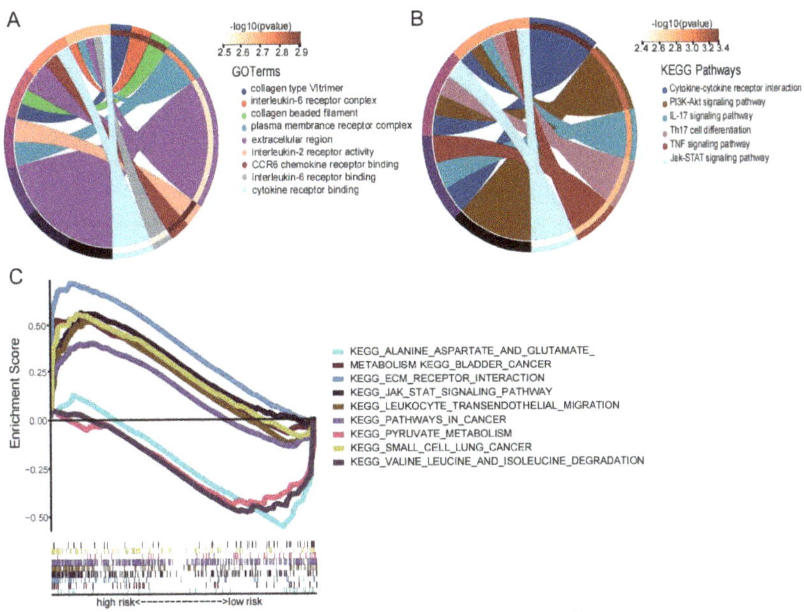


**Figure 10.** Enrichment analysis. (**A**) The KEGG diagram was made according to the DEGs between the high- and low-risk groups. (**B**) The GO diagram was made according to the DEGs between the high- and low-risk groups. (**C**) Gene set enrichment analysis of high- and low-risk groups.

### 3.8. MDK Is Elevated in GBM Patients and Related to Poor Prognosis

In previous studies, we found that hypoxia increased the expression of *MDK* in human umbilical vein endothelial cells [25] and *MDK* promoted the proliferation and migration of human breast cancer cells [26]. However, in GBM, the relationship between *MDK* and hypoxia and EMT has not been researched. Therefore, we chose *MDK* for further in vitro experiments to verify the efficacy of our model. First, in the TCGA database, we found that *MDK* was highly expressed in GBM (Figure 11A) and closely related to important clinical pathological features such as IDH wild type and 1p19q co-deletion (Figure 11B,C). At the same time, our clinical samples achieved the same results (Figure 11D,E). Finally, to verify the impact of *MDK* on the prognosis of GBM, the K-M curve demonstrated that the patients with high expression of *MDK* had a worse prognosis (Figure 11F–H). In conclusion, *MDK* was highly expressed in GBM and might be a potential prognostic marker.

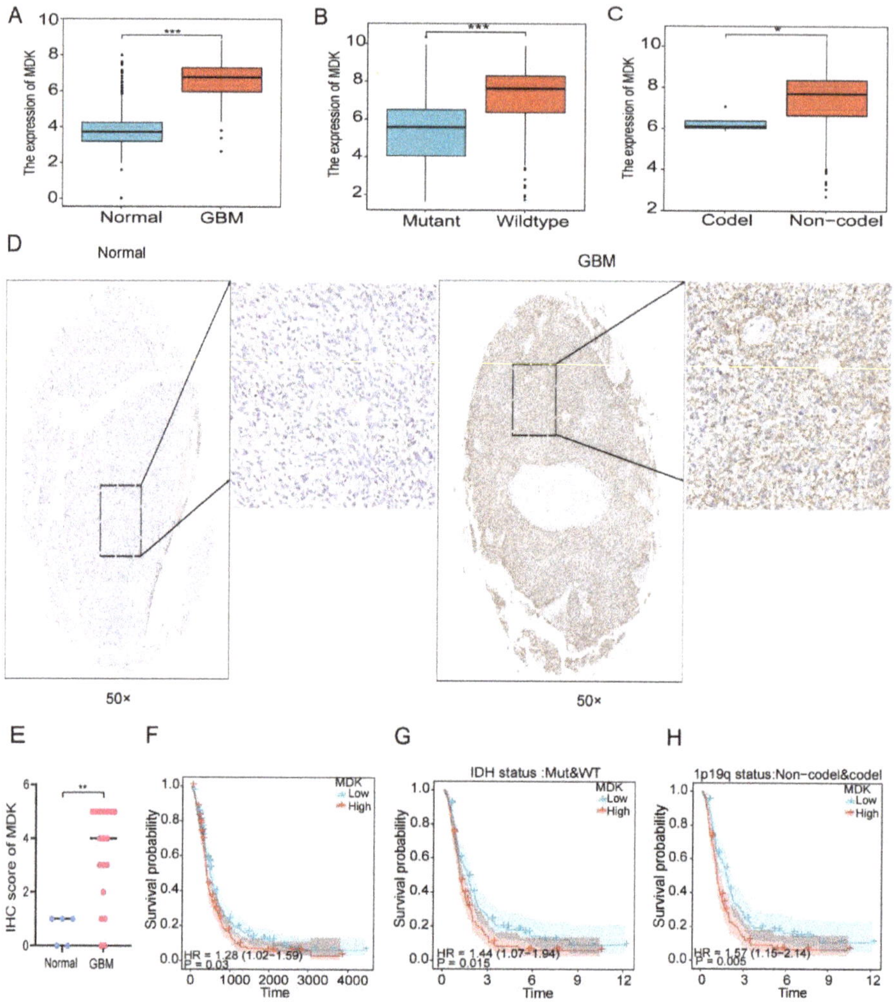


**Figure 11.** *MDK* is elevated in GBM patients and related to poor prognosis. (**A**) The gene expression of *MDK* in GBM in the TCGA database. (**B**,**C**) The gene expression of *MDK* in GBM with IDH status and 1p19q status in the TCGA database. (**D**,**E**) Representative IHC images of clinical samples, quantitative statistical analysis is shown below. Images on the right represent areas within the square magnified. (**F**–**H**) The K-M survival analysis of *MDK* in GBM in the TCGA database. (* $p < 0.05$, ** $p < 0.01$, *** $p < 0.001$, $p < 0.05$ was considered significant).

### *3.9. Hypoxia Increases MDK Expression in GBM Cells*

The growth of GBM is often accompanied by hypoxia, which causes GBM cells to alter the heredity and metabolism to adapt to this environment. In this article, we attempted to investigate the connection between *MDK* and hypoxia in GBM. First, in the TCGA database, we found that the expression of *MDK* was significantly positively correlated with hypoxia-related markers (HIF1A, HK2, VEGFA, and CA9) (Figure 12A). Then, GBM cells were cultured under 1% $O_2$ conditions to simulate a hypoxia environment. According to PCR findings, hypoxia significantly increased the expression of *MDK* mRNA (Figure 12B). As shown in Figure 12C,D, *MDK* protein expression peaked after 24 h of hypoxia. At the same time, the fluorescence intensity of *MDK* in the hypoxia group was significantly increased (Figure 12E–H). In conclusion, hypoxia significantly induced the expression of *MDK* in GBM cells.

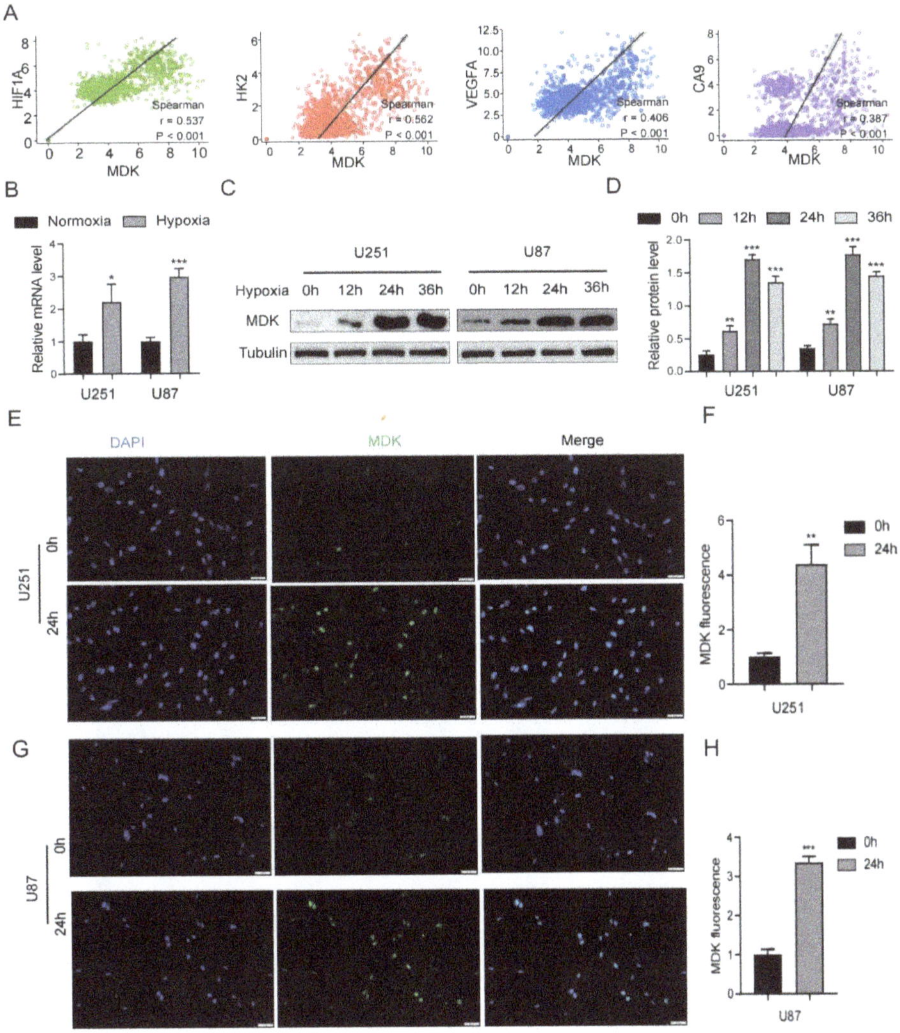

**Figure 12.** Hypoxia increases *MDK* expression in GBM cells. (**A**) Correlation between *MDK* and hypoxia-related markers such as HIF1F, HK2, VEGFA, and CA9 in the TCGA database. (**B**) The mRNA levels of *MDK* after GBM cells were cultured in hypoxia. (**C**,**D**) The protein levels of *MDK* after GBM cells were cultured in hypoxia, quantitative statistical analysis is shown on the right. (**E**–**H**) After GBM cells were cultured in hypoxia for 24 h, immunofluorescence was used to assess the *MDK* expression and relative fluorescence intensity was statistically analyzed. Scale bars: 20 μm. (* $p < 0.05$, ** $p < 0.01$, *** $p < 0.001$, $p < 0.05$ was considered significant).

### *3.10. Knockdown MDK Inhibits GBM Cells' Proliferation and EMT*

To analyze the role of *MDK* on the proliferation and EMT of GBM cells, we knocked down the expression of *MDK* by transient transfection and WB was utilized to verify the success of plasmid construction (Figure 13G,H). As shown in Figure 13A–C, the knockdown of *MDK* significantly inhibited the proliferation of GBM cells. In the TCGA database, the results revealed that *MDK* was significantly positively correlated with EMT-related markers

(ZEB1, Snail1, Vimentin, and N-cadherin) (Figure 13D). Additionally, the knockdown of *MDK* reduced the number of GBM cells invading (Figure 13E,F). According to WB results, the knockdown of *MDK* dramatically enhanced the expression of epithelial cell-related marker E-cadherin while significantly decreasing the expression of mesenchymal cell-related markers (Vimentin and Snail1) (Figure 13G,H). In short, the knockdown of *MDK* inhibited the proliferation and EMT of GBM cells.

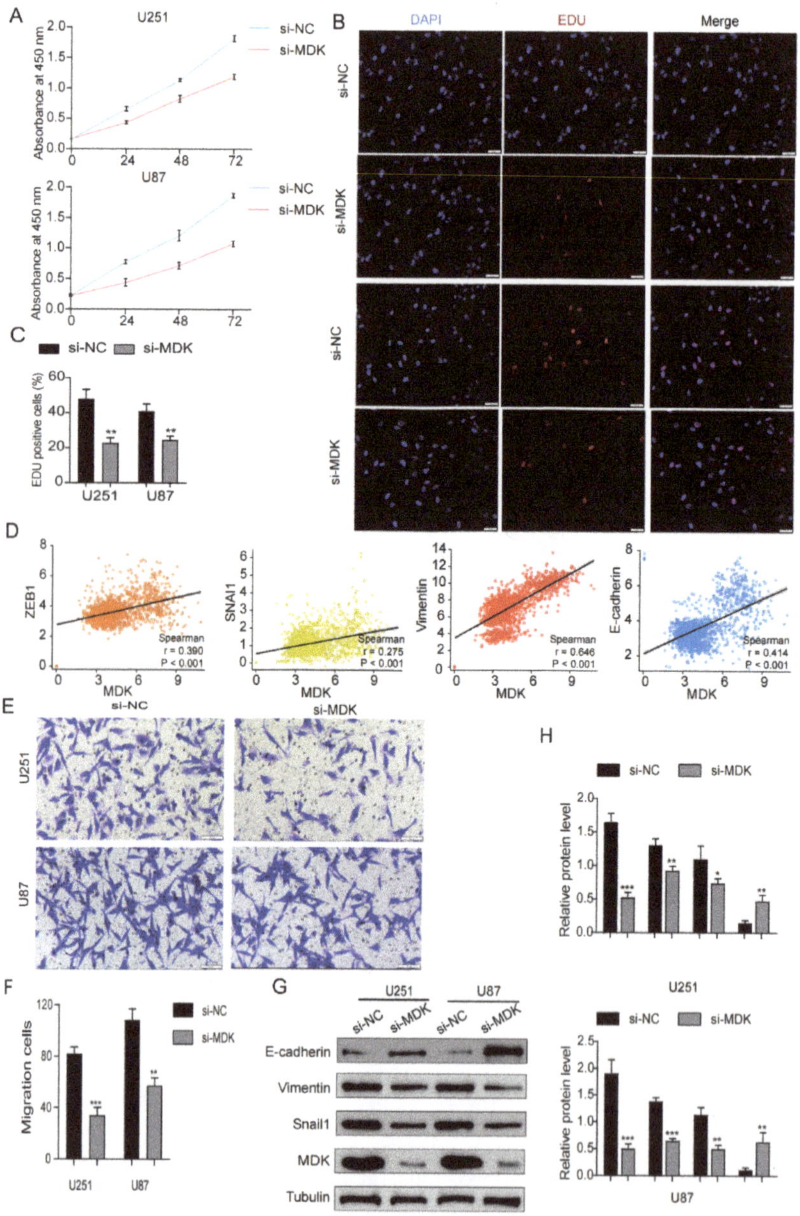


**Figure 13.** Knockdown *MDK* inhibits GBM cells' proliferation and EMT. (**A**) CCK8 assay was utilized to detect the viability of GBM cells transfected with *MDK* plasmid. (**B**,**C**) EdU assay was utilized to

detect the proliferation of GBM cells transfected with *MDK* plasmid, EdU-positive cells were defined as EdU cells/blue cells. (**D**) Correlation between *MDK* and EMT-related markers such as ZEB1, Snail1, Vimentin, and N-cadherin in the TCGA database. (**E**,**F**) The transwell assay was utilized to detect the invasive ability of GBM cells transfected with *MDK* plasmid and the results were statistically analyzed. Scale bars: 50 μm. (**G**,**H**) WB showing the protein levels of E-cadherin, Vimentin, Snail1, and *MDK* and quantitative statistical analysis is shown on the right. (* $p < 0.05$, ** $p < 0.01$, *** $p < 0.001$, $p < 0.05$ was considered significant).

### 3.11. Knockdown MDK Reverses Hypoxia-Induced EMT in GBM Cells

A hypoxic environment is regarded as the main driving force of the malignant behavior of tumors and stimulates EMT to enhance tumor metastasis and diffusion [27]. In our study, we found that hypoxia can induce the expression of *MDK* in GBM cells, and then we speculated whether *MDK* mediates hypoxia-induced EMT. As shown in Figure 14A–C, hypoxia significantly increased the migratory distance of GBM cells, but the knockdown of *MDK* reversed this result. In addition, the knockdown of *MDK* blocked the enhancement of the invasion of GBM cells by hypoxia (Figure 14D,E). Finally, the knockdown of *MDK* decreased hypoxia-induced Vimentin and Snaila expression and increased E-cadherin expression (Figure 14F,G). These results indicated that *MDK* mediates hypoxia-induced EMT.

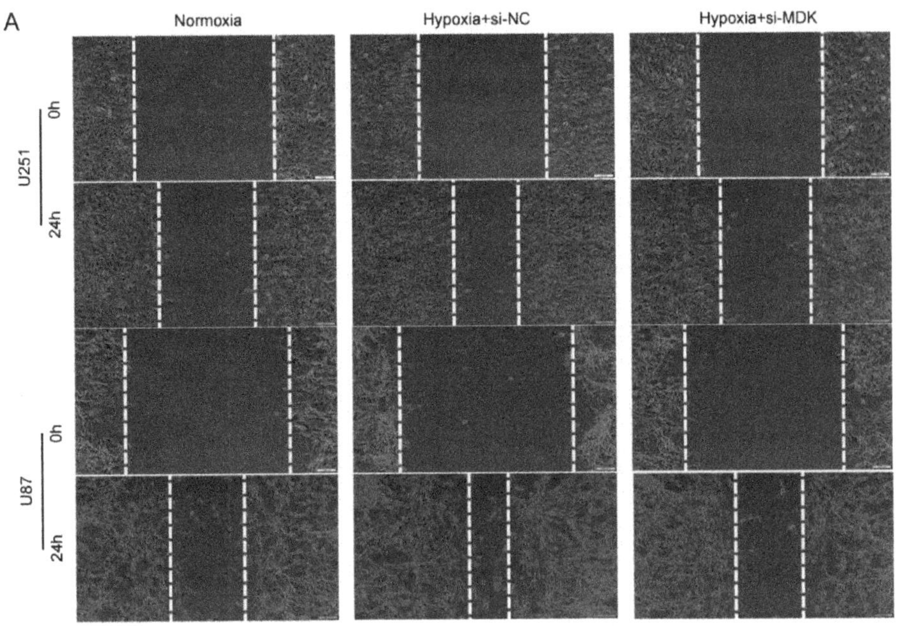


**Figure 14.** *Cont.*

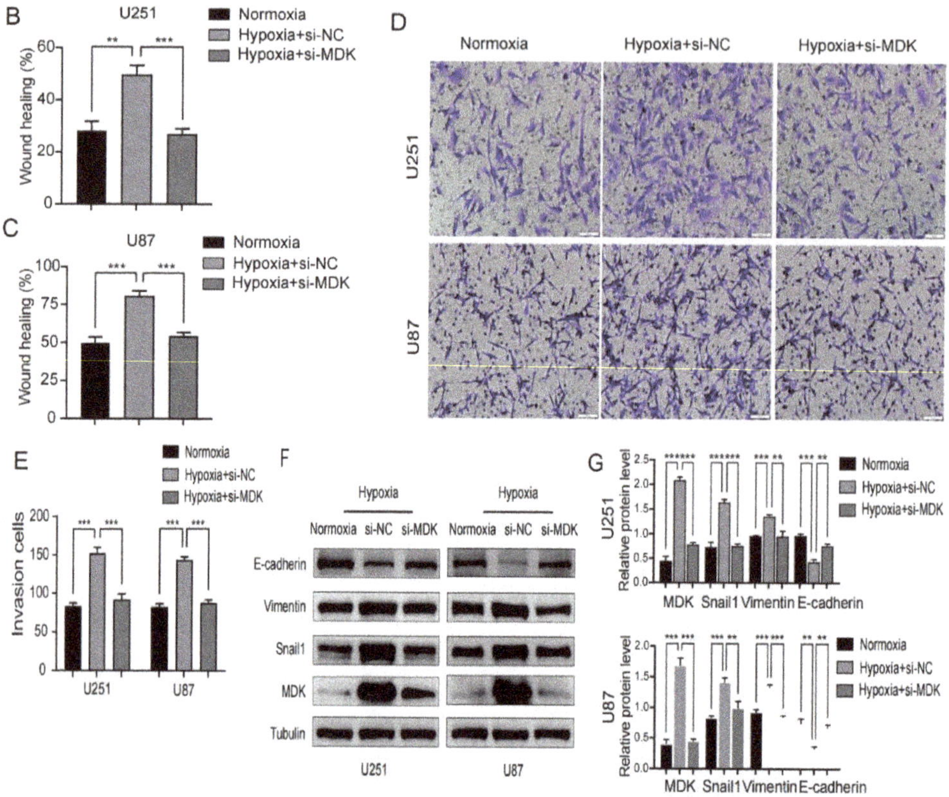


**Figure 14.** Knockdown *MDK* reverses hypoxia-induced EMT in GBM cells. (**A–C**) The wound healing assay was utilized to detect the migratory ability of MDA knockdown GBM cells in hypoxia and the results were statistically analyzed. Scale bars: 100 μm. (**D**,**E**) The transwell assay was utilized to detect the invasive ability of MDA knockdown GBM cells in hypoxia and quantitative statistical analysis is shown. Scale bars: 50 μm. (**F**,**G**) WB was performed to measure the expression of E-cadherin, Vimentin, Snail1, and *MDK*. Tubulin was utilized as internal control. ** $p < 0.01$, *** $p < 0.001$. Protein levels of E-cadherin, Vimentin, Snail1, and *MDK* and quantitative statistical analysis are shown on the right.

## 4. Discussion

GBM has the characteristics of high incidence, strong invasiveness, low survival, and a high recurrence rate. Although the traditional treatment methods have made great progress, they are still not enough to solve the problem of frequent recurrence in GBM patients, which prompt people to more actively study the pathophysiology and molecular targeted therapy of gliomas. Hypoxia can induce EMT, promote tumor migration, and promote invasion, anti-apoptosis, and the degradation of the extracellular matrix. As far as we are aware, this is the first paper combining bioinformatics techniques to construct a risk model and investigate the EMT induced by hypoxia in GBM patients. In this paper, we used multiple public datasets to investigate the predictive role of novel biomarkers in GBM patients. The differential expression of GBM and normal samples and WGCNA analysis were used to identify 59 differentially expressed genes related to both hypoxia and EMT. LASSO and Cox regression analyses were used to screen out the best two prognostic genes (*STC1* and *MDK*) and construct the risk model. The risk score could accurately predict the overall survival of GBM patients and serve as an independent risk factor. Based on the risk model, we further discussed the differences in high- and low-risk groups with common clinical traits, immune cell infiltration, tumor mutation, drug treatment, and enrichment analysis. The prognostic gene *MDK* was used for an in vitro experiment to verify the promoting effect of *MDK* on tumor proliferation, invasion, and metastasis in the hypoxic environment. This investigation marks the initial discovery that the downregulation of the *MDK* gene effectively diminishes the proliferation, migration, and epithelial–mesenchymal transition (EMT) of glioblastoma multiforme (GBM) cells in response to hypoxic conditions.

In this study, we constructed a risk model that included two genes: *STC1* and *MDK*. *STC1* is a protein-coding gene associated with diseases including colon [28] and breast cancers [29], it can enhance the metastasis potential of HCC through the JNK signaling pathway [30], and it reduces the immune function of macrophages [31–33]. *MDK* is a secreted protein that acts as a cytokine and growth factor, and it is mediated by cell surface proteoglycan and non-proteoglycan receptors [34–41]. It regulates the inflammatory response, cell proliferation, cell adhesion, cell growth, cell survival, tissue regeneration, cell differentiation, cell migration, and other processes [35–43]. For example, *MDK* can promote T-cell proliferation through the NFAT signaling pathway and Th1 cell differentiation, inhibit the emergence of drug-resistant dendritic cells, and thereby inhibit the differentiation of regulatory T cells [43]. It has also been reported that the *MDK* gene can promote the proliferation of human cancer cells, such as gastric cancer cells [44] and glioma cells [45]. As a result of binding to anaplastic lymphoma kinase (ALK), insulin receptor substrate 1(IRS1) is phosphorylated and mitogen-activated protein kinase (MAPK) and pi3 kinase are activated, which promotes cell growth [40]. The transition of epithelial cells to mesenchymal cells is promoted through interaction with NOTCH2 [34]. Then, we performed a clinical analysis based on this two-gene risk model, and the Kaplan–Meier survival curve showed that the survival probability of patients in the high-risk group was significantly lower ($p < 0.01$). Patients with an age $\geq 40$ years had a lower survival probability than those aged <40 years, and the IDH wild type group had a lower survival rate than the IDH mutation group. Clinical subgroup analyses showed that age $\geq 40$, IDH wild type, 1p19q non-codel, and MGMTp un-methylated people have a higher risk score in the high-risk group compared to the low-risk group. Age has been reported as a risk factor for glioma [46], patients with IDH mutations have a better prognosis [47], patients with the 1p19q codel have a better prognosis [48,49], and MGMT methylated patients are more sensitive to treatment with TMZ [50,51]. Meanwhile, the AUC value of the risk score reached 0.807, giving our risk model an advantage in terms of accuracy compared to models constructed in previous studies [52,53]. In addition, the nomogram constructed with clinical information in the TCGA and CGGA cohorts could be used to predict patient survival at 1, 3, and 5 years, while univariate and multivariate Cox analyses showed that age and risk score were independent risk factors for patient prognosis.

In addition to prognostic value, we discovered that the risk model was linked to tumor immunity and tumor mutations. A growing body of evidence suggests that tumor immunity is important in tumorigenesis and treatment response [54,55]. The high-risk group had a significantly higher Stromal Score and Estimate Score than the low-risk group, according to immunological analysis. The risk score was significantly inversely related to RNAss, implying that as the risk score increased, tumor stem cell differentiation decreased. The proportion of immune cells in the high-risk group was significantly higher than in the low-risk group. As immune checkpoints or chemokines, LIF, VEGFA, SPP1, CCL2, CXCL8, THBS1, and POST genes were significantly higher in the high- and low-risk groups, suggesting that they could be used as therapeutic targets. For example, studies have shown that infiltrating tumor-associated macrophages promote tumor growth partly by secreting SPP1, which can promote glioma cell survival and angiogenesis [56]. In the presence of the Spp1 protein, KPDC cell migration was significantly increased, whereas basal KPC cell migration was significantly decreased after Spp1 inhibition [57]. VEGFA is also upregulated in many tumors, and its expression is associated with tumor development, including ovarian carcinomas, and is a target in many cancer therapies under development [58,59]. Tumor mutation analysis revealed that the missense mutations in the TTN gene were the most common, and the difference in TMB between the high- and low-risk groups was not statistically significant. However, patients in the high-TMB group had a lower survival probability with the lowest survival probability in the high-risk group in combination with high TMB. Those with higher TMB were more likely to be recognized and attacked by immune cells, and they responded more significantly to immunotherapy [60,61]. There were significant differences in drug sensitivity between the high- and low-risk groups for idasanutlin, CGM097, AMG-232, CEP-32496, LY2090314, and 3-AQC.

Furthermore, pathway enrichment revealed that the extracellular matrix and cytokine pathways were the most different between the high- and low-risk groups. The high-risk group was concentrated on cancer signaling pathways and the extracellular matrix, whereas the low-risk group was concentrated on amino acid metabolism. Cancer cells have been shown to regulate the cytokine environment in order to change the cell composition of the microenvironment, which promotes tumor progression [62–64]. Extracellular matrix (ECM) components in particular have been identified as key regulators of cancer progression [65].

Finally, we selected *MDK*, a prognostic gene, for a further in vitro experiment, which was shown to be highly expressed in GBM and associated with a poorer prognosis. The expression of *MDK* was elevated in the hypoxic environment, and the proliferation and migration ability of GBM cells were reduced after knockdown of the *MDK* gene. The knockdown of the *MDK* gene in the hypoxic environment could reverse hypoxia-induced EMT, indicating that the hypoxia-induced EMT of GBM is mediated by *MDK*.

In conclusion, we used bioinformatics to effectively construct a risk model which performed admirably in survival and prognosis prediction in GBM patients. The model's predictions in training and validation cohorts verified its viability, and an in vitro experiment revealed that *MDK* mediates the EMT of tumor cells in a hypoxic environment. Nonetheless, our research has some limitations. On the one hand, our model is based on a retrospective analysis and will need to be validated in prospective studies. On the other hand, while this study is primarily based on bioinformatics techniques, the functional mechanisms and interactions of genes are important for validation, necessitating more complex and experimental data collection and evaluation.

**Supplementary Materials:** The following supporting information can be downloaded at https://www.mdpi.com/article/10.3390/biomedicines12010092/s1, Table S1: Demographic and Clinical Baseline Data of Glioma Patients.

**Author Contributions:** M.X. and S.T. performed the data analysis and wrote the manuscript. L.G. participated in the study design, figures, writing, and revised the manuscript. All authors have read and agreed to the published version of the manuscript.

**Funding:** This study was supported by National Science Foundation of China (Project # 82270861 to Gao), the Fundamental Research Funds for the Central Universities (Project # 2042020kf1079 to Gao), the Planned international development Project of Wuhan University (Project # WHU-GJZDZX TS03 to Gao), China UK joint training program for young scientist of Wuhan University.

**Institutional Review Board Statement:** This study was approved by the Ethics Committee of the Renmin Hospital of Wuhan University (approval number: 2012LKSZ (010) H). All methods were performed in accordance with the relevant guidelines and regulations. All patients had signed informed consent, and brain resections were collected from GBM patients.

**Data Availability Statement:** https://www.jianguoyun.com/p/DdCiP5UQiKCaCxj23vQEIAA.

**Conflicts of Interest:** The authors declare no conflict of interest.

## Abbreviations

EMT: Epithelial–mesenchymal transition; GBM: glioblastoma; WGCNA: weighted correlation network analysis; HIFs: hypoxia-inducible factors; DEGs: differential expression genes; TCGA: The Cancer Genome Atlas; GTEx: Genotype-Tissue Expression; LASSO: least absolute shrinkage and selection operator; CGGA: Chinese Glioma Genome Atlas; MsigDB: Molecular Signatures Database; NCBI: National Center for Biotechnology Information; TMB: tumor mutation burden; GDSC: Genomics of Drug Sensitivity in Cancer; CTRP: Cancer Therapeutics Response Portal; GO: Gene Ontology; KEGG: Kyoto Encyclopedia of Genes and Genomes; GSEA: Gene Set Enrichment Analysis; ALK: anaplastic lymphoma kinase; IRS1: insulin receptor substrate 1; MAPK: mitogen-activated protein kinase; ECM: extracellular matrix.

## References


1. Ostrom, Q.T.; Gittleman, H.; Fulop, J.; Liu, M.; Blanda, R.; Kromer, C.; Wolinsky, Y.; Kruchko, C.; Barnholtz-Sloan, J.S. CBTRUS Statistical Report: Primary Brain and Central Nervous System Tumors Diagnosed in the United States in 2008–2012. *Neuro Oncol.* **2015**, *17* (Suppl. S4), iv1–iv62. [CrossRef] [PubMed]
2. Dobes, M.; Khurana, V.G.; Shadbolt, B.; Jain, S.; Smith, S.F.; Smee, R.; Dexter, M.; Cook, R. Increasing incidence of glioblastoma multiforme and meningioma, and decreasing incidence of Schwannoma (2000–2008): Findings of a multicenter Australian study. *Surg. Neurol Int.* **2011**, *2*, 176. [PubMed]
3. Davis, M.E. Glioblastoma: Overview of Disease and Treatment. *Clin. J. Oncol. Nurs.* **2016**, *20* (Suppl. S5), S2–S8. [CrossRef] [PubMed]
4. Asthagiri, A.R.; Pouratian, N.; Sherman, J.; Ahmed, G.; Shaffrey, M.E. Advances in brain tumor surgery. *Neurol Clin.* **2007**, *25*, 975–1003. [CrossRef] [PubMed]
5. Choi, S.H.; Stuckey, D.W.; Pignatta, S.; Reinshagen, C.; Khalsa, J.K.; Roozendaal, N.; Martinez-Quintanilla, J.; Tamura, K.; Keles, E.; Shah, K. Tumor Resection Recruits Effector T Cells and Boosts Therapeutic Efficacy of Encapsulated Stem Cells Expressing IFNbeta in Glioblastomas. *Clin. Cancer Res.* **2017**, *23*, 7047–7058. [CrossRef]
6. Tsai, J.H.; Yang, J. Epithelial-mesenchymal plasticity in carcinoma metastasis. *Genes Dev.* **2013**, *27*, 2192–2206. [CrossRef]
7. Bertout, J.A.; Patel, S.A.; Simon, M.C. The impact of $O_2$ availability on human cancer. *Nat. Rev. Cancer.* **2008**, *8*, 967–975. [CrossRef]
8. Semenza, G.L. Hypoxia-inducible factors: Mediators of cancer progression and targets for cancer therapy. *Trends Pharmacol. Sci.* **2012**, *33*, 207–214. [CrossRef]
9. Lequeux, A.; Noman, M.Z.; Xiao, M.; Van Moer, K.; Hasmim, M.; Benoit, A.; Bosseler, M.; Viry, E.; Arakelian, T.; Berchem, G.; et al. Targeting HIF-1 alpha transcriptional activity drives cytotoxic immune effector cells into melanoma and improves combination immunotherapy. *Oncogene* **2021**, *40*, 4725–4735. [CrossRef]
10. Chae, Y.C.; Vaira, V.; Caino, M.C.; Tang, H.-Y.; Seo, J.H.; Kossenkov, A.V.; Ottobrini, L.; Martelli, C.; Lucignani, G.; Bertolini, I.; et al. Mitochondrial Akt Regulation of Hypoxic Tumor Reprogramming. *Cancer Cell* **2016**, *30*, 257–272. [CrossRef]
11. Kim, K.H.; Moon, M.; Yu, S.B.; Mook-Jung, I.; Kim, J.I. RNA-Seq analysis of frontal cortex and cerebellum from 5XFAD mice at early stage of disease pathology. *J. Alzheimers Dis.* **2012**, *29*, 793–808. [CrossRef] [PubMed]
12. Mirnics, K.; Korade, Z.; Arion, D.; Lazarov, O.; Unger, T.; Macioce, M.; Sabatini, M.; Terrano, D.; Douglass, K.C.; Schor, N.F.; et al. Presenilin-1-dependent transcriptome changes. *J. Neurosci.* **2005**, *25*, 1571–1578. [CrossRef] [PubMed]
13. Tang, K.; Ji, X.; Zhou, M.; Deng, Z.; Huang, Y.; Zheng, G.; Cao, Z. Rank-in: Enabling integrative analysis across microarray and RNA-seq for cancer. *Nucleic Acids Res.* **2021**, *49*, e99. [CrossRef] [PubMed]
14. The Cancer Genome Atlas Research Network. Comprehensive genomic characterization defines human glioblastoma genes and core pathways. *Nature* **2008**, *455*, 1061–1068. [CrossRef] [PubMed]
15. Huang, C.; Yi, H.; Zhou, Y.; Zhang, Q.; Yao, X. Pan-Cancer Analysis Reveals SH3TC2 as an Oncogene for Colorectal Cancer and Promotes Tumorigenesis via the MAPK Pathway. *Cancers* **2022**, *14*, 3735. [CrossRef] [PubMed]
16. Wang, Y.; Qian, T.; You, G.; Peng, X.; Chen, C.; You, Y.; Yao, K.; Wu, C.; Ma, J.; Sha, Z.; et al. Localizing seizure-susceptible brain regions associated with low-grade gliomas using voxel-based lesion-symptom mapping. *Neuro Oncol.* **2015**, *17*, 282–288. [CrossRef] [PubMed]
17. Zhao, Z.; Meng, F.; Wang, W.; Wang, Z.; Zhang, C.; Jiang, T. Comprehensive RNA-seq transcriptomic profiling in the malignant progression of gliomas. *Sci. Data* **2017**, *4*, 170024. [CrossRef] [PubMed]
18. Ritchie, M.E.; Phipson, B.; Wu, D.; Hu, Y.; Law, C.W.; Shi, W.; Smyth, G.K. limma powers differential expression analyses for RNA-sequencing and microarray studies. *Nucleic Acids Res.* **2015**, *43*, e47. [CrossRef]
19. Ning, W.; Acharya, A.; Li, S.; Schmalz, G.; Huang, S. Identification of Key Pyroptosis-Related Genes and Distinct Pyroptosis-Related Clusters in Periodontitis. *Front. Immunol.* **2022**, *13*, 862049. [CrossRef]

20. Nair, G.G.; Liu, J.S.; Russ, H.A.; Tran, S.; Saxton, M.S.; Chen, R.; Juang, C.; Li, M.-L.; Nguyen, V.Q.; Giacometti, S.; et al. Author Correction: Recapitulating endocrine cell clustering in culture promotes maturation of human stem-cell-derived beta cells. *Nat. Cell Biol.* **2019**, *21*, 792. [CrossRef]
21. Langfelder, P.; Horvath, S. WGCNA: An R package for weighted correlation network analysis. *BMC Bioinform.* **2008**, *9*, 559. [CrossRef] [PubMed]
22. Iorio, F.; Knijnenburg, T.A.; Vis, D.J.; Bignell, G.R.; Menden, M.P.; Schubert, M.; Aben, N.; Gonçalves, E.; Barthorpe, S.; Lightfoot, H.; et al. A Landscape of Pharmacogenomic Interactions in Cancer. *Cell* **2016**, *166*, 740–754. [CrossRef] [PubMed]
23. Zou, Y.; Palte, M.J.; Deik, A.A.; Li, H.; Eaton, J.K.; Wang, W.; Tseng, Y.-Y.; Deasy, R.; Kost-Alimova, M.; Dančík, V.; et al. A GPX4-dependent cancer cell state underlies the clear-cell morphology and confers sensitivity to ferroptosis. *Nat. Commun.* **2019**, *10*, 1617. [CrossRef] [PubMed]
24. Corsello, S.M.; Nagari, R.T.; Spangler, R.D.; Rossen, J.; Kocak, M.; Bryan, J.G.; Humeidi, R.; Peck, D.; Wu, X.; Tang, A.A.; et al. Discovering the anti-cancer potential of non-oncology drugs by systematic viability profiling. *Nat. Cancer* **2020**, *1*, 235–248. [CrossRef] [PubMed]
25. Weckbach, L.T.; Groesser, L.; Borgolte, J.; Pagel, J.I.; Pogoda, F.; Schymeinsky, J.; Müller-Höcker, J.; Shakibaei, M.; Muramatsu, T.; Deindl, E.; et al. Midkine acts as proangiogenic cytokine in hypoxia-induced angiogenesis. *Am. J. Physiol. Heart Circ. Physiol.* **2012**, *303*, H429–H438. [CrossRef] [PubMed]
26. Zhang, L.; Song, L.; Xu, Y.; Xu, Y.; Zheng, M.; Zhang, P.; Wang, Q. Midkine promotes breast cancer cell proliferation and migration by upregulating NR3C1 expression and activating the NF-kappaB pathway. *Mol. Biol. Rep.* **2022**, *49*, 2953–2961. [CrossRef]
27. Jiang, J.; Tang, Y.L.; Liang, X.H. EMT: A new vision of hypoxia promoting cancer progression. *Cancer Biol. Ther.* **2011**, *11*, 714–723. [CrossRef]
28. Pena, C.; Cespedes, M.V.; Lindh, M.B.; Kiflemariam, S.; Mezheyeuski, A.; Edqvist, P.H.; Hägglöf, C.; Birgisson, H.; Bojmar, L.; Jirström, K.; et al. *STC1* expression by cancer-associated fibroblasts drives metastasis of colorectal cancer. *Cancer Res.* **2013**, *73*, 1287–1297. [CrossRef]
29. Nakagawa, T.; Martinez, S.R.; Goto, Y.; Koyanagi, K.; Kitago, M.; Shingai, T.; Elashoff, D.A.; Ye, X.; Singer, F.R.; Giuliano, A.E.; et al. Detection of circulating tumor cells in early-stage breast cancer metastasis to axillary lymph nodes. *Clin. Cancer Res.* **2007**, *13*, 4105–4110. [CrossRef]
30. Chan, K.K.; Leung, C.O.; Wong, C.C.; Ho, D.W.; Chok, K.S.; Lai, C.L.; Ng, I.O.-L.; Lo, R.C.-L. Secretory Stanniocalcin 1 promotes metastasis of hepatocellular carcinoma through activation of JNK signaling pathway. *Cancer Lett.* **2017**, *403*, 330–338. [CrossRef]
31. Sheikh-Hamad, D. Mammalian stanniocalcin-1 activates mitochondrial antioxidant pathways: New paradigms for regulation of macrophages and endothelium. *Am. J. Physiol. Renal Physiol.* **2010**, *298*, F248–F254. [CrossRef] [PubMed]
32. Ma, S.; Xie, N.; Li, W.; Yuan, B.; Shi, Y.; Wang, Y. Immunobiology of mesenchymal stem cells. *Cell Death Differ.* **2014**, *21*, 216–225. [CrossRef] [PubMed]
33. Oh, J.Y.; Ko, J.H.; Lee, H.J.; Yu, J.M.; Choi, H.; Kim, M.K.; Wee, W.R.; Prockop, D.J. Mesenchymal stem/stromal cells inhibit the NLRP3 inflammasome by decreasing mitochondrial reactive oxygen species. *Stem Cells* **2014**, *32*, 1553–1563. [CrossRef] [PubMed]
34. Huang, Y.; Hoque, M.O.; Wu, F.; Trink, B.; Sidransky, D.; Ratovitski, E.A. Midkine induces epithelial-mesenchymal transition through Notch2/Jak2-Stat3 signaling in human keratinocytes. *Cell Cycle* **2008**, *7*, 1613–1622. [CrossRef]
35. Kurosawa, N.; Chen, G.Y.; Kadomatsu, K.; Ikematsu, S.; Sakuma, S.; Muramatsu, T. Glypican-2 binds to midkine: The role of glypican-2 in neuronal cell adhesion and neurite outgrowth. *Glycoconj J.* **2001**, *18*, 499–507. [CrossRef] [PubMed]
36. Maeda, N.; Ichihara-Tanaka, K.; Kimura, T.; Kadomatsu, K.; Muramatsu, T.; Noda, M. A receptor-like protein-tyrosine phosphatase PTPzeta/RPTPbeta binds a heparin-binding growth factor midkine. Involvement of arginine 78 of midkine in the high affinity binding to PTPzeta. *J. Biol. Chem.* **1999**, *274*, 12474–12479. [CrossRef] [PubMed]
37. Muramatsu, H.; Zou, K.; Sakaguchi, N.; Ikematsu, S.; Sakuma, S.; Muramatsu, T. LDL receptor-related protein as a component of the midkine receptor. *Biochem. Biophys. Res. Commun.* **2000**, *270*, 936–941. [CrossRef]
38. Muramatsu, H.; Zou, P.; Suzuki, H.; Oda, Y.; Chen, G.Y.; Sakaguchi, N.; Sakuma, S.; Maeda, N.; Noda, M.; Takada, Y.; et al. alpha4beta1- and alpha6beta1-integrins are functional receptors for midkine, a heparin-binding growth factor. *J. Cell Sci.* **2004**, *117 Pt 22*, 5405–5415. [CrossRef]
39. Sakaguchi, N.; Muramatsu, H.; Ichihara-Tanaka, K.; Maeda, N.; Noda, M.; Yamamoto, T.; Michikawa, M.; Ikematsu, S.; Sakuma, S.; Muramatsu, T. Receptor-type protein tyrosine phosphatase zeta as a component of the signaling receptor complex for midkine-dependent survival of embryonic neurons. *Neurosci. Res.* **2003**, *45*, 219–224. [CrossRef]
40. Stoica, G.E.; Kuo, A.; Powers, C.; Bowden, E.T.; Sale, E.B.; Riegel, A.T.; Wellstein, A. Midkine binds to anaplastic lymphoma kinase (ALK) and acts as a growth factor for different cell types. *J. Biol. Chem.* **2002**, *277*, 35990–35998. [CrossRef]
41. Weckbach, L.T.; Gola, A.; Winkelmann, M.; Jakob, S.M.; Groesser, L.; Borgolte, J.; Pogoda, F.; Pick, R.; Pruenster, M.; Müller-Höcker, J.; et al. The cytokine midkine supports neutrophil trafficking during acute inflammation by promoting adhesion via beta2 integrins (CD11/CD18). *Blood* **2014**, *123*, 1887–1896. [CrossRef] [PubMed]
42. Horiba, M.; Kadomatsu, K.; Nakamura, E.; Muramatsu, H.; Ikematsu, S.; Sakuma, S.; Hayashi, K.; Yuzawa, Y.; Matsuo, S.; Kuzuya, M.; et al. Neointima formation in a restenosis model is suppressed in midkine-deficient mice. *J. Clin. Invest.* **2000**, *105*, 489–495. [CrossRef] [PubMed]

43. Sonobe, Y.; Li, H.; Jin, S.; Kishida, S.; Kadomatsu, K.; Takeuchi, H.; Mizuno, T.; Suzumura, A. Midkine inhibits inducible regulatory T cell differentiation by suppressing the development of tolerogenic dendritic cells. *J. Immunol.* **2012**, *188*, 2602–2611. [CrossRef] [PubMed]
44. Xu, Y.; Qu, X.; Zhang, X.; Luo, Y.; Zhang, Y.; Luo, Y.; Hou, K.; Liu, Y.; Mizuno, T.; Suzumura, A. Midkine positively regulates the proliferation of human gastric cancer cells. *Cancer Lett.* **2009**, *279*, 137–144. [CrossRef] [PubMed]
45. Luo, J.; Wang, X.; Xia, Z.; Yang, L.; Ding, Z.; Chen, S.; Lai, B.; Zhang, N. Transcriptional factor specificity protein 1 (SP1) promotes the proliferation of glioma cells by up-regulating midkine (*MDK*). *Mol Biol. Cell.* **2015**, *26*, 430–439. [CrossRef] [PubMed]
46. Wang, Y.; Liu, X.; Guan, G.; Xiao, Z.; Zhao, W.; Zhuang, M. Identification of a Five-Pseudogene Signature for Predicting Survival and Its ceRNA Network in Glioma. *Front. Oncol.* **2019**, *9*, 1059. [CrossRef] [PubMed]
47. Liao, P.; Ostrom, Q.T.; Stetson, L.; Barnholtz-Sloan, J.S. Models of epigenetic age capture patterns of DNA methylation in glioma associated with molecular subtype, survival, and recurrence. *Neuro Oncol.* **2018**, *20*, 942–953. [CrossRef] [PubMed]
48. Eckel-Passow, J.E.; Lachance, D.H.; Molinaro, A.M.; Walsh, K.M.; Decker, P.A.; Sicotte, H.; Pekmezci, M.; Rice, T.; Kosel, M.L.; Smirnov, I.V.; et al. Glioma Groups Based on 1p/19q, IDH, and TERT Promoter Mutations in Tumors. *N. Engl. J. Med.* **2015**, *372*, 2499–2508. [CrossRef] [PubMed]
49. Weller, M.; Weber, R.G.; Willscher, E.; Riehmer, V.; Hentschel, B.; Kreuz, M.; Felsberg, J.; Beyer, U.; Löffler-Wirth, H.; Kaulich, K.; et al. Molecular classification of diffuse cerebral WHO grade II/III gliomas using genome- and transcriptome-wide profiling improves stratification of prognostically distinct patient groups. *Acta Neuropathol.* **2015**, *129*, 679–693. [CrossRef]
50. Taylor, J.W.; Schiff, D. Treatment considerations for MGMT-unmethylated glioblastoma. *Curr. Neurol. Neurosci. Rep.* **2015**, *15*, 507. [CrossRef]
51. Pandith, A.A.; Qasim, I.; Zahoor, W.; Shah, P.; Bhat, A.R.; Sanadhya, D.; Shah, Z.A.; Naikoo, N.A. Concordant association validates MGMT methylation and protein expression as favorable prognostic factors in glioma patients on alkylating chemotherapy (Temozolomide). *Sci. Rep.* **2018**, *8*, 6704. [CrossRef] [PubMed]
52. Liu, Y.Q.; Wu, F.; Li, J.J.; Li, Y.F.; Liu, X.; Wang, Z.; Chai, R.-C. Gene Expression Profiling Stratifies IDH-Wildtype Glioblastoma with Distinct Prognoses. *Front. Oncol.* **2019**, *9*, 1433. [CrossRef] [PubMed]
53. Qiu, H.; Li, Y.; Cheng, S.; Li, J.; He, C.; Li, J. A Prognostic Microenvironment-Related Immune Signature via ESTIMATE (PROMISE Model) Predicts Overall Survival of Patients with Glioma. *Front. Oncol.* **2020**, *10*, 580263. [CrossRef] [PubMed]
54. Martin-Orozco, N.; Muranski, P.; Chung, Y.; Yang, X.O.; Yamazaki, T.; Lu, S.; Hwu, P.; Restifo, N.P.; Overwijk, W.W.; Dong, C. T helper 17 cells promote cytotoxic T cell activation in tumor immunity. *Immunity* **2009**, *31*, 787–798. [CrossRef] [PubMed]
55. Wang, W.; Green, M.; Choi, J.E.; Gijon, M.; Kennedy, P.D.; Johnson, J.K.; Liao, P.; Lang, X.; Kryczek, I.; Sell, A.; et al. CD8(+) T cells regulate tumour ferroptosis during cancer immunotherapy. *Nature* **2019**, *569*, 270–274. [CrossRef] [PubMed]
56. Chen, P.; Zhao, D.; Li, J.; Liang, X.; Li, J.; Chang, A.; Henry, V.K.; Lan, Z.; Spring, D.J.; Rao, G.; et al. Symbiotic Macrophage-Glioma Cell Interactions Reveal Synthetic Lethality in PTEN-Null Glioma. *Cancer Cell* **2019**, *35*, 868–884.e6. [CrossRef] [PubMed]
57. Whittle, M.C.; Izeradjene, K.; Rani, P.G.; Feng, L.; Carlson, M.A.; DelGiorno, K.E.; Wood, L.D.; Goggins, M.; Hruban, R.H.; Chang, A.E.; et al. RUNX3 Controls a Metastatic Switch in Pancreatic Ductal Adenocarcinoma. *Cell* **2015**, *161*, 1345–1360. [CrossRef]
58. Sopo, M.; Anttila, M.; Hamalainen, K.; Kivela, A.; Yla-Herttuala, S.; Kosma, V.M.; Keski-Nisula, L.; Sallinen, H. Expression profiles of VEGF-A, VEGF-D and VEGFR1 are higher in distant metastases than in matched primary high grade epithelial ovarian cancer. *BMC Cancer.* **2019**, *19*, 584. [CrossRef]
59. Sayles, L.C.; Breese, M.R.; Koehne, A.L.; Leung, S.G.; Lee, A.G.; Liu, H.Y.; Spillinger, A.; Shah, A.T.; Tanasa, B.; Straessler, K.; et al. Genome-Informed Targeted Therapy for Osteosarcoma. *Cancer Discov.* **2019**, *9*, 46–63. [CrossRef]
60. Chan, T.A.; Yarchoan, M.; Jaffee, E.; Swanton, C.; Quezada, S.A.; Stenzinger, A.; Peters, S. Development of tumor mutation burden as an immunotherapy biomarker: Utility for the oncology clinic. *Ann. Oncol.* **2019**, *30*, 44–56. [CrossRef]
61. Stenzinger, A.; Allen, J.D.; Maas, J.; Stewart, M.D.; Merino, D.M.; Wempe, M.M.; Dietel, M. Tumor mutational burden standardization initiatives: Recommendations for consistent tumor mutational burden assessment in clinical samples to guide immunotherapy treatment decisions. *Genes Chromosomes Cancer* **2019**, *58*, 578–588. [CrossRef] [PubMed]
62. Qian, B.-Z.; Li, J.; Zhang, H.; Kitamura, T.; Zhang, J.; Campion, L.R.; Kaiser, E.A.; Snyder, L.A.; Pollard, J.W. CCL2 recruits inflammatory monocytes to facilitate breast-tumour metastasis. *Nature* **2011**, *475*, 222–225. [CrossRef] [PubMed]
63. Hoesel, B.; Schmid, J.A. The complexity of NF-kappaB signaling in inflammation and cancer. *Mol. Cancer* **2013**, *12*, 86. [CrossRef] [PubMed]
64. Hanahan, D.; Weinberg, R.A. Hallmarks of cancer: The next generation. *Cell* **2011**, *144*, 646–674. [CrossRef]
65. Lu, P.; Weaver, V.M.; Werb, Z. The extracellular matrix: A dynamic niche in cancer progression. *J. Cell Biol.* **2012**, *196*, 395–406. [CrossRef]

biomedicines

MDPI

*Article*

# Clinicopathological and Molecular Characteristics of *IDH*-Wildtype Glioblastoma with *FGFR3::TACC3* Fusion

**Hyunsik Bae [1], Boram Lee [2], Soohyun Hwang [2], Jiyeon Lee [3], Hyun-Soo Kim [2,*] and Yeon-Lim Suh [2,*]**

[1] Pathology Center, Seegene Medical Foundation, Seoul 04805, Republic of Korea; hbae0827@gmail.com
[2] Department of Pathology and Translational Genomics, Samsung Medical Center, Sungkyunkwan University School of Medicine, Seoul 06351, Republic of Korea; boram.lee@samsung.com (B.L.); juzzsaw@naver.com (S.H.)
[3] Department of Pathology, Guro Hospital, Korea University College of Medicine, Seoul 08308, Republic of Korea; jiyeon_l@naver.com
* Correspondence: hyun-soo_kim@naver.com (H.-S.K.); ylsuh76@skku.edu (Y.-L.S.)

**Abstract:** The World Health Organization Classification of Tumors of the Central Nervous System recently incorporated histological features, immunophenotypes, and molecular characteristics to improve the accuracy of glioblastoma (GBM) diagnosis. *FGFR3::TACC3* (F3T3) fusion has been identified as an oncogenic driver in *IDH*-wildtype GBMs. Recent studies have demonstrated the potential of using FGFR inhibitors in clinical trials and TACC3-targeting agents in preclinical models for GBM treatment. However, there is limited information on the clinicopathological and genetic features of *IDH*-wildtype GBMs with F3T3 fusion. The aim of this study was to comprehensively investigate the clinical manifestations, histological features, and mutational profiles of F3T3-positive GBMs. Between September 2017 and February 2023, 25 consecutive cases (5.0%) of F3T3-positive GBM were extracted from 504 cases of *IDH*-wildtype GBM. Clinicopathological information and targeted sequencing results obtained from 25 primary and 4 recurrent F3T3-positive GBMs were evaluated and compared with those from F3T3-negative GBMs. The provisional grades determined by histology only were distributed as follows: 4 (26/29; 89.7%), 3 (2/29; 6.9%), and 2 (1/29; 3.4%). Grade 2–3 tumors were ultimately diagnosed as grade 4 GBMs based on the identification of the *TERT* promoter mutation and the combined gain of chromosome 7 and loss of chromosome 10 (7+/10−). F3T3-positive GBMs predominantly affected women (2.6 females per male). The mean age of patients with an F3T3-positive GBM at initial diagnosis was 62 years. F3T3-positive GBMs occurred more frequently in the cortical locations compared to F3T3-negative GBMs. Imaging studies revealed that more than one-third (12/29; 41.4%) of F3T3-positive GBMs displayed a circumscribed tumor border. Seven of the seventeen patients (41.2%) whose follow-up periods exceeded 20 months died of the disease. Histologically, F3T3-positive GBMs more frequently showed curvilinear capillary proliferation, palisading nuclei, and calcification compared to F3T3-negative GBMs. Molecularly, the most common alterations observed in F3T3-positive GBMs were *TERT* promoter mutations and 7+/10−, whereas amplifications of *EGFR*, *PDGFRA*, and *KIT* were not detected at all. Other genetic alterations included *CDKN2A/B* deletion, *PTEN* mutation, *TP53* mutation, *CDK4* amplification, and *MDM2* amplification. Our observations suggest that F3T3-positive GBM is a distinct molecular subgroup of the *IDH*-wildtype GBM. Both clinicians and pathologists should consider this rare entity in the differential diagnosis of diffuse astrocytic glioma to make an accurate diagnosis and to ensure appropriate therapeutic management.

**Keywords:** adult-type diffuse glioma; glioblastoma; *IDH*-wildtype; *FGFR3::TACC3* fusion

## 1. Introduction

The 2016 World Health Organization (WHO) Classification of Tumors of the Central Nervous System (CNS) incorporated histological features and immunophenotypes to improve the accuracy of glioblastoma (GBM) diagnosis [1]. With the integration of the next-generation sequencing (NGS) technique into the diagnostic process, novel molecular characteristics of GBMs have been discovered. In the 2021 WHO Classification of Tumors of the CNS, the adoption of the Consortium to Inform Molecular and Practical Approaches to CNS Tumor Taxonomy (cIMPACT-NOW) further improved the classification of gliomas, especially aiding in prognosis prediction [2].

Isocitrate dehydrogenase (*IDH*)-wildtype glioblastoma (GBM) is a highly heterogeneous high-grade glioma [3]. The molecular diagnosis of *IDH*-wildtype GBM can be established based on the following three molecular parameters: the combined gain of chromosome 7 and loss of chromosome 10 (7+/10−), epidermal growth factor receptor (*EGFR*) amplification, and telomerase reverse transcriptase (*TERT*) promoter mutation [2]. Even though research on oncogenic driver mutations applicable to specific targeted therapies for GBMs is actively ongoing, their rarity has limited their use in treatment.

Gene fusion refers to the hybridization of coding or regulatory DNA sequences between genes resulting from genomic rearrangements such as translocations, deletions, duplications, and inversions [4]. The fusion between fibroblast growth factor receptor 3 (*FGFR3*) and transforming acidic coiled-coil-containing protein 3 (*TACC3*; F3T3 fusion) has been recently identified as an oncogenic driver in 3% of *IDH*-wildtype GBMs [5]. This fusion gene encodes a protein located at the poles of the mitotic spindle and exhibits hyperphosphorylation and constitutive activation of the kinase domain [6]. *IDH*-wildtype GBMs with F3T3 fusion (F3T3-positive GBMs) have been shown to exhibit distinct histological and molecular features [6–8], including the palisading of monotonous, ovoid nuclei; the curvilinear proliferation of thin capillaries; calcification; desmoplastic vessels; and amplifications of cyclin-dependent kinase 4 (*CDK4*) and mouse double minute 2 homolog (*MDM2*) [9–11]. It has been also reported that patients with an F3T3-positive GBM show better prognoses than those with an F3T3-negative GBM [5,9,11]. Some studies have highlighted differences in methylation profiles, tumor mutational burdens (TMBs), and copy number alterations between F3T3-positive and -negative GBMs [5,12].

Some previous studies have been conducted on F3T3-positive GBMs from multi-institutional cohorts [8,13], but their clinicopathological features remain to be clarified. In this study, we aimed to comprehensively analyze the clinical manifestations, imaging findings, histological features, and molecular characteristics of 25 consecutive cases of F3T3-positive GBM diagnosed at a single institution.

## 2. Materials and Methods

### 2.1. Patient Selection

Between September 2017 and February 2023, 25 cases of diffuse astrocytic gliomas with F3T3 fusion were extracted from a cohort of 504 *IDH*-wildtype GBM patients (5.0%) using next-generation sequencing (NGS) analysis. Twenty-five and four tumors were diagnosed as primary and recurrent F3T3-positive GBMs, respectively. Among the 479 patients with *IDH*-wildtype GBM without F3T3 fusion, 40 were randomly selected to establish an age-matched control group using the simultaneously conducted NGS analysis. Clinicopathological and demographic information was collected from the electronic medical records.

### 2.2. Histological and Neuroimaging Analysis

Two board-certified neuropathologists (H.B. and Y.-L.S.) independently reviewed all available hematoxylin and eosin-stained slides to ensure consistency and accuracy in the assessment of histological features. The following histological features were assessed: curvilinear capillary proliferation, nuclear palisading, calcification, palisading necrosis, desmoplastic vessels, tumor giant cells, and perivascular rosette formation. They also reviewed all available brain magnetic resonance images (MRIs) and the corresponding radiological interpretations.

### 2.3. NGS

Formalin-fixed, paraffin-embedded tissues were microdissected manually on unstained sections under a microscope. Genomic DNA was extracted using the QIAamp DNA FFPE Tissue Kit (Qiagen, Hilden, Germany). The DNA samples were fragmented using the Covaris S2 instrument (Covaris, Woburn, MA, USA), and a library was prepared using the SureSelect XT Automation Reagent Kit (Agilent Technologies, Santa Clara, CA, USA). To ensure the appropriate fragment size, the median size of the DNA fragments was determined using the Agilent 4200 Tape Station (Agilent, Santa Clara, CA, USA), with the desired range set between 300 and 400 bp. The concentration of the DNA was measured using the Qubit dsDNA High Sensitivity Assay Kit (Thermo Fisher Scientific, Waltham, MA, USA). Sequencing was performed on the NextSeq 500/550 platform using the TG NextSeq 500/550 High Output Kit v2.0 and the TG NextSeq 500/550 Mid Output Kit v2.0 (NextSeq 550 Dx, Illumina, San Diego, CA, USA). Data analysis and interpretation were conducted using BrainTumorSCAN v2.0 (Geninus, Seoul, Republic of Korea). The BrainTumorSCAN (Geninus) pipeline consisted of a sequencing panel targeting 233 glioma-related genes. The obtained sequence data were analyzed to identify clinically relevant genomic alterations, including single-nucleotide variants (SNVs), small insertions/deletions (indels), copy number alterations (CNAs), and rearrangements/fusions. SNVs and indels with a variant allele frequency (VAF) below 1% were excluded from the analysis. Copy number gains were defined as having an average copy number greater than four, while losses were defined as having a copy number less than one. Translocations were identified based on supporting reads ranging from four to twelve, depending on the sample quality. The NGS results were manually reviewed by four board-certified pathologists (H.B., B.L., S.H., and Y.-L.S.) and one bioinformatician (D.G.K.). Reported variants were cross-referenced with ClinVar (https://www.ncbi.nlm.nih.gov/clinvar/, accessed on 1 December 2023) to exclude variants that were reported as benign or likely benign. The genomic data analysis visualization was performed using the cBioPortal Oncoprinter (https://www.cbioportal.org/oncoprinter, accessed on 1 December 2023).

### 2.4. Statistical Analysis

Statistical analysis was performed using IBM SPSS Statistics for Windows, v27.0 (IBM Corporation, Armonk, NY, USA). Categorical variables, including clinical features, histological features, and imaging findings, were compared between F3T3-positive and -negative GBMs using Fisher's exact test or the chi-square test. Overall survival was defined as the time from initial diagnosis to death, with there being censored observations for patients who were still alive at the last follow-up. Survival curves were generated using the Kaplan–Meier method, and differences between the curves were assessed using the log-rank test. Statistical significance was defined as a $p < 0.05$.

## 3. Results

### 3.1. Demographics

Table 1 summarizes the association between F3T3 fusion and the clinicopathological characteristics of *IDH*-wildtype GBMs. Twenty-five primary and four recurrent tumor specimens were obtained from 25 patients. F3T3-positive GBMs predominantly affected women (2.6 females per male; $p = 0.003$). The mean age of patients with an F3T3-positive GBM at initial diagnosis was 62 years (median age, 60 years; range, 40–84 years), which was similar to that of the 479 patients with an F3T3-negative GBM (60 years; $p = 0.424$; range, 19–87 years). Seventeen of the twenty-nine tumors (58.6%) presented as a single lesion and were totally excised. The remaining 12 tumors (41.4%) showed multiple lesions and were diagnosed with navigation-guided stereotactic biopsies. Nineteen patients (76.0%) received concurrent chemoradiation therapy with temozolomide, while two (8.0%) received radiation therapy only. The mean follow-up period for the 25 patients with an F3T3-positive GBM was 19 months (median, 12 months; range, 3–69 months). Fifteen patients (60.0%) developed tumor recurrence or progression during the follow-up period. Six of the seventeen patients whose follow-up period exceeded 20 months died of disease. Eight patients with a follow-up period shorter than 20 months were alive at the last follow-up. Survival analysis revealed that patients with an F3T3-positive GBM had a longer overall survival than those with an F3T3-negative GBM, but the difference in survival rate between the two groups was not statistically significant ($p = 0.088$; Figure 1).

**Table 1.** Association between F3T3 fusion and the clinicopathological characteristics of *IDH*-wildtype GBMs.

| Characteristic | | Number of Cases (%) | | *p* Value |
|---|---|---|---|---|
| | | F3T3-Positive GBM | F3T3-Negative GBM | |
| Sex | Male | 7 (28.0) | 282 (58.9) | 0.003 * |
| | Female | 18 (72.0) | 197 (41.1) | |
| Mean age (range; years) | | 62 (40–84) | 60 (19–87) | 0.424 |
| Tumor laterality | Single | 17 (58.6) | 374 (78.1) | 0.006 * |
| | Multiple | 12 (41.4) | 105 (21.9) | |
| Type of procedure | Resection | 17 (58.6) | 380 (79.3) | 0.004 * |
| | Biopsy | 12 (41.4) | 99 (20.7) | |
| Overall survival | Mean (months) | 23 | 19 | 0.088 |
| | Alive | 19 (76.0) | 214 (44.7) | |
| | Dead | 6 (24.0) | 265 (55.3) | |

* Statistically significant.

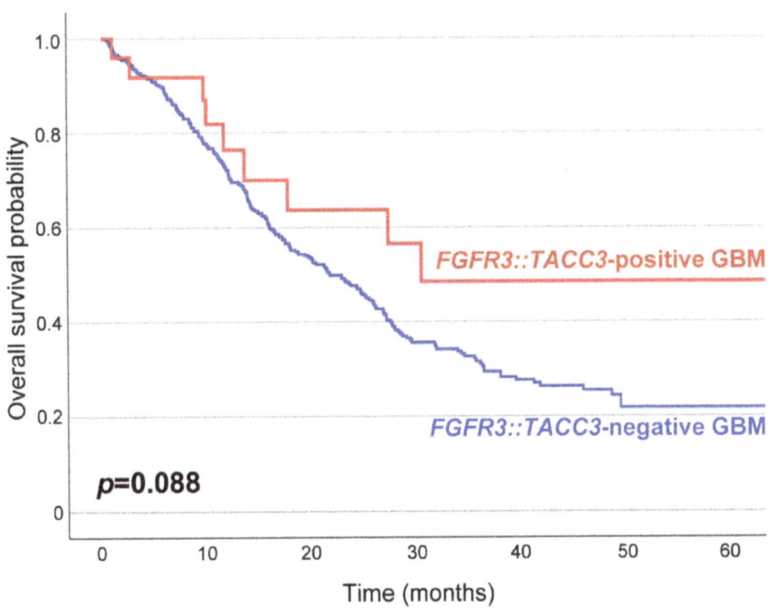


**Figure 1.** Kaplan–Meier plots showing the overall survival probabilities of patients with F3T3-positive GBM (blue line) and those with F3T3-negative GBM (red line). The overall survival rate of F3T3-positive GBM patients was higher than that of F3T3-negative GBM patients, but the difference was not statistically significant ($p = 0.088$).

### 3.2. Histological Analysis

The histological grades of 29 tumor tissues were distributed as follows: 26 were grade 4 (89.7%), 2 were grade 3 (6.9%), and 1 was grade 2 (3.4%). Since the lower-grade tumors displayed neither tumor cell necrosis nor microvascular proliferation, their provisional grade was grade 2 or 3. However, they were finally confirmed as WHO grade 4 GBMs based on their NGS results showing either a *TERT* promoter mutation (1/3), 7+/10− (1/3), or both (1/3; Table 2). Figure 2 depicts representative photomicrographs showing the histological features of an F3T3-positive GBM.

**Table 2.** Histological and molecular grades of F3T3-positive GBMs.

| Histological Grade | Number of Cases (%) | Molecular Feature | | | |
|---|---|---|---|---|---|
| | | *TERT* Promoter Mutation Only | 7+/10− Only | Both *TERT* Promoter Mutation and 7+/10− | Absent |
| 2 | 1 (3.4) | 1 | 0 | 0 | 0 |
| 3 | 2 (6.9) | 0 | 1 | 1 | 0 |
| 4 | 26 (89.7) | 3 | 6 | 16 | 1 |

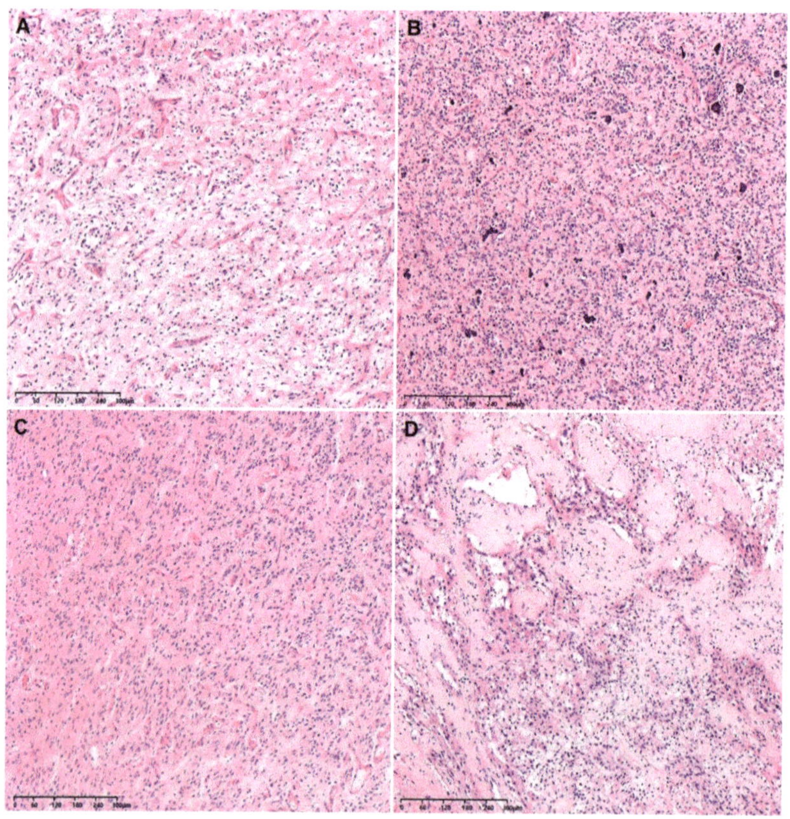


**Figure 2.** Representative photomicrographs showing the histological features of an F3T3-positive GBM. The tumor tissue exhibits curvilinear capillary proliferation (**A**), scattered calcification within the tumor (**B**), nuclear palisading (**C**), and desmoplastic vessels (**D**).

As shown in Table 3, the most common histological feature observed in the F3T3-positive group was curvilinear capillary proliferation (25/29; 86.2%), followed by desmoplastic vessels (18/29; 62.1%), palisading nuclei (16/29; 55.2%), palisading necrosis (14/29; 48.3%), and calcification within the tumor (13/29; 44.8%). Tumor giant cells and a perivascular rosette pattern were observed in 10 (34.5%) and 8 (27.6%) cases of F3T3-positive GBM, respectively. Compared to the F3T3-negative GBMs, F3T3-positive tumors showed curvilinear capillary proliferation ($p < 0.001$), nuclear palisading ($p = 0.001$), and calcification ($p = 0.006$) more frequently. The frequencies of desmoplastic vessels ($p = 0.429$), palisading necrosis ($p = 0.429$), tumor giant cells ($p = 0.271$), and perivascular rosette formation ($p = 0.114$) were not significantly different between F3T3-positive and -negative GBMs.

**Table 3.** Association between F3T3 and the histological features of IDH-wildtype GBMs.

| Characteristic | Number of Cases (%) | | p Value |
|---|---|---|---|
| | **F3T3-Positive GBM ($n = 29$)** | **F3T3-Negative GBM ($n = 40$)** | |
| Curvilinear capillary proliferation | 25 (86.2) | 6 (15.0) | <0.001 * |
| Desmoplastic vessels | 18 (62.1) | 21 (52.5) | 0.429 |
| Nuclear palisading | 16 (55.2) | 6 (15.0) | 0.001 * |
| Palisading necrosis | 14 (48.3) | 16 (40.0) | 0.494 |
| Calcification | 13 (44.8) | 6 (15.0) | 0.006 * |
| Tumor giant cells | 10 (34.5) | 9 (22.5) | 0.271 |
| Perivascular rosettes | 8 (27.6) | 5 (12.5) | 0.114 |

* Statistically significant.

### 3.3. Neuroimaging Analysis

As shown in Table 4, F3T3-positive GBMs were most frequently found in the right cerebral hemisphere (11/29; 37.9%) and the frontal lobe (16/29; 55.2%). These tumors were not found in the deep nuclei or cerebellum. Ten tumors (34.5%) were found in the bilateral cerebral hemispheres. In 12 cases (41.4%), the tumors had multiple anatomical locations. In contrast, F3T3-negative GBMs occurred most commonly in the left cerebral hemisphere (19/40; 47.5%) and the frontal lobe (20/40; 50.0%). While the tumor laterality and location were not significantly different according to the status of F3T3 fusion, cortical localization of tumors was significantly more frequent in F3T3-positive GBMs than in tumors without F3T3 ($p = 0.006$). Twelve cases of F3T3-positive GBM (41.4%) showed relatively well-circumscribed tumor borders, but there was no statistical significance in tumor border configuration between the two groups ($p = 0.093$).

### 3.4. Molecular Characteristics

Twenty-seven tumor tissue samples were available for NGS. As shown in Figure 3, the most common fusion type was identified between *FGFR3* exon 17 and *TACC3* exon 11 (7/27; 25.9%), followed by the fusion between *FGFR3* exon 18 and *TACC3* exon 11 (6/27; 22.2%). The most common exon was exon 18 in *FGFR3* (15/27; 55.6%) and exon 11 in *TACC3* (13/27; 48.1%). All fusions were in-frame fusions, encoding a structural protein with consistent tyrosine kinase activity. The mean TMB of F3T3-positive GBMs was 7.08 per megabase (range of 1.51–19.67 per megabase). The most common genetic alterations detected in F3T3-positive GBMs were 7+/10− (22/27; 81.5%), followed by the *TERT* promoter mutation (19/27; 70.4%). None of the F3T3-positive GBMs demonstrated amplifications in *EGFR*, platelet-derived growth factor receptor A (*PDGFRA*), or *KIT*. Other alterations included CDK inhibitor 2A (*CDKN2A*) deletion (13/27; 48.1%), CDK inhibitor 2B (*CDKN2B*) deletion (10/27; 37.0%), phosphatase and tensin homolog deletion in chromosome 10 (*PTEN*) mutation (9/27; 33.3%), *CDK4* amplification (6/27; 22.2%), *TP53* mutation (5/27; 18.5%), and murine double minute 2 (*MDM2*) amplification (4/27; 14.8%). The information on O-6-methylguanine-DNA methyltransferase (*MGMT*) promoter methylation was available in 25 F3T3-positive GBMs. Thirteen F3T3-positive tumors (52.0%) had a methylated *MGMT* promoter.

**Table 4.** Association between F3T3 fusion and the locational information of *IDH*-wildtype GBMs.

| Characteristic | | Number of Cases (%) | | $p$ Value |
|---|---|---|---|---|
| | | **F3T3-Positive GBM ($n$ = 29)** | **F3T3-Negative GBM ($n$ = 40)** | |
| Laterality | Right | 11 (37.9) | 15 (37.5) | 0.971 |
| | Left | 8 (27.6) | 19 (47.5) | 0.094 |
| | Bilateral | 10 (34.5) | 6 (15.0) | 0.058 |
| Location | Frontal lobe | 16 (55.2) | 20 (50.0) | 0.671 |
| | Parietal lobe | 9 (31.0) | 7 (17.5) | 0.189 |
| | Temporal lobe | 8 (27.6) | 15 (37.5) | 0.389 |
| | Occipital lobe | 2 (6.9) | 2 (5.0) | 0.739 |
| Epicenter | Cortex | 5 (17.2) | 0 (0.0) | 0.006 * |
| | Subcortex | 7 (24.1) | 3 (7.5) | 0.053 |
| | White matter | 17 (58.7) | 37 (92.5) | 0.001 * |
| Border | Circumscribed | 12 (41.4) | 9 (22.5) | 0.093 |
| | Infiltrative | 17 (58.6) | 31 (77.5) | 0.179 |

* Statistically significant.

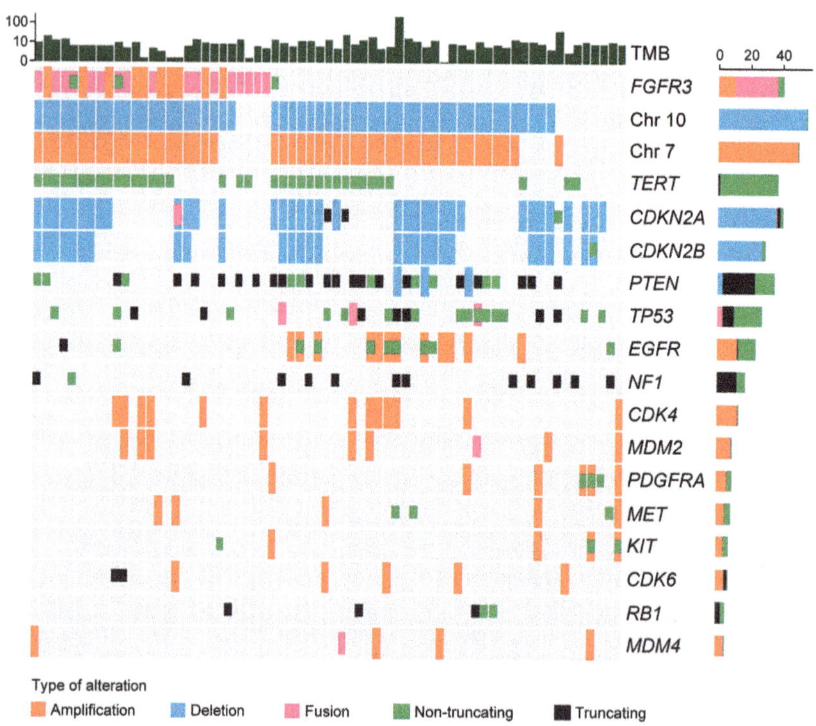


**Figure 3.** Molecular characteristics of F3T3-positive GBMs.

Table 5 summarizes the difference in the frequencies of genetic alterations between F3T3-positive and -negative GBMs. In total, 28 and 17 of the 40 F3T3-negative GBMs had 7+/10− and the *TERT* promoter mutation, respectively. There was no significant difference in the alteration frequencies according to the presence of F3T3 fusion ($p = 0.842$ and $p = 0.773$). The *EGFR* and *PDGFRA* amplifications were detected in 12 ($p = 0.002$) and 6 ($p = 0.035$) F3T3-negative GBMs, respectively. There were no significant differences in the frequencies of other mutations and the degree of *MGMT* promoter methylation between the two groups.

**Table 5.** Differences in the frequencies of genetic alterations between F3T3-positive and -negative GBMs.

| Alteration | Number of Cases (%) | | *p* Value |
|---|---|---|---|
| | F3T3-Positive GBM ($n = 27$) | F3T3-Negative GBM ($n = 40$) | |
| 7+/10− | 22 (81.5) | 28 (70.0) | 0.842 |
| *TERT* promoter mutation | 19 (70.4) | 17 (42.5) | 0.773 |
| *EGFR* amplification | 0 (0.0) | 12 (30.0) | 0.002 * |
| *PDGFRA* amplification | 0 (0.0) | 6 (15.0) | 0.035 * |
| *KIT* amplification | 0 (0.0) | 4 (10.0) | 0.142 |
| *CDKN2A* deletion | 13 (48.1) | 23 (57.5) | 0.211 |
| *CDKN2B* deletion | 10 (37.0) | 18 (45.0) | 0.211 |
| *PTEN* mutation | 9 (33.3) | 24 (60.0) | 0.046 * |
| *CDK4* amplification | 6 (22.2) | 7 (17.5) | 0.755 |
| *TP53* mutation | 5 (18.5) | 17 (42.5) | 0.105 |
| *MDM2* amplification | 4 (14.8) | 4 (10.0) | 0.705 |
| *MGMT* promoter methylation | 13 (52.0) | 12 (30.0) | 0.310 |

* Statistically significant.

### 3.5. Brief Presentation of a Case of F3T3-Positive GBM Showing Unusual Histology

We encountered a case of an F3T3-positive GBM displaying unusual histology. A 40-year-old woman presented with a 6.7 cm sized multiloculated mass in the left temporal lobe (Figure 4A). She underwent gross total resection. The tumor exhibited a relatively well-circumscribed border. Histologically, the tumor consisted of round-to-oval cells arranged in an alveolar or micropapillary pattern (Figure 4B). The tumor cells possessed pleomorphism nuclei with vesicular chromatin, conspicuous nucleoli, and an eosinophilic cytoplasm. The peritumoral area showed significant inflammatory infiltrates with eosinophils. The tumor lacked the characteristic histological features typically seen in F3T3-positive GBMs. Instead, tumor cell necrosis, frequent mitotic figures, and a cell-in-cell phenomenon were observed (Figure 4C). Immunostaining revealed that the tumor cells were negative for glial fibrillary acidic protein and oligodendrocyte transcription factor 2, but they were diffusely positive for CD56 and S100 protein (Figure 4D). NGS analysis revealed the presence of 7+/10−, F3T3 fusion, and amplifications of *CDK4* and *MDM2*, compatible with the diagnosis of *IDH*-wildtype GBM with F3T3 fusion.

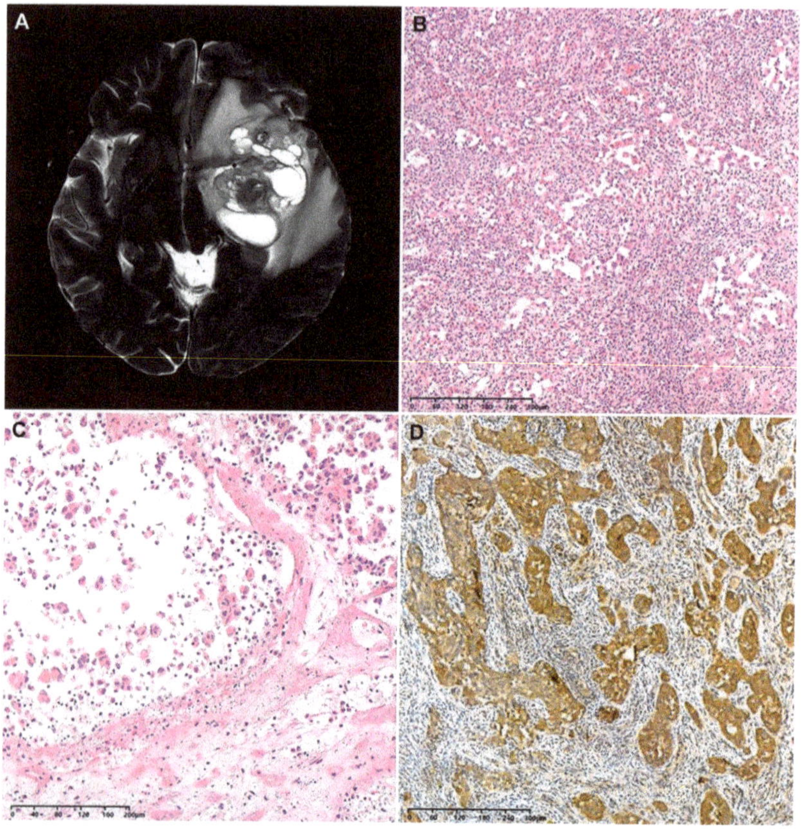


**Figure 4.** A case of F3T3-positive GBM showing unusual histology. (**A**) The brain MRI revealed a relatively well-circumscribed, multiloculated cystic mass in the left temporal lobe. (**B**) The tumor cells were arranged in an alveolar or micropapillary pattern. The peritumoral area exhibited mixed inflammatory infiltrates, including eosinophils. (**C**) In some microscopic foci, individually dispersed tumor cells displayed a cell-in-cell phenomenon. (**D**) Immunostaining revealed that the tumor cells diffusely and strongly expressed the S100 protein.

## 4. Discussion

Diffuse gliomas are categorized into three groups: adult-type diffuse gliomas, pediatric-type diffuse low-grade gliomas, and pediatric-type diffuse high-grade gliomas [8]. Extensive molecular research has been conducted in the field of brain tumors to achieve an integrated diagnosis, leading to a comprehensive understanding of the molecular landscape [14–16]. Even though there is a pressing need for novel drug research specific to brain tumors, patients with a GBM still present dismal prognoses and often develop resistance to conventional therapy [5]. Clinical trials targeting several oncogenic mutations and cellular signaling pathways frequently observed in GBMs, such as *EGFR* mutation or amplification, the phosphoinositide 3-kinase/Akt/mammalian target of the rapamycin pathway, the p53 pathway, and the retinoblastoma (RB) pathway, have failed to improve outcomes due to challenges related to blood–brain barrier penetration, drug stability, and safety concerns [17]. While *FGFR* alterations are frequently observed in GBMs, their clinical significance may be limited since therapeutically meaningful alterations, such as F3T3 fusion, are rare [18].

FGFR proteins modulate various signaling pathways through molecular alterations and have significant importance in cancer biology. Multiple FGFR inhibitors currently in clinical development demonstrate therapeutic relevance and scalability [19]. The *TACC3* gene has been identified as an important partner for *FGFR* fusions and is associated with the pathogenesis of several solid tumors [19]. The TACC3 protein possesses a coiled-coil domain at its C terminus, which promotes stability and organization of the mitotic spindle [20]. Since F3T3 fusion was first reported in GBMs and urothelial carcinomas of the urinary bladder [21,22], it has now been observed in pulmonary adenocarcinoma and squamous cell carcinoma, nasopharyngeal carcinoma, and uterine cervical squamous cell carcinoma [23–27]. The mechanism of action of this fusion protein involves delaying mitotic progression and inducing aneuploidy by increasing downstream signaling through the mitogen-activated protein kinase pathway, activating mitochondrial biogenesis and metabolism, and recruiting endogenous TACC3 from the mitotic spindle. These processes are believed to enhance malignant transformation [19,28,29]. The discovery of oncogenic gene fusions has significant clinical implications, suggesting the possibility of targeted therapy. A personalized approach utilizing *FGFR* fusion-targeted inhibitors may improve the prognosis of patients with gliomas [30]. Tabernero et al. [31] reported partial responses among patients with F3T3-positive tumors who were treated with the *FGFR* inhibitor erdafitinib. Several targeted therapies have been evaluated in clinical trials for patients with gliomas with *FGFR* fusion, including erdafitinib, ponatinib, and infigratinib [1]. These studies indicated that F3T3 fusion is a potential therapeutic target. Particularly, tumors with F3T3 fusion appear to be more sensitive to *FGFR*-targeted therapy compared to those with other *FGFR* abnormalities. This finding is meaningful because treatment options for patients with aggressive tumors with F3T3 fusion, such as a GBM and urothelial carcinoma, are limited [32].

F3T3 fusion has been detected in 3–12% of adult GBM cases [5,11,12,33]. Consistent with previous data, in this study, we detected F3T3 fusion in 5.0% (25/504) of the *IDH*-wildtype GBMs. Notably, we observed a higher frequency of F3T3 fusion in women (72.0%) compared to men (28.0%). When compared to F3T3-negative GBMs, a significant female predilection was observed (2.6 females per male; $p = 0.003$). In contrast, three previous studies reported either equal rates of F3T3 fusion in men and women [9] or slightly higher rates in men (1.47 male per female and 1.2 male per female) [5,8]. Since the previous studies showing equal gender distribution were conducted in Western populations, we considered that the difference in gender distribution may reflect racial difference. It is reasonable to assume that the relative incidence of F3T3-positive GBM may vary by ethnicity. Both domestic multi-institutional and global multinational studies are necessary to confirm or disprove this assumption and to explore the underlying factors contributing to this gender-based difference in the occurrence of F3T3 fusion.

Our study provides the first evidence that GBMs with F3T3 fusion are more frequently located in the cerebral cortical area compared to F3T3-negative tumors. Twelve of the twenty-nine F3T3-positive GBMs (41.3%) were located in the cerebral cortex or subcortical area. Consistent with our results, Di Stefano et al. [30] investigated the radiological profiles of gliomas with F3T3 fusion and found they more often affected the insula and temporal lobe. In contrast, Roux et al. [34] analyzed MRI scans of 392 GBM patients with the *IDH* wildtype and reported 63.3% (248/392) of GBMs were located in the subcortical white matter of the subventricular zones of both hemispheres. Since they did not mention the presence of absence of F3T3 fusion in their cases, we cannot determine whether they examined F3T3-positive tumors, F3T3-negative tumors, or a mixture of both types.

Bielle et al. [9] reported that F3T3-positive GBMs are histologically characterized by monomorphic, ovoid nuclei; nuclear palisading; a curvilinear capillary network; calcification; and desmoplastic vessels. We found that curvilinear capillary networks ($p < 0.001$), nuclear palisading ($p = 0.001$), and calcification ($p = 0.004$) were more frequently identified in F3T3-positive GBMs compared to the control group. In cases where these three recurrent characteristic histological findings are encountered in the diagnosis of *IDH*-wildtype GBM, pathologists should consider the possibility of an F3T3 fusion and perform a molecular test. Additionally, FGFR3 immunostaining has been recommended as a screening tool for detecting F3T3 fusion, with a sensitivity of 100% and a specificity of 92% [9]. In F3T3-positive cases, FGFR3 immunostaining consistently revealed a positive expression, although heterogeneous staining patterns were occasionally observed. Particularly, areas exhibiting the characteristic histological features exhibited a stronger staining intensity [9]. Therefore, FGFR3 immunostaining can facilitate the efficient prediction of F3T3 fusion, making the diagnostic process easier.

We found that none of the *IDH*-wildtype GBMs with F3T3 fusion had *EGFR* amplification, which is consistent with the findings of a previous study by Di Stefano et al. [11]. These observations suggest that F3T3 fusion is mutually exclusive with *IDH* mutation and *EGFR* amplification. Similarly, none of the F3T3-positive GBMs presented with *KIT* or *PDGFRA* amplifications, while 83 (17.4%) and 63 (13.2%) of the 479 F3T3-negative GBMs had *PDGFRA* and *KIT* amplifications, respectively. The absence of the amplification of these genes in our F3T3-positive GBM cases raises the possibility of mutual exclusivity between *KIT/PDGFRA* and F3T3 fusion, similar to that of *IDH* mutation and *EGFR* amplification. Consistent with our data, Mata et al. [5] also reported a lack of *PDGFRA* and *KIT* amplifications in their F3T3-positive GBMs, demonstrating significant differences from that of F3T3-negative GBMs. In contrast, according to the 2021 WHO Classification of Tumors of the CNS, *PDGFRA* alteration has been reported in 10–15% of GBM cases [14,21,35]. Nobusawa et al. [36] observed the amplifications of *PDGFRA*, *KIT*, and *KDR* in 33 (8.5%), 17 (4.4%), and 13 (3.3%) of 390 GBMs, respectively. Further studies are warranted to investigate whether the absence of *EGFR*, *PDGFRA*, and *KIT* amplifications is due to the rarity of reported GBM cases with F3T3 fusion or is due to these alterations and F3T3 fusion being mutually exclusive events. We found no significant difference in the frequency of TP53 alteration between the F3T3-positive and -negative groups. Even though we further analyzed whether the difference exists according to the type of *TP53* mutation, Fisher's exact test revealed that non-truncating mutation ($p = 0.102$), truncating mutation ($p = 0.583$), and fusion ($p = 0.372$) of *TP53* did not show any statistical significance. Further investigations using a larger cohort are necessary to compare both the mutational profiles of *TP53* and the expression patterns of p53 protein between F3T3-positive and -negative GBMs.

Unlike the previous reports by Bielle et al. [9] and Di Stefano et al. [11], this study represents a large series of 25 consecutive F3T3-positive GBM cases diagnosed in a single institution. We comprehensively analyzed a wide range of clinicopathological characteristics, including clinical features, radiological characteristics, histological findings, and molecular characteristics. Bielle et al. [9] described a multi-institutional series of 30 patients with diffuse gliomas, including low-grade (grade 2) tumors. They demonstrated recurrent morphological features and molecular characteristics. Similarly, Di Stefano et al. [11] collected 80 F3T3-positive glioma cases from multiple institutions encompassing low-grade (grades 2 and 3) gliomas. They investigated the radiological features and molecular characteristics. In this study, by combining the clinical, pathological, and imaging features, we provided strong, unique insights into understanding the implications of F3T3 fusion in *IDH*-wildtype GBMs.

In summary, our findings suggest that F3T3-positive GBM is a distinct molecular subgroup within the *IDH*-wildtype GBM group. Both clinicians and pathologists should consider this rare entity in the differential diagnosis of diffuse astrocytic glioma to make an accurate diagnosis and to ensure appropriate therapeutic management. Identifying this rare entity based on its distinctive imaging findings and histological features allows for the potential utilization of FGFR3 inhibitors or TACC3-targeting agents, opening up additional treatment options for patients. As clinical trials investigating novel therapeutic agents progress, it will be meaningful to assess their efficacy in F3T3-positive GBMs and to compare their impact on survival rates with those in the non-fusion group.

**Author Contributions:** Conceptualization, H.B. and Y.-L.S.; methodology, H.B. and B.L.; software, H.B.; validation, J.L. and H.-S.K.; formal analysis, H.B.; investigation, S.H.; resources, Y.-L.S.; data curation, B.L.; writing—original draft preparation, H.B. and Y.-L.S.; writing—review and editing, H.-S.K. and Y.-L.S.; visualization, H.B. and H.-S.K.; supervision, Y.-L.S.; funding acquisition, H.-S.K. All authors have read and agreed to the published version of the manuscript.

**Funding:** This research was supported by the National Research Foundation of Korea (NRF) grant funded by the Korean Government (MSIT) (2023R1A2C2006223).

**Institutional Review Board Statement:** This study was conducted in accordance with the Declaration of Helsinki and approved by the Institutional Review Board of the Samsung Medical Center (protocol code: 2022-09-065-002; date of approval: 27 September 2022).

**Informed Consent Statement:** Regarding the retrospective nature of this study, the Institutional Review Board waived the requirement for the investigators to obtain signed informed consent.

**Data Availability Statement:** No new data were created or analyzed in this study. Data sharing is not applicable to this article.

**Acknowledgments:** We would like to thank Deok Geun Kim for their technical support.

**Conflicts of Interest:** The authors declare no conflicts of interest.

## References

1. *WHO Classification of Tumours of the Central Nervous System*, 4th ed.; Louis, D.N.; Ohgaki, H.; Wiestler, O.D.; Cavenee, W.K. (Eds.) WHO: Lyon, France, 2016.
2. WHO Classification of Tumors Editorial Board. (Ed.) *WHO Classification of Tumours: Central Nervous System Tumours*, 5th ed.; WHO: Lyon, France, 2021.
3. Polivka, J.; Svajdler, M.; Priban, V.; Mracek, J.; Kasik, P.; Martinek, P.; Ptakova, N.; Sharif Bagheri, M.; Shetti, D.; Pesta, M.; et al. Oncogenic fusions in gliomas: An institutional experience. *Anticancer Res.* **2022**, *42*, 1933–1939. [CrossRef] [PubMed]
4. Xu, T.; Wang, H.; Huang, X.; Li, W.; Huang, Q.; Yan, Y.; Chen, J. Gene fusion in malignant glioma: An emerging target for next-generation personalized treatment. *Transl. Oncol.* **2018**, *11*, 609–618. [CrossRef] [PubMed]
5. Mata, D.A.; Benhamida, J.K.; Lin, A.L.; Vanderbilt, C.M.; Yang, S.R.; Villafania, L.B.; Ferguson, D.C.; Jonsson, P.; Miller, A.M.; Tabar, V.; et al. Genetic and epigenetic landscape of *IDH*-wildtype glioblastomas with *FGFR3-TACC3* fusions. *Acta Neuropathol. Commun.* **2020**, *8*, 186. [CrossRef]
6. Gott, H.; Uhl, E. *FGFR3-TACCs3* fusions and their clinical relevance in human glioblastoma. *Int. J. Mol. Sci.* **2022**, *23*, 8675. [CrossRef] [PubMed]
7. Lasorella, A.; Sanson, M.; Iavarone, A. *FGFR-TACC* gene fusions in human glioma. *Neuro Oncol.* **2017**, *19*, 475–483. [CrossRef] [PubMed]
8. Metais, A.; Tauziede-Espariat, A.; Garcia, J.; Appay, R.; Uro-Coste, E.; Meyronet, D.; Maurage, C.A.; Vandenbos, F.; Rigau, V.; Chiforeanu, D.C.; et al. Clinico-pathological and epigenetic heterogeneity of diffuse gliomas with *FGFR3::TACC3* fusion. *Acta Neuropathol. Commun.* **2023**, *11*, 14. [CrossRef]
9. Bielle, F.; Di Stefano, A.L.; Meyronet, D.; Picca, A.; Villa, C.; Bernier, M.; Schmitt, Y.; Giry, M.; Rousseau, A.; Figarella-Branger, D.; et al. Diffuse gliomas with *FGFR3-TACC3* fusion have characteristic histopathological and molecular features. *Brain Pathol.* **2018**, *28*, 674–683. [CrossRef] [PubMed]
10. Broggi, G.; Piombino, E.; Altieri, R.; Romano, C.; Certo, F.; Barbagallo, G.M.V.; Vigneri, P.; Condorelli, D.; Colarossi, L.; Colarossi, C.; et al. Glioblastoma, *IDH*-Wild type with *FGFR3-TACC3* fusion: When morphology may reliably predict the molecular profile of a tumor. A case report and literature review. *Front. Neurol.* **2022**, *13*, 823015. [CrossRef] [PubMed]
11. Di Stefano, A.L.; Picca, A.; Saragoussi, E.; Bielle, F.; Ducray, F.; Villa, C.; Eoli, M.; Paterra, R.; Bellu, L.; Mathon, B.; et al. Clinical, molecular, and radiomic profile of gliomas with *FGFR3-TACC3* fusions. *Neuro Oncol.* **2020**, *22*, 1614–1624. [CrossRef]
12. Wu, Z.; Lopes Abath Neto, O.; Bale, T.A.; Benhamida, J.; Mata, D.; Turakulov, R.; Abdullaev, Z.; Marker, D.; Ketchum, C.; Chung, H.J.; et al. DNA methylation analysis of glioblastomas harboring *FGFR3-TACC3* fusions identifies a methylation subclass with better patient survival. *Acta Neuropathol.* **2022**, *144*, 155–157. [CrossRef]
13. Dono, A.; El Achi, H.; Bundrant, B.E.; Goli, P.S.; Zhu, P.; Ozkizilkaya, H.I.; Esquenazi, Y.; Ballester, L.Y. Infiltrating gliomas with *FGFR* alterations: Histologic features, genetic alterations, and potential clinical implications. *Cancer Biomark.* **2023**, *36*, 117–131. [CrossRef]
14. Brennan, C.W.; Verhaak, R.G.; McKenna, A.; Campos, B.; Noushmehr, H.; Salama, S.R.; Zheng, S.; Chakravarty, D.; Sanborn, J.Z.; Berman, S.H.; et al. The somatic genomic landscape of glioblastoma. *Cell* **2013**, *155*, 462–477. [CrossRef]
15. Cancer Genome Atlas Research Network. Comprehensive genomic characterization defines human glioblastoma genes and core pathways. *Nature* **2008**, *455*, 1061–1068. [CrossRef] [PubMed]
16. Frattini, V.; Trifonov, V.; Chan, J.M.; Castano, A.; Lia, M.; Abate, F.; Keir, S.T.; Ji, A.X.; Zoppoli, P.; Niola, F.; et al. The integrated landscape of driver genomic alterations in glioblastoma. *Nat. Genet.* **2013**, *45*, 1141–1149. [CrossRef] [PubMed]

17. Le Rhun, E.; Preusser, M.; Roth, P.; Reardon, D.A.; van den Bent, M.; Wen, P.; Reifenberger, G.; Weller, M. Molecular targeted therapy of glioblastoma. *Cancer Treat Rev.* **2019**, *80*, 101896. [CrossRef] [PubMed]
18. Yang, K.; Wu, Z.; Zhang, H.; Zhang, N.; Wu, W.; Wang, Z.; Dai, Z.; Zhang, X.; Zhang, L.; Peng, Y.; et al. Glioma targeted therapy: Insight into future of molecular approaches. *Mol. Cancer* **2022**, *21*, 39. [CrossRef]
19. Costa, R.; Carneiro, B.A.; Taxter, T.; Tavora, F.A.; Kalyan, A.; Pai, S.A.; Chae, Y.K.; Giles, F.J. *FGFR3-TACC3* fusion in solid tumors: Mini review. *Oncotarget* **2016**, *7*, 55924–55938. [CrossRef]
20. Hood, F.E.; Royle, S.J. Pulling it together: The mitotic function of TACC3. *Bioarchitecture* **2011**, *1*, 105–109. [CrossRef]
21. Singh, D.; Chan, J.M.; Zoppoli, P.; Niola, F.; Sullivan, R.; Castano, A.; Liu, E.M.; Reichel, J.; Porrati, P.; Pellegatta, S.; et al. Transforming fusions of *FGFR* and *TACC* genes in human glioblastoma. *Science* **2012**, *337*, 1231–1235. [CrossRef]
22. Williams, S.V.; Hurst, C.D.; Knowles, M.A. Oncogenic *FGFR3* gene fusions in bladder cancer. *Hum. Mol. Genet.* **2013**, *22*, 795–803. [CrossRef]
23. Capelletti, M.; Dodge, M.E.; Ercan, D.; Hammerman, P.S.; Park, S.I.; Kim, J.; Sasaki, H.; Jablons, D.M.; Lipson, D.; Young, L.; et al. Identification of recurrent *FGFR3-TACC3* fusion oncogenes from lung adenocarcinoma. *Clin. Cancer Res.* **2014**, *20*, 6551–6558. [CrossRef] [PubMed]
24. Carneiro, B.A.; Elvin, J.A.; Kamath, S.D.; Ali, S.M.; Paintal, A.S.; Restrepo, A.; Berry, E.; Giles, F.J.; Johnson, M.L. *FGFR3-TACC3*: A novel gene fusion in cervical cancer. *Gynecol. Oncol. Rep.* **2015**, *13*, 53–56. [CrossRef] [PubMed]
25. Shinmura, K.; Kato, H.; Matsuura, S.; Inoue, Y.; Igarashi, H.; Nagura, K.; Nakamura, S.; Maruyama, K.; Tajima, M.; Funai, K.; et al. A novel somatic *FGFR3* mutation in primary lung cancer. *Oncol. Rep.* **2014**, *31*, 1219–1224. [CrossRef]
26. Tamura, R.; Yoshihara, K.; Saito, T.; Ishimura, R.; Martinez-Ledesma, J.E.; Xin, H.; Ishiguro, T.; Mori, Y.; Yamawaki, K.; Suda, K.; et al. Novel therapeutic strategy for cervical cancer harboring *FGFR3-TACC3* fusions. *Oncogenesis* **2018**, *7*, 4. [CrossRef] [PubMed]
27. Yuan, L.; Liu, Z.H.; Lin, Z.R.; Xu, L.H.; Zhong, Q.; Zeng, M.S. Recurrent *FGFR3-TACC3* fusion gene in nasopharyngeal carcinoma. *Cancer Biol. Ther.* **2014**, *15*, 1613–1621. [CrossRef]
28. Frattini, V.; Pagnotta, S.M.; Tala; Fan, J.J.; Russo, M.V.; Lee, S.B.; Garofano, L.; Zhang, J.; Shi, P.; Lewis, G.; et al. A metabolic function of *FGFR3-TACC3* gene fusions in cancer. *Nature* **2018**, *553*, 222–227. [CrossRef] [PubMed]
29. Sarkar, S.; Ryan, E.L.; Royle, S.J. *FGFR3-TACC3* cancer gene fusions cause mitotic defects by removal of endogenous TACC3 from the mitotic spindle. *Open Biol.* **2017**, *7*, 170080. [CrossRef] [PubMed]
30. Di Stefano, A.L.; Fucci, A.; Frattini, V.; Labussiere, M.; Mokhtari, K.; Zoppoli, P.; Marie, Y.; Bruno, A.; Boisselier, B.; Giry, M.; et al. Detection, characterization, and inhibition of *FGFR-TACC* Fusions in *IDH* wild-type glioma. *Clin. Cancer Res.* **2015**, *21*, 3307–3317. [CrossRef]
31. Tabernero, J.; Bahleda, R.; Dienstmann, R.; Infante, J.R.; Mita, A.; Italiano, A.; Calvo, E.; Moreno, V.; Adamo, B.; Gazzah, A.; et al. Phase I dose-escalation study of JNJ-42756493, an oral pan-fibroblast growth factor receptor inhibitor, in patients with advanced solid tumors. *J. Clin. Oncol.* **2015**, *33*, 3401–3408. [CrossRef]
32. Parish, A.; Schwaederle, M.; Daniels, G.; Piccioni, D.; Fanta, P.; Schwab, R.; Shimabukuro, K.; Parker, B.A.; Helsten, T.; Kurzrock, R. Fibroblast growth factor family aberrations in cancers: Clinical and molecular characteristics. *Cell Cycle* **2015**, *14*, 2121–2128. [CrossRef]
33. Asif, S.; Fatima, R.; Krc, R.; Bennett, J.; Raza, S. Comparative proteogenomic characterization of glioblastoma. *CNS Oncol.* **2019**, *8*, CNS37. [CrossRef] [PubMed]
34. Roux, A.; Roca, P.; Edjlali, M.; Sato, K.; Zanello, M.; Dezamis, E.; Gori, P.; Lion, S.; Fleury, A.; Dhermain, F.; et al. MRI atlas of *IDH* wild-type supratentorial glioblastoma: Probabilistic maps of phenotype, management, and outcomes. *Radiology* **2019**, *293*, 633–643. [CrossRef] [PubMed]
35. Verhaak, R.G.; Hoadley, K.A.; Purdom, E.; Wang, V.; Qi, Y.; Wilkerson, M.D.; Miller, C.R.; Ding, L.; Golub, T.; Mesirov, J.P.; et al. Integrated genomic analysis identifies clinically relevant subtypes of glioblastoma characterized by abnormalities in *PDGFRA*, *IDH1*, *EGFR*, and *NF1*. *Cancer Cell* **2010**, *17*, 98–110. [CrossRef] [PubMed]
36. Nobusawa, S.; Stawski, R.; Kim, Y.H.; Nakazato, Y.; Ohgaki, H. Amplification of the *PDGFRA*, *KIT* and *KDR* genes in glioblastoma: A population-based study. *Neuropathology* **2011**, *31*, 583–588. [CrossRef]

*biomedicines*

MDPI

*Article*

# Adult IDH Wild-Type Glioblastoma Ultrastructural Investigation Suggests a Possible Correlation between Morphological Biomarkers and Ki-67 Index

**Pietro Familiari [1,†], Michela Relucenti [2,*,†], Pierfrancesco Lapolla [1,3], Mauro Palmieri [1], Manila Antonelli [4], Loredana Cristiano [5], Claudio Barbaranelli [6], Myriam Catalano [7], Luca D'Angelo [1], Giuseppe Familiari [2], Antonio Santoro [1], Alessandro Frati [1,8,†] and Placido Bruzzaniti [1,9,†]**

1 Department of Human Neurosciences, Division of Neurosurgery, Policlinico Umberto I University Hospital, Sapienza University of Rome, 00185 Rome, Italy; pietro.familiari@uniroma1.it (P.F.); pierfrancesco.lapolla@nds.ox.ac.uk (P.L.); mauro.palmieri@uniroma1.it (M.P.); lucadangelo_80@hotmail.com (L.D.); antonio.santoro@uniroma1.it (A.S.); alessandro.frati@uniroma1.it (A.F.); placido.bruzzaniti@uniroma1.it (P.B.)

2 Department of Anatomy, Histology, Forensic Medicine, and Orthopedics, Sapienza University of Rome, 00185 Rome, Italy; giuseppe.familiari@uniroma1.it

3 Nuffield Department of Surgical Sciences, University of Oxford, Oxford OX3 9DU, UK

4 Department of Radiological, Oncological and Anatomo-Pathological Sciences, Sapienza University of Rome, 00185 Rome, Italy; manila.antonelli@uniroma1.it

5 Department of Life, Health and Environmental Sciences, University of L'Aquila, 67100 L'Aquila, Italy; loredana.cristiano@univaq.it

6 Department of Psychology, Sapienza University of Rome, 00185 Rome, Italy; claudio.barbaranelli@uniroma1.it

7 Department of Physiology and Pharmacology "Vittorio Erspamer", Sapienza University of Rome, 00185 Rome, Italy; myriam.catalano@uniroma1.it

8 Department of Neurosurgery, Istituto di Ricovero e Cura a Carattere Scientifico Neuromed, 86077 Pozzilli, Italy

9 Fabrizio Spaziani Hospital, 03100 Frosinone, Italy

* Correspondence: michela.relucenti@uniroma1.it

† These authors contributed equally to this study.

**Abstract:** Glioblastoma is an aggressive brain tumor with an average life expectancy between 14 and 16 months after diagnosis. The Ki-67 labeling index (LI), a measure of cellular proliferation, is emerging as a prognostic marker in GBM. In this study, we investigated the ultrastructure of glioblastoma tissue from 9 patients with the same molecular profile (adult IDH wild-type glioblastoma, wild-type ATRX, and positive for TP53 expression, GFAP expression, and EGFR overexpression) to find possible ultrastructural features to be used as biomarkers and correlated with the only parameter that differs among our samples, the Ki-67 LI. Our main results were the visualization of the anatomical basis of astrocyte-endothelial cells crosstalk; the ultrastructural in situ imaging of clusters of hyperactivated microglia cells (MsEVs); the ultrastructural in situ imaging of microglia cells storing lipid vesicles (MsLVs); the ultrastructural in situ imaging of neoplastic cells mitophagy (NCsM). The statistical analysis of our data indicated that MsEVs and MsLVs correlate with the Ki-67 LI value. We can thus assume they are good candidates to be considered morphological biomarkers correlating to Ki-67 LI. The role of NCsM instead must be further evaluated. Our study findings demonstrate that by combining ultrastructural characteristics with molecular information, we can discover biomarkers that have the potential to enhance diagnostic precision, aid in treatment decision-making, identify targets for therapy, and enable personalized treatment plans tailored to each patient. However, further research with larger sample sizes is needed to validate these findings and fully utilize the potential of ultrastructural analysis in managing glioblastoma.

**Keywords:** microglia; glioblastoma; EVs; biomarker; Ki67; lipid vesicles

## 1. Introduction

Glioblastoma is a type of brain cancer that is highly malignant; it represents up to 48.6% of the primary malignant tumor in the central nervous system. Despite the advances in neurosurgery, radiation therapy, and chemotherapy, the median survival time of glioblastoma patients still ranges from 9 to 16 months [1]. Although individual differences exist, with some patients surviving only for a few months and others for years [2]. Wild-type isocitrate dehydrogenase (IDH) glioblastoma is mostly a primary or de novo tumor that occurs in people over the age of 50 who have a brief clinical history (less than three months before diagnosis) and no previously present lower-grade glioma. [3]. IDH-wildtype refers to an associated genetic profile characteristically altered. [4]. Genetic/molecular markers related to Glioblastoma are ATRX mutation, TP53 expression, GFAP, and EGFR.

ATRX [5] is a chromatin remodeler protein, recurrently mutated in WHO grade II/III astrocytic glioma and secondary glioblastoma [6–8], where 75% of astrocytic gliomas with IDH1 and TP53 mutations also carry ATRX mutations [6]).

TP53 is one of the most commonly deregulated genes in cancer. The p53-ARF-MDM2 pathway is deregulated in 84% of glioblastoma patients and 94% of glioblastoma cell lines. Deregulated p53 pathway components have been implicated in glioblastoma cell invasion, migration, proliferation, evasion of apoptosis, and cancer cell stemness [9].

Glial fibrillary acidic protein (GFAP) is an intermediate filament protein found in gliomas that astrocytes and neural progenitor cells generate. Changes in GFAP alternative splicing are linked to glioma malignancy. The canonical isoform GFAP is downregulated in higher-grade cancers, resulting in greater GFAP isoform dominance in the network [10].

Glioblastoma frequently has epidermal growth factor receptor (EGFR) mutations. EGFR activation can be increased in a variety of ligand-dependent and ligand-independent ways. Evidence suggests that EGFR is overexpressed in most primary glioblastomas and some of the secondary glioblastomas and is characteristic of more aggressive glioblastoma phenotypes. Mutations of EGFR occur in roughly one-third of all classical tumors and often in mesenchymal, proneural, and neural glioblastomas [11].

Ki-67 is a nuclear protein; it reflects the physiological proliferative state of the cells in which this marker is expressed [12]. Different studies have investigated the prognostic potential of the Ki-67 labeling index (LI) in gliomas [13]. The Ki-67 increases with increasing WHO grade [14,15], it was demonstrated that there was no difference in median Ki-67 LI between IDH-mutated and IDH-wildtype glioblastomas ($p = 0.9$) and Ki-67 LI was not associated with survival in glioblastomas in either univariate ($p = 0.9$) or multivariate analysis including MGMT promoter methylation status and excluding IDH-mutated glioblastomas ($p = 0.2$). Ki-67 LI may be of value in the differential diagnostic setting, but it must not be over-interpreted in the clinicopathological context. In the innovative paper [16], the authors provide an initial study on the predictive value using multiple MRI characteristics for Ki-67 LI in glioma. Their results demonstrate that the degree of peritumoral edema was more severe in the group with high Ki-67 LI, which also suggested that there is a certain correlation between peritumoral edema and Ki-67 LI, both of which are related to the higher malignant degree of glioma.

Recent studies focus on the glioblastoma microenvironment, which consists of an extracellular matrix, interstitial fluid, and non-neoplastic cells compartmentalized in anatomically-distinct regions called "tumor niches", in which glioma stem cells are localized [17–19]. Multiple elements influence intercellular communication in this intricate microenvironment. Extracellular vesicles (EVs), small membrane-bound particles produced by cells, have emerged as an ultrastructural morphological biomarker in brain malignancies in recent years [20–25].

Mitophagy is an additional ultrastructural morphological biomarker that could relate to biomolecular data, the process by which dysfunctional or damaged mitochondria are selectively removed from the cell. Mitophagy is an essential mechanism for maintaining cellular homeostasis that prevents the accumulation of damaged mitochondria, which increases cell death or senescence [26]. The role of mitophagy in glioblastoma appears to be complex and context-dependent. Some studies indicate that inhibition of mitophagy in glioblastoma cells results in decreased proliferation and increased sensitivity to radiation therapy, and another study found that activation of mitophagy in glioblastoma cells resulted in decreased tumor growth and increased survival in a mouse model [27]. More research is needed to fully understand the mechanisms underlying the effects of mitophagy in glioblastoma.

The presence of cells containing a discrete amount of lipid vesicles is another ultrastructural feature that may be related to biomolecular markers. Lipid vesicles are considered dynamic organelles that store and mobilize lipids in cells. It has been proven that glioblastoma cells with high amounts of lipid droplets are more resistant to chemotherapy and radiation therapy than cells with low lipid droplets. In addition, it was observed that in glioblastoma cells, the inhibition of lipid droplet formation can sensitize these cells to chemotherapy [28]; thus, their presence promotes therapy resistance. On the other hand, lipid droplets have also been implicated in promoting tumor growth and invasion in glioblastoma, being possible that lipid droplets can be transferred from glioblastoma cells to endothelial cells, promoting angiogenesis and tumor growth [29,30].

Ultrastructural studies on glioblastoma architecture exist, but no one focuses on the possible relation between ultrastructural characteristics and molecular biomarkers. We investigated the ultrastructure of glioblastoma tissue from 9 patients with the same molecular profile (adult IDH wild-type glioblastoma, wild-type ATRX, and positive for TP53 expression, GFAP expression, and EGFR overexpression) to find possible ultrastructural features to be used as biomarkers and correlated with the only parameter that differs among our samples, the Ki-67 labeling index.

## 2. Materials and Methods

### 2.1. Patients Characteristics

Four males (age 58 ± 7.4) and five females (age 54.8 ± 6.9) were diagnosed with Glioblastoma Grade IV CNS-WHO2021, IDH wild type. A neuropathologist reviewed hematoxylin and eosin-stained formalin-fixed paraffin-embedded sections for confirmation of a diagnosis of glioblastoma WHO grade IV-wild type as per 2021 World Health Organization criteria (Supplementary Materials Table S1).

### 2.2. Histopathological Procedures and Molecular Characterization of Glioblastoma Tumors

Hematoxylin and eosin staining: FFPE 3-mm-thick sections from tissue samples were cut, deparaffination with xylol, and stained with hematoxylin and eosin.

Immunohistochemistry: an immunohistochemical panel in which GFAP, ATRX, P53, IDH1 R132H, and Ki67 were utilized. This panel is routinely performed in Policlinico Umberto 1 Hospital according to the key diagnostic genes, molecules, pathways, and/or combinations in the 2021 WHO Classification of CNS Tumors [3,4].

The procedure was performed on Leica BOND-III fully automated IHC stainer (Leica Biosystems, Nussloch, Germany). The protocol was executed: tissue sections were cut at 5 µm, dried at 70 °C for 30 min, and then dewaxed. The Bond Rx system with Epitope Retrieval Solution 1 (pH 6) for 30 min was used for antigen retrieval. Sections were incubated for 30 min with GFAP (Leica Novocastra, Milan Italy, clone GA5, mouse monoclonal, 1:400 dilution), ATRX (NBP1-83077, rabbit polyclonal, 1:1000 dilution; Novus Biologicals, Milan, Italy), p53 (NCL-L-p53-DO7, mouse polyclonal, 0.875 µg/mL; Leica Novocastra, Milan, Italy), IDH1 R132H (DIA H09, mouse monoclonal antibody, Dianova, BIOZOL, Hamburg, Germany), Ki64 (NCL-L-Ki67-MM1—mouse monoclonal, 84 mg/L, 1:200 dilution, Leica Novocastra, Milan, Italy), Ki67 Clone MIB-1 and anti-EGFR Antibody (clone RM240, 1:150; RevMAb Bioscences, Burlingame, CA, USA).

### 2.3. Light and Transmission Electron Microscopy Protocols

Samples were fixed in 2.5% glutaraldehyde in PB 0.1 M pH 7.4 for at least 48 h at 4 °C and then rinsed with PB [31]. Samples were then post-fixed with $OsO_4$ 1.33%

(Agar Scientific, Stansted, UK) for 2 h, washed with PB for 20 min, and incubated with 1% tannic acid (Sigma-Aldrich, St. Louis, MO, USA) in $dH_2O$ for 30 min. Dehydration in ascending ethanol series (30%, 70%, 95%, 100% $v/v \times 3$) was performed. Ethanol substitution with propylene oxide was performed (BDH Italia, Milan, Italy), and then samples were embedded in a mixture of 50:50 propylene oxide and epoxy resin Agar 100 (SIC, Rome, Italy) overnight at 25 °C under a chemical fume hood. Finally, samples were embedded in Agar 100 resin and put in a stove at 60 °C for 48 h. Semithin sections (1 µm thick) were collected on glass slides, stained blue by Azur II, to perform light microscopy observations in a light microscope (Carl Zeiss Axioskop-40, Zeiss, Oberkochen, Germany). Light microscopy observation of 1 µm thick epoxy resin semithin sections allows imaging at high magnification [32]. Then, ultrathin sections (80–90 nm), for transmission electron microscopy (TEM) observation, were cut using an ultramicrotome (Leica EM UC6, Vienna, Austria). Ultrathin sections were collected on 100-mesh copper grids (Assing, Rome, Italy) stained with Uranyless© solution (Electron Microscopy Sciences, Hatfield, PA, USA) and led citrate [33,34]. Imaging was performed using a transmission electron microscope set with an accelerating voltage of 60 kV (Carl Zeiss EM10, Thornwood, NY, USA). Images were acquired with a CCD digital camera (AMT CCD, Deben UK Ltd., Suffolk, UK) [35].

### 2.4. Semiquantitative Evaluation of Ultrastructural Characteristics

The semiquantitative evaluation of ultrastructural morphological characteristics was performed by examining pictures from 30 microscopic fields captured at 2000× for each sample. Microglia cells showing EVS secretion (MsEVs), Microglia cells storing lipid vesicles (MsLVs), and neoplastic cells mitophagy (NCsM) were evaluated as absent or present in each image. The overall percentage of positivity was reported for each parameter and each patient in Table S4 of Supplementary Materials. Statistical analysis of data was performed by MedCalc© software (version 20.218, MedCalc Software ltd, Ostend, Belgium) and SPSS© statistical software (version 29, IBM©-SPSS© Statistics, Milan, Italy) [36,37].

## 3. Results

### 3.1. Histopathological and Immunohistochemical Analysis

The results of the molecular characterization of glioblastoma samples are summarized in Table S2 of the Supplementary Materials section. All the patients were negative for ATRX mutation and positive for TP53 and GFAP expression and showed EGFR hyperexpression. The value of Ki-67 ranges from 10% to 40%. The histopathological examination showed high cellularity, the presence of pleomorphic nuclei, and areas of necrosis and diffuse neovascularization (Table S3 Supplementary Materials). The images reported in Figures 1 and 2 depict the characteristics of glioblastoma IDH wild type. Figure 1A shows a highly cellular tumor in which microvascular proliferation is evident in the center of the microscopic field. The image in Figure 1B shows the immunohistochemical staining results for IDHR232H, a genetic marker commonly used to identify glioblastomas with IDH mutations. The neoplastic cells are negative for this marker, indicating that they do not carry this specific mutation. Figure 1C,D provide the control, representing a glioblastoma IDH mutant with the same staining as Figure 1A,B.

Figure 2A illustrates the immunohistochemical staining for EGFR. As it is visible from the orange staining, the neoplastic cells are positive.

The images in Figure 2B,C show the results of the immunohistochemical staining for Ki67 in two different samples: B shows a sample of our patient group, and many orange-stained cells are visible, indicating high proliferative activity. In Figure 2C, a control sample with few and scattered, orange-stained cells showed low proliferative activity.

### 3.2. Light Microscopy Analysis of Glioblastoma Tissue

Figure 3 shows our samples' comprehensive glioblastoma architecture overview. The numerous visible blood vessel sections show a high level of vascularization. In particular, two transverse and round-shaped sections, two "Y" shaped sections indicative of branching vessels, and one longitudinal section are present. The histological variability of glioblastoma is clear from the left and right parts of the image, with a range of neoplastic cells displayed. The right side of the image contains neoplastic cells with large oval nuclei, dispersed chromatin, one prominent nucleolus, and pale cytoplasm. On the left side of the image, perivascular neoplastic cells are visible, arranged around small blood vessels with darker cytoplasm, irregularly shaped nuclei, and less dispersed chromatin. A neuron and its axon interspersed among fibrillar material and neoplastic cells were also present, as well as small and large intercellular spaces.

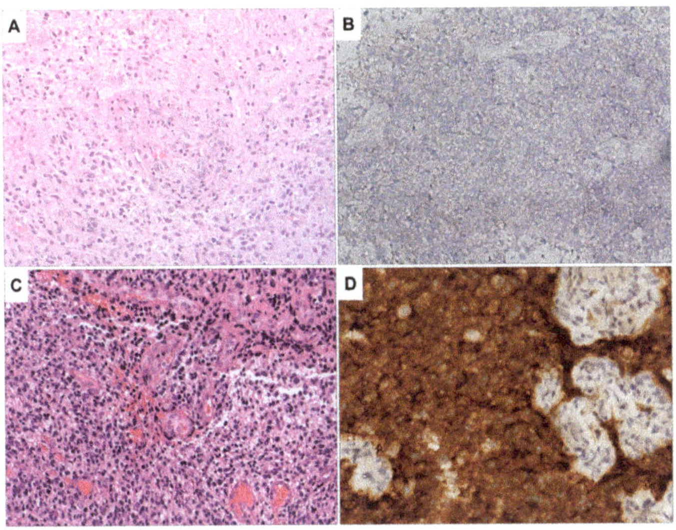


**Figure 1.** (**A**) Glioblastoma, *IDH* wild type. H&E staining, the tumor is highly cellular; microvascular proliferation is evident in the center of the image (100×). (**B**) Immunohistochemistry for $IDH^{R232H}$ is negative in neoplastic cells (200×). No orange staining is visible. (**C**) As a control sample, a glioblastoma, IDH mutant is shown. H&E staining, this sample shows pleomorphic and hyperchromatic neoplastic cells. Microvascular proliferation is present (100×). (**D**) As a control sample for positivity for $IDH^{R232H}$, a glioblastoma IDH mutant is presented. This sample is heavily stained in orange, indicating that $IDH^{R232H}$ is positive in neoplastic cells (200×).

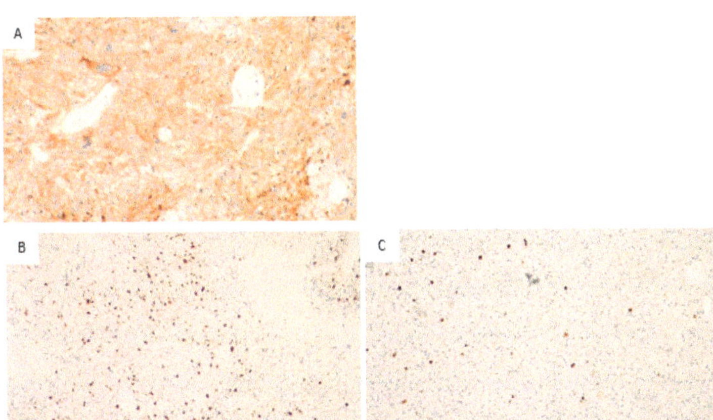


**Figure 2.** Glioblastoma, IDH wild type. (**A**) Immunohistochemical staining for EGFR, the neoplastic

cells positive for EGFR are stained in orange (200×), so the sample is positive for EGFR. (**B**) Immunohistochemical staining for Ki67, numerous cells positive for Ki67 (stained in orange) are visible. Thus, a high proliferative activity is present (200×). (**C**) In the control sample, immunohistochemical staining for Ki67, in this sample is evident the presence of a few proliferative cells with orange staining. This indicates a low proliferation index.

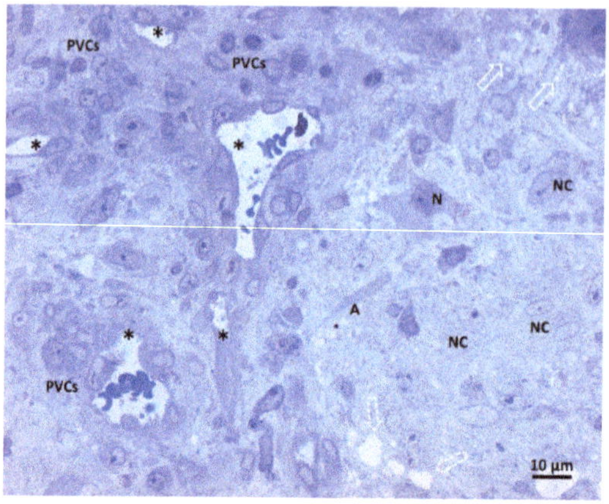


**Figure 3.** Light microscopy image of methylene blue stained semithin section. Magnification 400×, bar 10 microns. This picture shows a panoramic view of glioblastoma tissue. Neoplastic cells of glial origin (NC), vessels sections (*). A neuron (N) and its axon (A) are interspersed among fibrillar material and neoplastic cells. Small intercellular spaces (empty arrow), more extensive areas (open dotted arrow), and perivascular neoplastic cells (PVCs).

Under higher magnification, the small glioblastoma vessels appeared irregular, and the triangular-shaped section in Figure 4A is a sign of vessel branching. They showed endothelial cell proliferation with many hyperplastic endothelial cells bulging into the lumen with oval or round shapes. The perivascular area contains mitotic figures (Figure 4A,B), sometimes close to the endothelium. Neoplastic cells moved towards the vessel and intravasation into its lumen (dotted lines). The above images visually represent the so-called perivascular niche, the microenvironment in which the molecular cross-talk between endothelial cells, astrocytes, and glioblastoma stem cells fuels the production of cells with invasive properties [38–40].

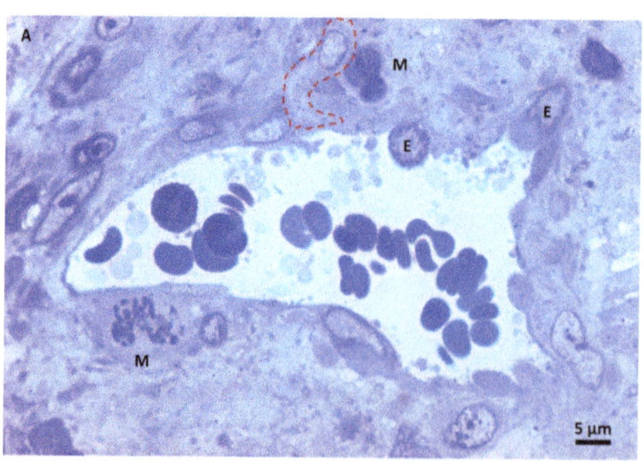


**Figure 4.** *Cont.*

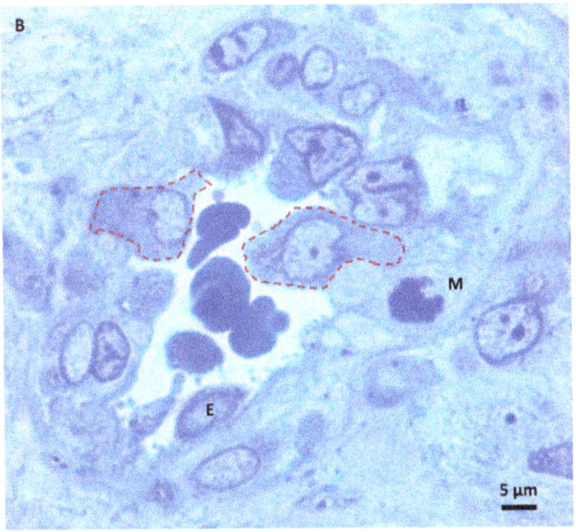


**Figure 4.** (**A**,**B**) Light microscopy images of methylene blue stained semithin section. Magnification 1000×, bar 5 microns. Hyperplastic endothelial cells (E). Mitotic figures (M) are sometimes very close to the endothelium—intravasating neoplastic cells (dotted lines).

### 3.3. *Transmission Electron Microscopy Analysis of Glioblastoma Tissue*

#### 3.3.1. Small Vessels

Transmission electron microscopy images allow more detailed visualization of the wall of glioblastoma small vessels (Figure 5A–C). They were lined by hyperplastic endothelial cells lying on a markedly thickened basement membrane. The luminal surface of many endothelial cells presents various bulbous cytoplasmic projections with wide stalks (cauliflower-like) and thin and long arms, which was visible projecting into the vessel lumen (Figure 5B). According to [41], the increased villous projections from the luminal surface of endothelial cells of glioblastoma vessels increase the surface area that actively resorbs the intravascular fluid.

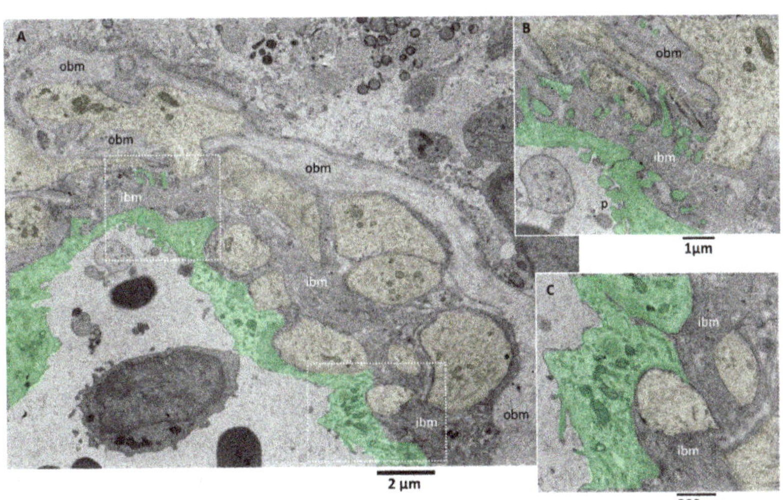


**Figure 5.** TEM, Glioblastoma small vessel wall. (**A**) Panoramic view of the vessel wall; note that a lymphocyte is visible in the lumen's endothelial cells (green). The vessel basement membrane

appears thickened, divided into two layers (inner basement membrane, ibm; outer basement membrane, obm), and penetrated by foot processes of astrocytes (yellow). (**B**) Endothelial cells (green) present cauliflower-like projections on their luminal surface (p), whereas the basal surface expresses long and deep finger-like projections (arrowheads) that penetrate the basement membrane to reach the astrocytes foot processes (yellow). (**C**) Astrocytes' foot processes (yellow) appear separated by the endothelial cells (green) by an excellent layer of ibm fibers.

The tight junctions between endothelial cells are well preserved. Interestingly, long projections develop from the abluminal aspect of endothelial cells, crossing the basement membrane's total thickness. They head towards astrocytes' footplates, contacting and wrapping around them. The basement membrane appeared formed by two layers, a more compact and darker one near the endothelium (inner basement membrane) and a looser and paler one distant from the endothelium (outer basement membrane). The basement membrane is discontinuous, and through its fenestrations, astrocytes' foot plates arrive very near the basal aspect of the endothelial cells (Figure 5A–C). A lymphocyte is visible in the vessel lumen.

Pictures captured at higher magnifications show the different layers of the basement membrane (Figure 6A,B). Astrocytes' foot plates were separated by the basal aspect of the endothelial cells through a thin layer of ibm (Figure 6A). The ibm is formed by thicker filaments circularly arranged and surrounded by a dark amorphous matrix. The ibm consisted of thinner and loosely organized filaments, predominantly oriented longitudinal, following the vessel length (Figure 6B).

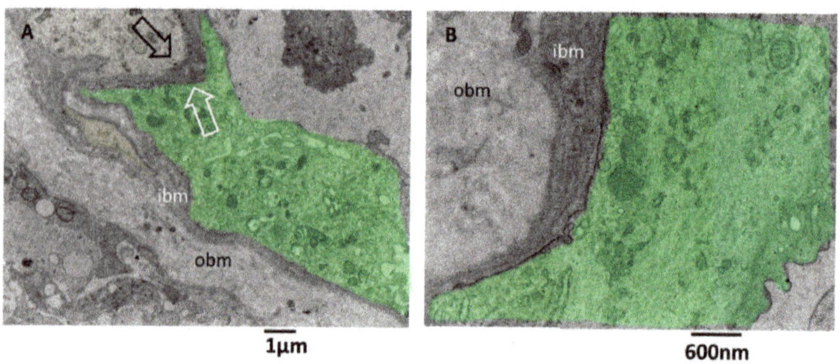


**Figure 6.** TEM, Glioblastoma small vessel wall, the basement membrane. (**A**) The endothelial cell (green) is separated by a skinny layer (black and white arrows) of ibm from the astrocyte foot process (yellow). (**B**) Endothelial cell (green) lies on a two-layered basement membrane. The outer layer (obm) filaments appear loosely arranged and longitudinally oriented. In the inner layer (ibm), filaments appear compact and circularly oriented.

There is still a debate in the literature on the alteration of vascular basement membrane in glioblastoma; our images confirm the findings of [42] and add further ultrastructural details of the basement membrane in alteration in GMB. The vascular basement membrane appears structured in two layers, an innermost and an outermost, with different filament arrangements. From our images, the basement membrane fenestrations across which the astrocyte foot processes arrive very near the inferior aspect of endothelial cells; this proximity is needed to realize the cross-talk with endothelial cells. In addition, the close contact between endothelial cell processes and astrocytes is presented. These characteristics allow the passage of lymphocytes, as demonstrated by [42], and as is visible in our images.

3.3.2. Neoplastic Cells

Glioblastoma is characterized by significant variability in its neoplastic cell population; our samples mainly consist of round astroglia-derived cells with different and altered structural characteristics. We observed neoplastic cells with an oval-shaped nucleus occupying the cell center, with one or two evident nucleoli. The cytoplasm contains a variable amount of endoplasmic reticulum and mitochondria (Figure 7A). At higher magnification (Figure 7B) is apparent that mitochondria were small and had a dark matrix, and the endoplasmic reticulum is sometimes enlarged. Many cells present a large cell body with a well-formed Golgi apparatus, sometimes more than one (Figure 7C), and variable content of glial filaments (type of intermediate filaments, glial filaments have been morphologically identified on TEM images according to [43]). In particular, the organelles' content decreases as the filaments amount increases. Glial filaments are interspersed in the neoplastic cell cytoplasm or can group in bundles (Figure 7D); vacuoles and enlarged endoplasmic reticulum were also present.

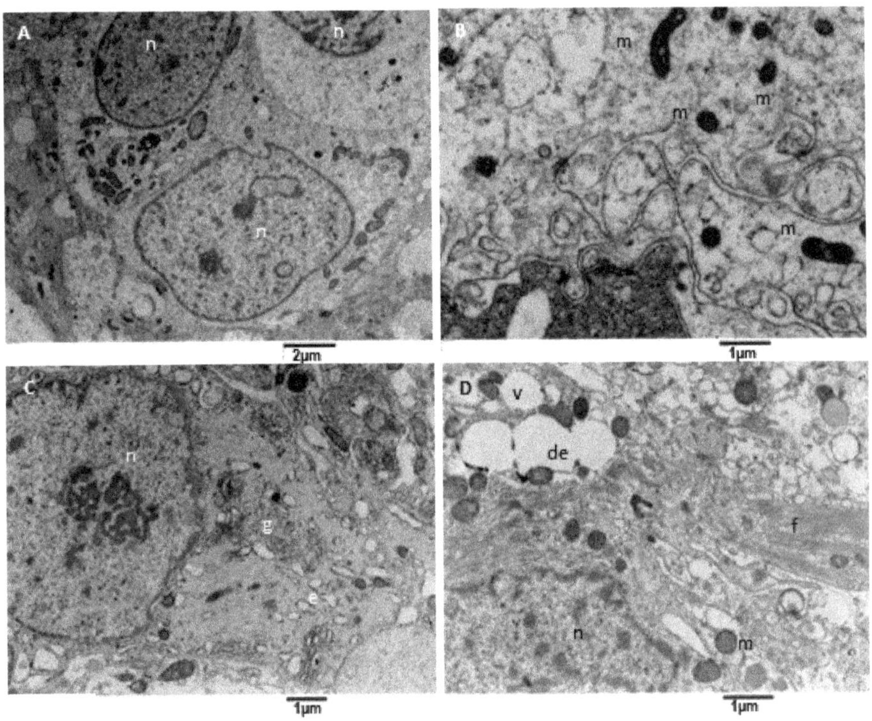


**Figure 7.** TEM, Neoplastic Cells. (**A**) The figure shows neoplastic cells (n) with a centrally located oval nucleus and cytoplasm rich in organelles, such as mitochondria. (**B**) Mitochondria (m) appear small and with a dense and dark matrix. (**C**) Some neoplastic cells show a well-developed Golgi apparatus (g) and moderately developed endoplasmic reticulum (e). (**D**) Some other cells contain glial filaments in the cytoplasm (f), mitochondria (m), vacuoles (v), and dilated endoplasmic reticulum (de).

### 3.3.3. Neoplastic Cells Mitophagy (NCsM)

Mitochondrial fission and mitophagy were observed in our samples, but with a variable degree since mitophagy was almost absent in one sample. This result is consistent with what happens in glioblastoma cells, in which an impairment of mitochondrial activity and autophagy suppression occurs; this drives stemness and invasion and indicates poor prognosis [44–46]. Our observation showed that neoplastic cells with active mitophagy had numerous cytoplasm organelles and sometimes glial filaments. Often, duplicated Golgi organs involved in active vesicle trafficking were present, the endoplasmic reticulum appeared to have a low/mild level of dilatation, mitochondria were actively dividing, and numerous mitophagy figures at different stages were also observed (Figure 8A–D).

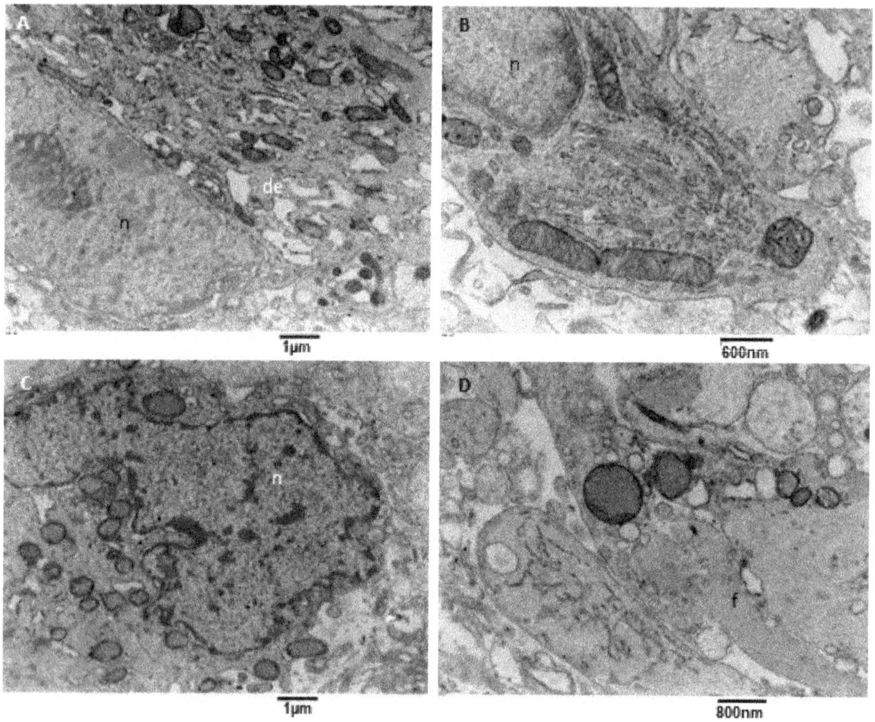


**Figure 8.** TEM, mitophagy in neoplastic cells. (**A**) A neoplastic cell is depicted. It is visible in the nucleus (n) and a prominent nucleolus. The cytoplasm contains dilated endoplasmic reticulum (de), and active mitochondrial fission (orange) can be observed. (**B**) This neoplastic cell shows a mitochondrion (orange) in the final stages of fission. A duplicated Golgi organ (yellow) is present. (**C**) A neoplastic cell with multiple mitochondria in advanced stages of mitophagy (violet) is represented. (**D**) At high magnification, two mitophagy figures (violet) are presented; note the cristae remnants inside the left mitophagy mitochondrion. Glial filaments (f) are also present in the cell cytoplasm.

3.3.4. Microglia Cells

Microglia cells are considered the tissue-resident macrophages of the central nervous system; they help maintain brain homeostasis, acting as the first line of defence against pathogens or threats [47–49]. Microglia are motile and dynamic cells whose constant expansion and retraction of processes allow them to monitor the surrounding parenchyma. Microglia are essential in neuroinflammation; they are the principal source of inflammatory mediators in the nervous system (cytokines, ROS, and soluble lipids, as reported in [50]). We observed microglia cells characterized by a soma with an oval/elongated nucleus, a small ring of cytoplasm surrounding the heart, and multiple long and tortuous cellular processes (hyper-ramified/hypertrophic according to [51], Figure 9A,B). Their endoplasmic reticulum appeared moderately dilated (a sign of moderate stress), and often mitochondria in tethering were observed (Figure 9D). Their mitochondria showed mild signs of stress (cristae disorganization) and a dark matrix. The abnormal mitochondrial ultrastructure could underlie the shift in energy-generating pathways in glioblastoma for cell survival and progression. Mitochondria entrapped in a network of intermediate filaments were also observed (Figure 9C). A close examination evidenced that intermediate filaments in a part bind to the mitochondria, in part, are arranged concentrically in a confinement scenario, as described by [52]. Rare dark microglia cells were observed, according to [53,54].

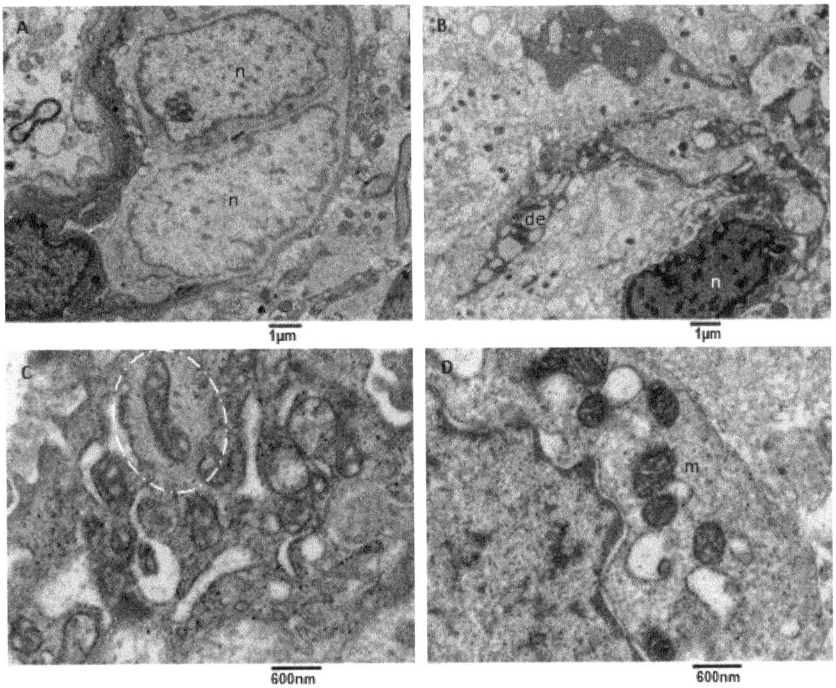


**Figure 9.** TEM, Microglia Cells. (**A**) A microglia cell (brown) surrounds two neoplastic cells (n) with two long processes. Bar 1 micron, (**B**) A microglia cell (brown) presenting multiple long processes in which a dilated endoplasmic reticulum (de) is visible. (**C**) Detail of a microglia cell cytoplasm showing mitochondria entrapped in a network of filaments (oval dotted line). (**D**) Detail of a microglia cell cytoplasm showing mitochondria (m) tethering with the dilated endoplasmic reticulum (de).

#### 3.3.5. Microglia Cells Secreting sEVs (MsEVs)

In literature, microglia cells are involved in glioblastoma [18], and it is also reported that sometimes they arrange in clusters [55] whose formation is due to the activity of glioblastoma cells. Briefly, glioblastoma cells initially secrete low levels of the chemokine CCL2 to attract microglia cells chemotactically, and then also microglia cells start to release EVs containing CCL2 [56], attracting even more microglial cells into the tumor, promoting tumor progression and development, as stated in [55,57,58]. The value of our ultrastructural observations is that our images show ex vivo and in situ the EVs secretion by microglia cells. Usually, this activity is demonstrated in vitro by analyzing the culture medium with methods like Nanosight tracking analysis. Our images show the presence of clusters of ramified microglia cells connected by tight junctions (Figure 10A) and actively secreting massive amounts of EVs with diameters less than 200 nm. Those cells showed dilated endoplasmic reticulum and mitochondria with moderate stress (Figure 10B,C).

#### 3.3.6. Microglia Cells Storing Lipid Vesicles (MsLVs)

Lipid droplet formation is undetectable in low-grade gliomas. The studies [59,60] reveal that their presence matches with progression to advanced-stage glioblastomas, which makes them a prospective biomarker and metabolic target in glioblastoma. We observed microglia cells storing lipid vesicles in the cytoplasm; sometimes, the lipid vesicle was single and large. More often, multiple lipid vesicles in a different emptying stage were present (Figure 11A). Lipid vesicles were in contact (tethering) with the nucleus but more often with the mitochondria (Figure 11B,C). The endoplasmic reticulum showed moderate dilatation; mitochondria appeared with mild signs of cristae disorganization and dark matrix (Figure 11D). This kind of microglia cells didn't show EVs secretion activity.

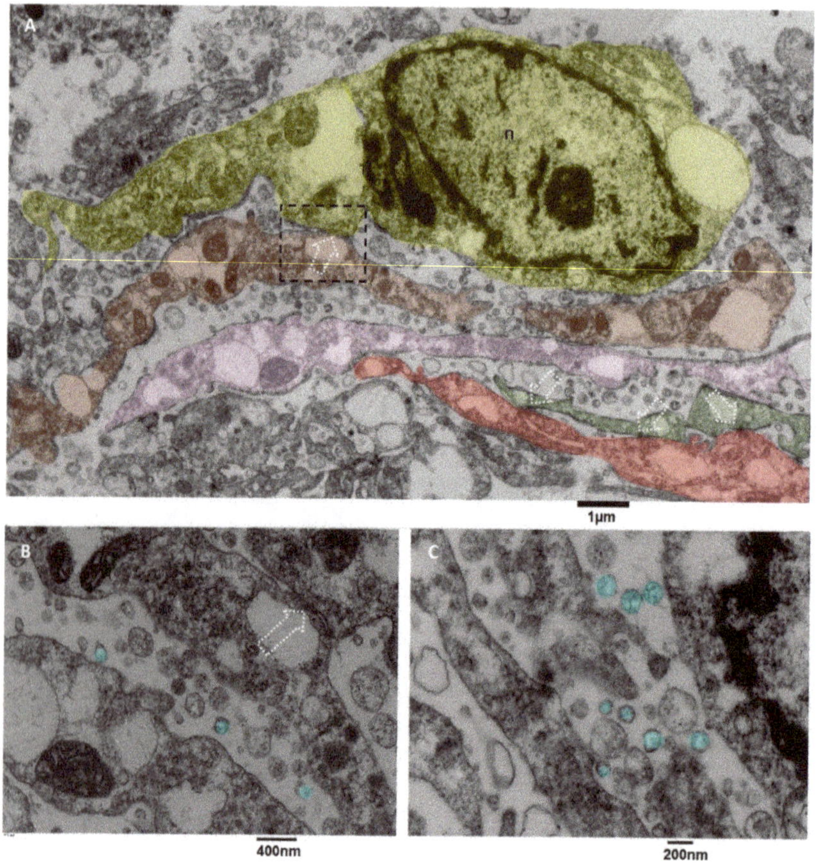


**Figure 10.** TEM, Microglia Cells. (**A**) panoramic view of at least five different microglia cells, represented in different colors (yellow, orange, pink, green and red), the dotted square representing a contact zone between two different microglia cells are enlarged in B, dotted arrows indicate the tight junctions between different cells, the endoplasmic reticulum appears dilated, sometimes severely. (**B**) The tight junction that connects two microglia cells is indicated with a dotted arrow; EVs are colored in light blue. Mitochondria show moderate signs of stress, with dilated cristae and a dark matrix; the endoplasmic reticulum is dilated. (**C**) the massive EVs secretion (light blue) is represented, and small vesicles are also visible in the cell's cytoplasm.

#### 3.3.7. Statistical Analysis

The percentage of positivity for microglia cells secreting sEVs (MsEVs), microglia cells storing lipid vesicles (MsLVs), and neoplastic cells with mitophagy (NCsM) were reported for each patient in Table S4 of the Supplementary Materials section and in Figure 12. Mean $\pm$ SD for MsEVs was 17.11 $\pm$ 6.17 with 95%CI = 12.36–21.85; Mean $\pm$ SD for MsLVs was 4.11 $\pm$ 5.15 with 95% CI = 0.14–8.07; Mean $\pm$ SD for neoplastic cells mitophagy was 16.22 $\pm$ 2.30 with 95% CI = 12.14–20.30.

Being Ki-67, the only molecular parameter differing among our patients, we aimed to verify the existence of any correlation between this parameter and the ultrastructural ones. So, we plotted Ki-67 values (see Table S4 of Supplementary Materials section) together with the values of MsEVs (blue line), MsLVs (orange line), and NCsM (gray line). and the plot is shown in Figure 13.

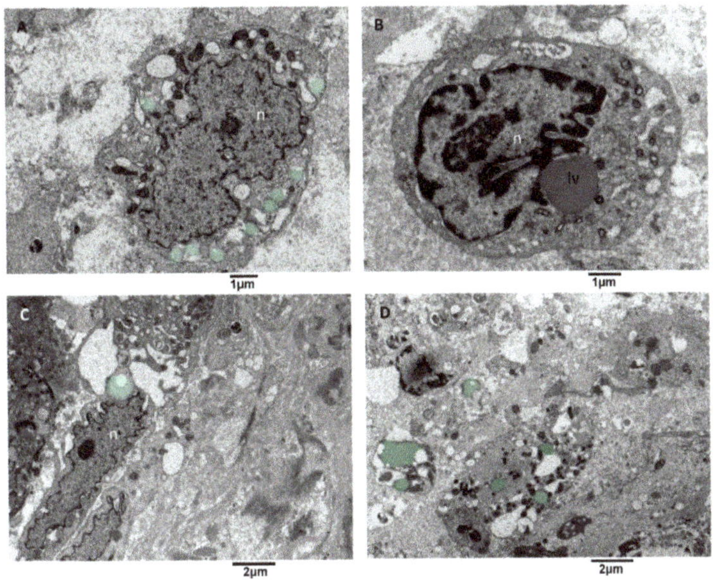


**Figure 11.** TEM, Microglia cells with lipid vesicles. (**A**) in this cell, multiple small lipid vesicles (green) surround the nucleus (n), and the endoplasmic reticulum appears enlarged. (**B**) this cell presents a single large lipid vesicle (lv) in tethering with the nucleus (n). (**C**) microglia cell with an elongated nucleus and severely dilated endoplasmic reticulum is shown; note the lipid vesicles in green near the nucleus (n). (**D**) This panoramic view shows the cytoplasm of different cells containing lipid vesicles (green).

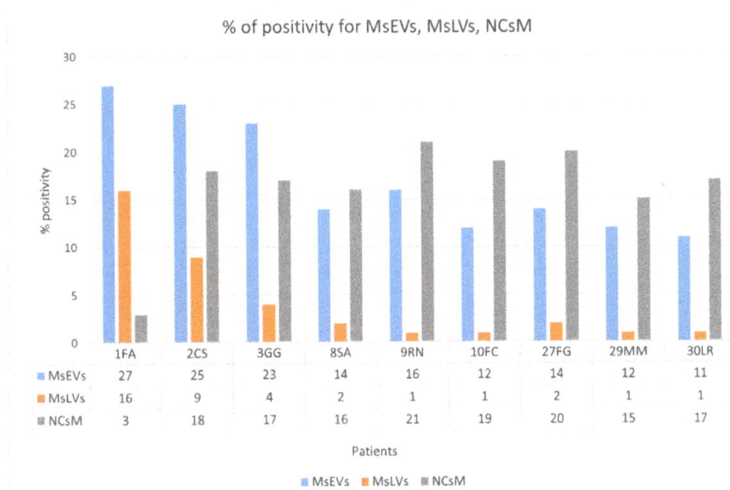


**Figure 12.** Percent of positivity for MsEVs (blue), MsLVs (orange), and NCsM (gray) in each patient. Note that high values of MsLVs are present in the same patient with high MsEVs. Values of NCsM appear instead almost constant among patients.

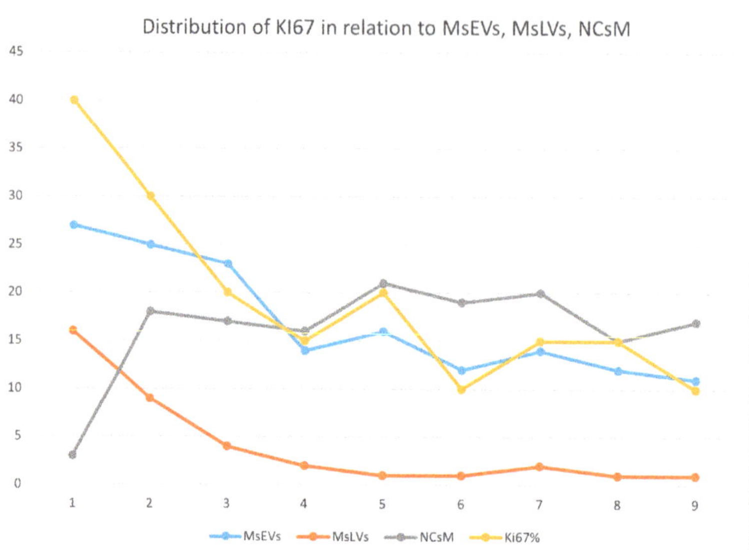


**Figure 13.** This figure illustrates the distribution of Ki-67 (yellow), MsEVs (blue line), MsLVs (orange line), and NCsM (gray line). High values of Ki-67 are related to high values for both MsEVs and MsLVs. Values of NCsM appear not related to Ki-67, having its value distribution line a different trend than that of Ki-67.

The relationship between KI-67 and the other parameters was investigated using correlation coefficients and graphically represented using the scatterplot. Being the sample particularly small, to have a more solid estimate of the correlation, bootstrap estimates were obtained using SPSS facilities, setting the number of replicated bootstrap samples to 10,000. Bootstrapping is particularly useful relative to smaller samples, especially in the presence of outliers [61]. The results of correlation analyses are precise (Figure 14A–C). MsEVs are highly correlated with Ki67, r = 0.918, $p < 0.0001$ (the 95% confidence interval limits for the bootstrap estimates are 0.776; 0.996). Figure 14A confirms the almost perfect linear relation between the two variables. MsLVs are also strongly correlated with Ki67, r = 0.951, $p < 0.0001$ (the 95% confidence interval limits for the bootstrap estimates are 0.290; 0.994). As shown in Figure 14B, the linearity of this relation is self-evident, albeit the presence of a subject whose score on Ki67 is somewhat higher concerning what expect from his very low score on MsLVs: this is probably the cause of a much wider confidence interval. Finally, although the negative correlation of NCsM is solid and significant, being r = −0.704, $p < 0.05$, the 95% confidence interval limits for the bootstrap estimates are −0.958; 0.468, being 0 a possible value with 95% of confidence, we cannot reject the null hypothesis that this correlation is 0. Scrutiny of the scatterplot (Figure 14C) shows that the negative correlation is almost uniquely due to a single subject with very high scores in Ki67 and very low scores on mitophagy: the other subjects are scattered without evidencing a clear pattern of association between the two variables.

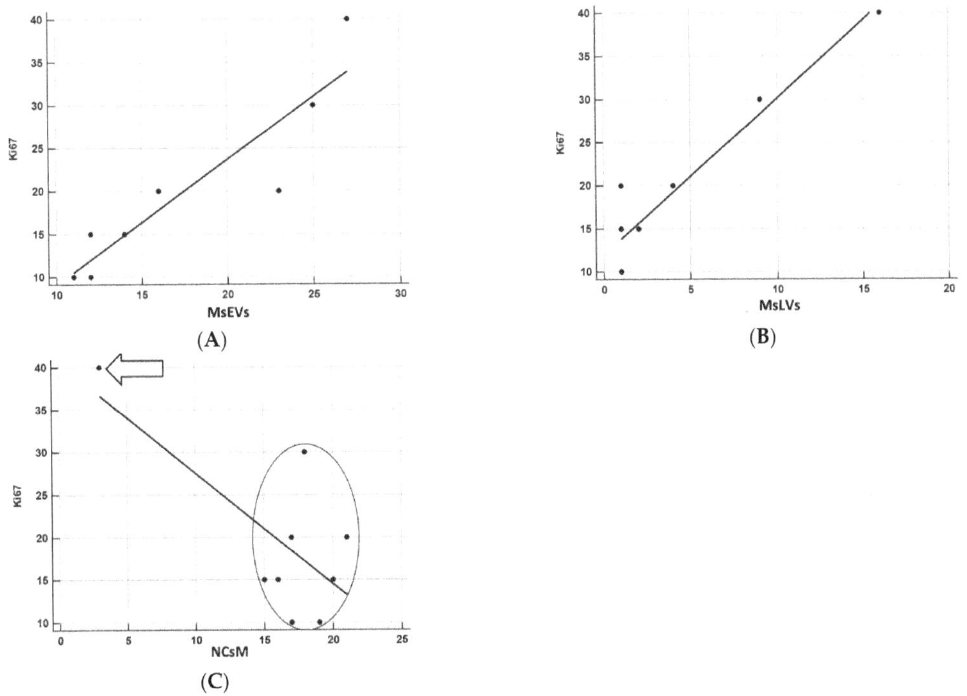


**Figure 14.** (**A**) Correlation analysis between Ki67 and MsEVs. r = 0.918, $p < 0.0001$; this result demonstrates the strong association between these two parameters, as it also appears clear by the regression line orientation the amount of MsEVs increases with increasing Ki-67. (**B**) Correlation analysis between Ki67 and MsLVs. r = 0.951, $p < 0.0001$; even in this case demonstrates the strong association between these two parameters, the regression line indicates that as the amount of MsLVs increase, the same is for Ki-67. (**C**) Correlation analysis between Ki67 and NCsM. r = −0.704, $p < 0.05$. Arrow: a single subject with very high scores in KI67 and very low scores in mitophagy. In the circle, all other subjects were comprised. A clear pattern of association between the two variables is not present in this case.

## 4. Discussion

The classification of CNS tumors has traditionally relied on histological observations and tests performed on tissue samples. However, the most recent version of the WHO Classification of Tumors of the Central Nervous System introduced a new approach. It now includes a wide range of molecular biomarkers that have proven clinically helpful and are crucial in accurately classifying CNS neoplasms.

In this new nomenclature, the presence in glioblastoma of a mutation in the enzyme IDH is introduced, and the once-named glioblastoma grade IV is now defined as Glioblastoma IDH-wild type [62–66].

Additional genetic and molecular indicators have recently been discovered to connect with glioblastoma. These include ATRX mutation [67,68], TP53 expression [69–71], GFAP [72,73], and EGFR [74–77]. According to the Supplementary Data in Table S2, our samples exhibit the wild type for IDH and ATRX, but they are positive for TP53, GFAP, and EGFR. The only molecular aspect that varies among our samples is the KI-67 LI, which represents a measure of cell proliferation.

KI-67 is evaluated using the monoclonal antibody Ki-67 (that stains nuclei reacting with nuclear proteins expressed in the cell cycle's GI, S, G2, and M phases), and the nuclear positivity found for Ki-67 is expressed in percentage, KI-67 labeling index

(LI). Ki-67 expression ranges from 15% to 40% in most glioblastomas, and high Ki-67 expression is related to lesion volume, an increased risk of recurrence, and a poor prognosis. [78]. According to our previous work [79], there is a correlation between the percentage staining of Ki-67 and overall survival in patients with IDH-WT glioblastoma; specifically, the percentage of Ki-67 staining >20% predicts poorer progression-free survival. However, there is still a debate on this in the literature and existing papers that report different results [15].

Our work aimed to perform an ultrastructural analysis of glioblastoma specimens with the same histological and molecular profile but different Ki-67 to find possible morphological characteristics related to Ki-67, but not to express any relationship with the overall survival.

We summarize and discuss the results of our study in five points:

1. Visualization of the anatomical basis of astrocyte-endothelial cells' crosstalk

During our observations, we identified micro vessels whose basement membrane (consisting of two layers) presented small openings called fenestrations. We observed that these fenestrations are crossed by astrocytes' foot processes, facilitating their proximity to the neighboring endothelial cells. We also observed that the endothelial cells extended through these fenestrations long processes towards the astrocytes. This feature represents the anatomical basis of their molecular crosstalk that, as demonstrated in a 3D coculture spheroids, significantly enhances tumor cells' stemness marker expression [80]. The signaling pathway between astrocytes and the endothelial cell barrier in mice in an orthotopic model of human glioblastoma [81] demonstrated the protection of cancer tissue from chemotherapeutic agents given by these cells. Moreover, altering vessels' basement membranes provides pathways for tumor growth and invasion and enables the entrance of immune cells that activate the astrocytes and promote neuroinflammation [82]. The alteration in the basement membrane could be exploited to develop new therapies.

2. Ultrastructural in situ imaging of clusters of hyperactivated microglia cells (MsEVs)

Our study of ex vivo samples demonstrates in situ the presence of clusters of hyperactivated microglia cells, contacting each other by tight junctions and secreting massive amounts of EVs.

In the context of glioblastoma, EVs released by tumor cells have been under study for the last ten years; neoplastic cells release EVs that contain high levels of the protein EG-FRvIII, which promotes tumor growth and invasion [83], EVs released by glioblastoma cells can induce angiogenesis, the process by which new blood vessels are formed, which is essential for tumor growth and progression [84,85]. EVs have also been implicated in therapy resistance in glioblastoma; EVs released by glioblastoma cells can transfer resistance to the chemotherapy drug temozolomide to recipient cells) and can induce resistance to radiation therapy [86]. Instead, EVs secreted by microglia are only recently investigated. They have been shown to contain a variety of inflammatory mediators, including cytokines and chemokines, which can be transferred to neighboring cells and modulate their activity. Some studies report that EVs from microglia cells attenuate glioblastoma proliferation. Still, the study of [87] clarified that only the spinal cord-derived EVs microglia under LPS stimulation significantly attenuated glioma proliferation, whether the cortical microglia were accounted for exerted anti-inflammatory effect. Neurogenesis/tumorigenesis properties are consistent with the relationship we found in our study. Our previous findings [53,54] show that microglia-secreted EVs influence the glioma microglia phenotype and modify tumor growth via transporting miR-124 to astrocytes and tumor cells. This miRNA modifies tumor metabolism by acting on glutamate homeostasis. More studies are needed to fully understand the genesis, the release mechanisms, and the effects of EVs in glioblastoma to develop targeted therapies that can modulate their formation and release.

3. Ultrastructural in situ imaging of microglia cells storing lipid vesicles (MsLVs)

We observed microglia cells storing lipid vesicles, active organelles that provide metabolic fuel, inflammatory intermediates, and signaling mediators to cells. The value of our images consists in the representation of microglia cells storing lipid vesicles in situ, in their in vivo tumor microenvironment, in which microglia cells express feels conditions and express a behavior (thus have a morphology) that it is not possible to recreate in vitro. Lipid vesicle biogenesis increases in neoplastic and inflammatory conditions. In literature is reported that microglia cells can accumulate lipid droplets in their cytoplasm, assuming a distinct phenotype called lipid droplet accumulating microglia [87]. The literature [88] reports evidence of a defective phagocytic phenotype in microglia cells storing lipid vesicles, in that enhanced phagocytic up-take of lipid exacerbates the lipid droplet accumulation burden and promotes chronic and self-sustained microglial activation [89]. In turn, sustained inflammation further pushes microglia into a hyperactivity state, creating feedback that exacerbates neuroinflammation and damages the integrity of the blood-brain barrier [90]. As our images demonstrate, the above-described overall scenario perfectly fits with our results. It reinforces our hypothesis of a relationship between the presence of microglia cells storing lipid vesicles and high levels of Ki-67LI.

4. Ultrastructural in situ imaging of neoplastic cells mitophagy (NCsM)

Our pictures evidenced the presence of mitophagy in neoplastic cells. In literature, this process is frequently triggered by oxidative stress or an increased need for energy, as it occurs during cancer growth and invasion [91]. Some research [92,93] investigated the involvement of mitophagy in gliomas. Mitophagy inhibition was shown to partially reverse cannabidiol-induced glioma cell death, indicating a favorable role for mitophagy in anti-tumor therapy [94]. Two investigations suggest that inducing mitophagy via FOXO3a protects gliomas against temozolomide-induced cytotoxicity, demonstrating that mitophagy has a double-edged sword effect on gliomas [95–97]. Thus, whether mitophagy is related to glioblastoma prognosis and the likely involvement of mitophagy-related genes need to be determined.

5. Statistical analysis of ultrastructural data

The statistical analysis of our semiquantitative data of the morphological biomarker (MsEVs and MsLVs, and NCsM) expression and their relationship with Ki67-LI was performed by bootstrapping, a procedure particularly useful to manage small samples [57] as it was in our case. The results indicate that the MsEVs and MsLVs strongly correlate with the Ki-67 LI value. We can thus assume they are good candidates to be considered morphological biomarkers correlating to Ki-67 LI (however, we are collecting specimens to increase the sample size).

The role of NCsM instead must be further evaluated, being absent a clear pattern of association. We underline the fact that observations on a larger sample size are needed, but based on the analysis we conducted, we can reasonably candidate MsEVs and MsLVs as morphological biomarkers related to the Ki67 index in adult glioblastoma IDH1.

The presence of such a correlation may be used to improve the diagnostic procedure by combining ultrastructural features with molecular information to provide a single, "integrated" diagnosis (in view of personalized medicine). It may be helpful to assess the aggressiveness and prognosis of the tumor (as a guide to treatment decisions and patient management). One further use of such biomarkers is that of potential therapeutic targets (revealing specific cellular structures and molecular pathways that are unique to the single individual tumor). Finally, their utilization may help develop personalized treatment plans; studying the ultrastructural characteristics of the tumor tissue from every single patient will allow the development of tailored therapies for the specific characteristics of the single and specific patient's tumor.

## 5. Conclusions

Our study of adult IDH Wild-Type Glioblastoma explored its ultrastructural characteristics and their correlation with the Ki-67 labeling index (LI). The results of our research show that integrating ultrastructural features with molecular information allows us to identify biomarkers that may improve diagnostic accuracy, guide treatment decisions, identify therapeutic targets, and facilitate personalized treatment plans tailored to individual patients. Future investigations with larger sample sizes are necessary to validate these findings and harness the potential of ultrastructural analysis in glioblastoma management.

**Supplementary Materials:** The following supporting information can be downloaded at: https://www.mdpi.com/article/10.3390/biomedicines11071968/s1. Table S1. Patients' data and diagnosis. Table S2. Molecular data. Table S3. Histological data. Table S4. Morphological data from picture analysis and relation with Ki-67 index.

**Author Contributions:** Conceptualization, P.F., M.R., A.F. and P.B.; methodology, P.L., M.P., M.A., L.C. and C.B.; software, C.B., L.C. and L.D.; validation, C.B., M.C. and A.S.; formal analysis, A.S., M.C. and A.F.; investigation, M.R., G.F., L.D. and C.B.; resources, A.S., G.F. and M.R.; data curation, M.A., M.C. and A.F.; writing—original draft preparation, P.F., M.R., A.F. and P.B.; writing—review and editing, P.F., M.R., A.F. and P.B.; visualization, A.S.; supervision, G.F.; project administration, G.F.; funding acquisition, M.R. All authors have read and agreed to the published version of the manuscript.

**Funding:** This research was funded by Sapienza Ateneo Funds 2022 project number RM122181672B66D0.

**Institutional Review Board Statement:** Anonymized human glioblastoma tissue samples and primary clinicopathologic data were obtained previous patient consent was granted. Tissue was used according to the guidelines of the Declaration of Helsinki and according to the guidelines of the Institutional Review Board of the Sapienza University of Rome (Authoriz. Number 0296/2023, the date of approval 29 March 2023).

**Informed Consent Statement:** Informed consent was obtained from all subjects involved in the study.

**Data Availability Statement:** Data are stored in The Electron microscopy unit of the Sapienza SAIMLAL department and are available upon written request.

**Acknowledgments:** We thank Ezio Battaglione for his skillful assistance.

**Conflicts of Interest:** The authors declare no conflict of interest.

## References

1. Grochans, S.; Cybulska, A.M.; Simińska, D.; Korbecki, J.; Kojder, K.; Chlubek, D.; Baranowska-Bosiacka, I. Epidemiology of Glioblastoma Multiforme-Literature Review. *Cancers* **2022**, *14*, 2412. [CrossRef] [PubMed]
2. Gerber, N.K.; Goenka, A.; Turcan, S.; Reyngold, M.; Makarov, V.; Kannan, K.; Beal, K.; Omuro, A.; Yamada, Y.; Gutin, P.; et al. Transcriptional diversity of long-term glioblastoma survivors. *Neuro Oncol.* **2014**, *16*, 1186–1195. [CrossRef] [PubMed]
3. Louis, D.N.; Perry, A.; Reifenberger, G.; von Deimling, A.; Figarella-Branger, D.; Cavenee, W.K.; Ohgaki, H.; Wiestler, O.D.; Kleihues, P.; Ellison, D.W. The 2016 World Health Organization Classification of Tumors of the Central Nervous System: A summary. *Acta Neuropathol.* **2016**, *131*, 803–820. [CrossRef] [PubMed]
4. Louis, D.N.; Perry, A.; Wesseling, P.; Brat, D.J.; Cree, I.A.; Figarella-Branger, D.; Hawkins, C.; Ng, H.K.; Pfister, S.M.; Reifenberger, G.; et al. The 2021 WHO Classification of Tumors of the Central Nervous System: A summary. *Neuro Oncol.* **2021**, *23*, 1231–1251. [CrossRef]
5. Qin, T.; Mullan, B.; Ravindran, R.; Messinger, D.; Siada, R.; Cummings, J.R.; Harris, M.; Muruganand, A.; Pyaram, K.; Miklja, Z.; et al. ATRX loss in glioma results in dysregulation of cell-cycle phase transition and ATM inhibitor radio-sensitization. *Cell Rep.* **2022**, *38*, 110216. [CrossRef]
6. Eckel-Passow, J.E.; Lachance, D.H.; Molinaro, A.M.; Walsh, K.M.; Decker, P.A.; Sicotte, H.; Pekmezci, M.; Rice, T.; Kosel, M.L.; Smirnov, I.V.; et al. Glioma Groups Based on 1p/19q, IDH, and TERT Promoter Mutations in Tumors. *N. Engl. J. Med.* **2015**, *372*, 2499–2508. [CrossRef]
7. Jiao, Y.; Killela, P.J.; Reitman, Z.J.; Rasheed, A.B.; Heaphy, C.M.; de Wilde, R.F.; Rodriguez, F.J.; Rosemberg, S.; Oba-Shinjo, S.M.; Nagahashi, M.; et al. Frequent ATRX, CIC, FUBP1 and IDH1 mutations refine the classification of malignant gliomas. *Oncotarget* **2012**, *3*, 709–722. [CrossRef]
8. Suzuki, H.; Aoki, K.; Chiba, K.; Sato, Y.; Shiozawa, Y.; Shiraishi, Y.; Shimamura, T.; Niida, A.; Motomura, K.; Ohka, F.; et al. Mutational landscape and clonal architecture in grade II and III gliomas. *Nat. Genet.* **2015**, *47*, 458–468. [CrossRef]
9. Zhang, Y.; Dube, C.; Gibert, M., Jr.; Cruickshanks, N.; Wang, B.; Coughlan, M.; Yang, Y.; Setiady, I.; Deveau, C.; Saoud, K.; et al. The p53 Pathway in Glioblastoma. *Cancers* **2018**, *10*, 297. [CrossRef]
10. Uceda-Castro, R.; van Asperen, J.V.; Vennin, C.; Sluijs, J.A.; van Bodegraven, E.J.; Margarido, A.S.; Robe, P.A.J.; van Rheenen, J.; Hol, E.M. GFAP splice variants fine-tune glioma cell invasion and tumour dynamics by modulating migration persistence. *Sci. Rep.* **2022**, *12*, 424. [CrossRef]

11. Xu, H.; Zong, H.; Ma, C.; Ming, X.; Shang, M.; Li, K.; He, X.; Du, H.; Cao, L. Epidermal growth factor receptor in glioblastoma. *Oncol. Lett.* **2017**, *14*, 512–516. [CrossRef] [PubMed]
12. Gerdes, J.; Schwab, U.; Lemke, H.; Stein, H. Production of a mouse monoclonal antibody reactive with a human nuclear antigen associated with cell proliferation. *Int. J. Cancer* **1983**, *31*, 13–20. [CrossRef]
13. Chen, W.J.; He, D.S.; Tang, R.X.; Ren, F.H.; Chen, G. Ki-67 is a valuable prognostic factor in gliomas: Evidence from a systematic review and meta-analysis. Asian Pac. *J. Cancer Prev.* **2015**, *16*, 411–420. [CrossRef] [PubMed]
14. Arshad, H.; Ahmad, Z.; Hasan, S.H. Gliomas: Correlation of histologic grade, Ki67 and p53 expression with patient survival. Asian Pac. *J. Cancer Prev.* **2010**, *11*, 1637–1640.
15. Dahlrot, R.H.; Bangsø, J.A.; Petersen, J.K.; Rosager, A.M.; Sørensen, M.D.; Reifenberger, G.; Hansen, S.; Kristensen, B.W. Prognostic role of Ki-67 in glioblastomas excluding contribution from non-neoplastic cells. *Sci. Rep.* **2021**, *11*, 17918. [CrossRef] [PubMed]
16. Du, N.; Shu, W.; Li, K.; Deng, Y.; Xu, X.; Ye, Y.; Tang, F.; Mao, R.; Lin, G.; Li, S.; et al. An initial study on the predictive value using multiple MRI characteristics for Ki-67 labeling index in glioma. *J. Transl. Med.* **2023**, *21*, 119. [CrossRef]
17. Hambardzumyan, D.; Bergers, G. Glioblastoma: Defining Tumor Niches. *Trends Cancer* **2015**, *1*, 252–265. [CrossRef]
18. Geribaldi-Doldán, N.; Fernández-Ponce, C.; Quiroz, R.N.; Sánchez-Gomar, I.; Escorcia, L.G.; Velásquez, E.P.; Quiroz, E.N. The Role of Microglia in Glioblastoma. *Front. Oncol.* **2021**, *10*, 603495. [CrossRef]
19. Buonfiglioli, A.; Hambardzumyan, D. Macrophages and microglia: The cerberus of glioblastoma. *Acta Neuropathol. Commun.* **2021**, *9*, 54. [CrossRef]
20. Rajendran, L.; Bali, J.; Barr, M.M.; Court, F.A.; Krämer-Albers, E.M.; Picou, F.; Raposo, G.; van der Vos, K.E.; van Niel, G.; Wang, J.; et al. Emerging roles of extracellular vesicles in the nervous system. *J. Neurosci.* **2014**, *34*, 15482–15489. [CrossRef]
21. Ciregia, F.; Urbani, A.; Palmisano, G. Extracellular Vesicles in Brain Tumors and Neurodegenerative Diseases. *Front. Mol. Neurosci.* **2017**, *10*, 276. [CrossRef] [PubMed]
22. Whitehead, C.A.; Kaye, A.H.; Drummond, K.J.; Widodo, S.S.; Mantamadiotis, T.; Vella, L.J.; Stylli, S.S. Extracellular vesicles and their role in glioblastoma. *Crit. Rev. Clin. Lab. Sci.* **2019**, *57*, 227–252. [CrossRef] [PubMed]
23. Gassama, Y.; Favereaux, A. Emerging Roles of Extracellular Vesicles in the Central Nervous System: Physiology, Pathology, and Therapeutic Perspectives. *Front. Cell. Neurosci.* **2021**, *15*, 626043. [CrossRef] [PubMed]
24. Balakrishnan, A.; Roy, S.; Fleming, T.; Leong, H.S.; Schuurmans, C. The Emerging Role of Extracellular Vesicles in the Glioma Microenvironment: Biogenesis and Clinical Relevance. *Cancers* **2020**, *12*, 1964. [CrossRef] [PubMed]
25. Pancholi, S.; Tripathi, A.; Bhan, A.; Acharya, M.M.; Pillai, P. Emerging Concepts on the Role of Extracellular Vesicles and Its Cargo Contents in Glioblastoma-Microglial Crosstalk. *Mol. Neurobiol.* **2022**, *59*, 2822–2837. [CrossRef]
26. Wang, Y.; Tang, C.; Cai, J.; Chen, G.; Zhang, D.; Zhang, Z.; Dong, Z. PINK1/Parkin-mediated mitophagy is activated in cisplatin nephrotoxicity to protect against kidney injury. *Cell Death Dis.* **2018**, *4*, 1113. [CrossRef]
27. Wang, J.; Qiu, X.; Huang, J.; Zhuo, Z.; Chen, H.; Zeng, R.; Wu, H.; Guo, K.; Yang, Q.; Ye, H.; et al. Development and validation of a novel mitophagy-related gene prognostic signature for glioblastoma multiforme. *BMC Cancer* **2022**, *22*, 644. [CrossRef]
28. Petan, T.; Jarc, E.; Jusović, M. Lipid Droplets in Cancer: Guardians of Fat in a Stressful World. *Molecules* **2018**, *23*, 1941. [CrossRef]
29. Leblond, M.M.; Gerault, A.N.; Corroyer-Dulmont, A.; MacKenzie, E.T.; Petit, E.; Bernaudin, M.; Valable, S. Hypoxia induces macrophage polarization and re-education toward an angiogenic phenotype in U87 and U251 glioblastoma models. *Oncoimmunology* **2018**, *7*, e1438106.
30. Cruz, A.L.; Barreto, E.D.A.; Fazolini, N.P.; Viola, J.P.; Bozza, P.T. Lipid droplets: Platforms with multiple functions in cancer hallmarks. *Cell Death Dis.* **2020**, *11*, 105. [CrossRef]
31. Cottignoli, V.; Relucenti, M.; Agrosì, G.; Cavarretta, E.; Familiari, G.; Salvador, L.; Maras, A. Biological Niches within Human Calcified Aortic Valves: Towards Understanding of the Pathological Biomineralization Process. *BioMed Res. Int.* **2015**, *2015*, 542687. [CrossRef] [PubMed]
32. Nardoni, M.; Della Valle, E.; Liberti, M.; Relucenti, M.; Casadei, M.A.; Paolicelli, P.; Apollonio, F.; Petralito, S. Can Pulsed Electromagnetic Fields Trigger On-Demand Drug Release from High-Tm Magnetoliposomes? *Nanomaterials* **2018**, *8*, 196. [CrossRef]
33. Relucenti, M.; Francescangeli, F.; De Angelis, M.L.; D'Andrea, V.; Miglietta, S.; Pilozzi, E.; Li, X.; Boe, A.; Chen, R.; Zeuner, A.; et al. The Ultrastructural Analysis of Human Colorectal Cancer Stem Cell-Derived Spheroids and Their Mouse Xenograft Shows That the Same Cells Types Have Different Ratios. *Biology* **2021**, *10*, 929. [CrossRef] [PubMed]
34. Relucenti, M.; Francescangeli, F.; De Angelis, M.L.; D'Andrea, V.; Miglietta, S.; Donfrancesco, O.; Li, X.; Chen, R.; Zeuner, A.; Familiari, G. A Different Exosome Secretion Pattern Characterizes Patient-Derived Colorectal Cancer Multicellular Spheroids and Their Mouse Xenografts. *Biology* **2022**, *11*, 1427. [CrossRef] [PubMed]
35. Petralito, S.; Paolicelli, P.; Nardoni, M.; Trilli, J.; Di Muzio, L.; Cesa, S.; Relucenti, M.; Matassa, R.; Vitalone, A.; Adrover, A.; et al. Gelation of the internal core of liposomes as a strategy for stabilization and modified drug delivery I. Physico-chemistry study. *Int. J. Pharm.* **2020**, *585*, 119467. [CrossRef] [PubMed]
36. Relucenti, M.; Miglietta, S.; Bove, G.; Donfrancesco, O.; Battaglione, E.; Familiari, P.; Barbaranelli, C.; Covelli, E.; Barbara, M.; Familiari, G. SEM BSE 3D Image Analysis of Human Incus Bone Affected by Cholesteatoma Ascribes to Osteoclasts the Bone Erosion and VpSEM dEDX Analysis Reveals New Bone Formation. *Scanning* **2020**, *2020*, 9371516. [CrossRef]

37. Lo Torto, F.; Relucenti, M.; Familiari, G.; Vaia, N.; Casella, D.; Matassa, R.; Miglietta, S.; Marinozzi, F.; Bini, F.; Fratoddi, I.; et al. The Effect of Postmastectomy Radiation Therapy on Breast Implants: Material Analysis on Silicone and Polyurethane Prosthesis. *Ann. Plast. Surg.* **2018**, *81*, 228–234. [CrossRef]
38. Adjei-Sowah, E.A.; O'Connor, S.A.; Veldhuizen, J.; Lo Cascio, C.; Plaisier, C.; Mehta, S.; Nikkhah, M. Investigating the Interactions of Glioma Stem Cells in the Perivascular Niche at Single-Cell Resolution using a Microfluidic Tumor Microenvironment Model. *Adv. Sci.* **2022**, *9*, e2201436. [CrossRef]
39. McCoy, M.G.; Nyanyo, D.; Hung, C.K.; Goerger, J.P.; Zipfel, R.W.; Williams, R.M.; Nishimura, N.; Fischbach, C. Endothelial cells promote 3D invasion of GLIOBLASTOMA by IL-8-dependent induction of cancer stem cell properties. *Sci. Rep.* **2019**, *9*, 9069. [CrossRef]
40. Akindona, F.A.; Frederico, S.C.; Hancock, J.C.; Gilbert, M.R. Exploring the origin of the cancer stem cell niche and its role in anti-angiogenic treatment for glioblastoma. *Front. Oncol.* **2022**, *12*, 947634. [CrossRef]
41. Dinda, A.K.; Sarkar, C.; Roy, S.; Kharbanda, K.; Mathur, M.; Khosla, A.K.; Banerji, A.K. A transmission and scanning electron microscopic study of tumoral and peritumoral microblood vessels in human gliomas. *J. Neuro Oncol.* **1993**, *16*, 149–158. [CrossRef] [PubMed]
42. Cribaro, G.P.; Saavedra-López, E.; Romarate, L.; Mitxitorena, I.; Díaz, L.R.; Casanova, P.V.; Roig-Martínez, M.; Gallego, J.M.; Perez-Vallés, A.; Barcia, C. Three-dimensional vascular microenvironment landscape in human glioblastoma. *Acta Neuropathol. Commun.* **2021**, *9*, 24. [CrossRef] [PubMed]
43. Sipe, J.C.; Herman, M.M.; Rubinstein, L.J. Electron microscopic observations on human glioblastomas and astrocytomas maintained in organ culture systems. *Am. J. Pathol.* **1973**, *73*, 589–606. [PubMed]
44. Ferese, R.; Lenzi, P.; Fulceri, F.; Biagioni, F.; Fabrizi, C.; Gambardella, S.; Familiari, P.; Frati, A.; Limanaqi, F.; Fornai, F. Quantitative Ultrastructural Morphometry and Gene Expression of mTOR-Related Mitochondriogenesis within Glioblastoma Cells. *Int. J. Mol. Sci.* **2020**, *21*, 4570. [CrossRef] [PubMed]
45. Lenzi, P.; Ferese, R.; Biagioni, F.; Fulceri, F.; Busceti, C.L.; Falleni, A.; Gambardella, S.; Frati, A.; Fornai, F. Rapamycin Ameliorates Defects in Mitochondrial Fission and Mitophagy in Glioblastoma Cells. *Int. J. Mol. Sci.* **2021**, *22*, 5379. [CrossRef] [PubMed]
46. Catalano, M.; D'Alessandro, G.; Lepore, F.; Corazzari, M.; Caldarola, S.; Valacca, C.; Faienza, F.; Esposito, V.; Limatola, C.; Cecconi, F.; et al. Autophagy induction impairs migration and invasion by reversing EMT in glioblastoma cells. *Mol. Oncol.* **2015**, *9*, 1612–1625. [CrossRef]
47. Li, Q.; Barres, B.A. Microglia and macrophages in brain homeostasis and disease. Nature reviews. *Immunology* **2018**, *18*, 225–242.
48. Wolf, S.A.; Boddeke, H.W.; Kettenmann, H. Microglia in Physiology and Disease. *Annu. Rev. Physiol.* **2017**, *79*, 619–643. [CrossRef]
49. Colonna, M.; Butovsky, O. Microglia Function in the Central Nervous System During Health and Neurodegeneration. *Annu. Rev. Immunol.* **2017**, *35*, 441–468. [CrossRef]
50. Graeber, M.B.; Li, W.; Rodriguez, M.L. Role of microglia in CNS inflammation. *FEBS Lett.* **2011**, *585*, 3798–3805. [CrossRef]
51. Augusto-Oliveira, M.; Arrifano, G.P.; Delage, C.I.; Tremblay, M.È.; Crespo-Lopez, M.E.; Verkhratsky, A. Plasticity of microglia. *Biol. Rev. Camb. Philos. Soc.* **2022**, *97*, 217–250. [CrossRef] [PubMed]
52. Schwarz, N.; Leube, R.E. Intermediate Filaments as Organizers of Cellular Space: How They Affect Mitochondrial Structure and Function. *Cells* **2016**, *5*, 30. [CrossRef] [PubMed]
53. St-Pierre, M.-K.; Carrier, M.; Ibáñez, F.G.; Šimončičová, E.; Wallman, M.-J.; Vallières, L.; Parent, M.; Tremblay, M.-È. Ultrastructural characterization of dark microglia during aging in a mouse model of Alzheimer's disease pathology and in human post-mortem brain samples. *J. Neuroinflamm.* **2022**, *19*, 235. [CrossRef] [PubMed]
54. Bisht, K.; Sharma, K.P.; Lecours, C.; Sánchez, M.G.; El Hajj, H.; Milior, G.; Olmos-Alonso, A.; Gómez-Nicola, D.; Luheshi, G.; Vallières, L.; et al. Dark microglia: A new phenotype predominantly associated with pathological states. *Glia* **2016**, *64*, 826–839. [CrossRef]
55. Yoshimura, T. The production of monocyte chemoattractant protein-1 (MCP$^{-1}$)/CCL2 in tumor microenvironments. *Cytokine* **2017**, *98*, 71–78. [CrossRef]
56. Rong, Y.; Ji, C.; Wang, Z.; Ge, X.; Wang, J.; Ye, W.; Tang, P.; Jiang, D.; Fan, J.; Yin, G.; et al. Small extracellular vesicles encapsulating CCL2 from activated astrocytes induce microglial activation and neuronal apoptosis after traumatic spinal cord injury. *J. Neuroinflamm.* **2021**, *18*, 196. [CrossRef]
57. Grimaldi, A.; Serpe, C.; Chece, G.; Nigro, V.; Sarra, A.; Ruzicka, B.; Relucenti, M.; Familiari, G.; Ruocco, G.; Pascucci, G.R.; et al. Microglia-Derived Microvesicles Affect Microglia Phenotype in Glioma. *Front. Cell. Neurosci.* **2019**, *13*, 41. [CrossRef]
58. Serpe, C.; Monaco, L.; Relucenti, M.; Iovino, L.; Familiari, P.; Scavizzi, F.; Raspa, M.; Familiari, G.; Civiero, L.; D'Agnano, I.; et al. Microglia-Derived Small Extracellular Vesicles Reduce Glioma Growth by Modifying Tumor Cell Metabolism and Enhancing Glutamate Clearance through miR-124. *Cells* **2021**, *10*, 2066. [CrossRef]
59. Fitzner, D.; Bader, J.M.; Penkert, H.; Bergner, C.G.; Su, M.; Weil, M.T.; Surma, M.A.; Mann, M.; Klose, C.; Simons, M. Cell-Type- and Brain-Region-Resolved Mouse Brain Lipidome. *Cell Rep.* **2020**, *32*, 108132. [CrossRef]
60. Geng, F.; Cheng, X.; Wu, X.; Yoo, J.Y.; Cheng, C.; Guo, J.Y.; Mo, X.; Ru, P.; Hurwitz, B.; Kim, S.H.; et al. Inhibition of SOAT1 Suppresses Glioblastoma Growth via Blocking SREBP-1-Mediated Lipogenesis. *Clin. Cancer Res.* **2016**, *22*, 5337–5348. [CrossRef]
61. Hayes, A.F. *Introduction to Mediation, Moderation, and Conditional Process Analysis: A Regression-Based Approach*; The Guilford Press: New York, NY, USA, 2022.

62. Reiter-Brennan, C.; Semmler, L.; Klein, A. The effects of 2-hydroxyglutarate on the tumorigenesis of gliomas. *Contemp. Oncol.* **2018**, *22*, 215–222. [CrossRef] [PubMed]
63. Fu, Y.; Zheng, S.; Zheng, Y.; Huang, R.; An, N.; Liang, A.; Hu, C. Glioma-derived isocitrate dehydrogenase-2 mutations induced up-regulation of HIF-1$\alpha$ and $\beta$-catenin signaling: Possible impact on glioma cell metastasis and chemo-resistance. *Int. J. Biochem. Cell Biol.* **2012**, *44*, 770–775. [CrossRef] [PubMed]
64. Lee, J.; Kim, E.; Chong, K.; Ryu, S.W.; Kim, C.; Choi, K.; Kim, J.H.; Choi, C. Atypical induction of HIF-1$\alpha$ expression by pericellular Notch1 signaling suffices for the malignancy of glioblastoma multiforme cells. *Cell. Mol. Life Sci.* **2022**, *79*, 537. [CrossRef] [PubMed]
65. Nicolas, S.; Abdellatef, S.; Haddad, M.A.; Fakhoury, I.; El-Sibai, M. Hypoxia and EGF Stimulation Regulate VEGF Expression in Human Glioblastoma Multiforme (GLIOBLASTOMA) Cells by Differential Regulation of the PI3K/Rho-GTPase and MAPK Pathways. *Cells* **2019**, *8*, 1397. [CrossRef]
66. Domènech, M.; Hernández, A.; Plaja, A.; Martínez-Balibrea, E.; Balañà, C. Hypoxia: The Cornerstone of Glioblastoma. *Int. J. Mol. Sci.* **2021**, *22*, 12608. [CrossRef]
67. He, J.; Mansouri, A.; Das, S. Alpha Thalassemia/Mental Retardation Syndrome X-Linked, the Alternative Lengthening of Telomere Phenotype, and Gliomagenesis: Current Understandings and Future Potential. *Front. Oncol.* **2018**, *7*, 322. [CrossRef]
68. Danussi, C.; Bose, P.; Parthasarathy, P.T.; Silberman, P.C.; Van Arnam, J.S.; Vitucci, M.; Tang, O.Y.; Heguy, A.; Wang, Y.; Chan, T.A.; et al. Atrx inactivation drives disease-defining phenotypes in glioma cells of origin through global epigenomic remodeling. *Nat. Commun.* **2018**, *9*, 1057. [CrossRef]
69. Olivier, M.; Hollstein, M.; Hainaut, P. TP53 mutations in human cancers: Origins, consequences, and clinical use. *Cold Spring Harb. Perspect. Biol.* **2010**, *2*, a001008. [CrossRef]
70. Mao, H.; Lebrun, D.G.; Yang, J.; Zhu, V.F.; Li, M. Deregulated signaling pathways in glioblastoma multiforme: Molecular mechanisms and therapeutic targets. *Cancer Investig.* **2012**, *30*, 48–56. [CrossRef]
71. Pearson, J.R.D.; Regad, T. Targeting cellular pathways in glioblastoma multiforme. *Signal Transduct. Target. Ther.* **2017**, *2*, 17040. [CrossRef]
72. Restrepo, A.; Smith, C.A.; Agnihotri, S.; Shekarforoush, M.; Kongkham, P.N.; Seol, H.J.; Northcott, P.; Rutka, J.T. Epigenetic regulation of glial fibrillary acidic protein by DNA methylation in human malignant gliomas. *Neuro Oncol.* **2011**, *13*, 42–50. [CrossRef]
73. Bignami, A.; Eng, L.F.; Dahl, D.; Uyeda, C.T. Localization of the glial fibrillary acidic protein in astrocytes by immunofluorescence. *Brain Res.* **1972**, *43*, 429–435. [CrossRef] [PubMed]
74. Oprita, A.; Baloi, S.C.; Staicu, G.A.; Alexandru, O.; Tache, D.E.; Danoiu, S.; Micu, E.S.; Sevastre, A.S. Updated Insights on EGFR Signaling Pathways in Glioma. *Int. J. Mol. Sci.* **2021**, *22*, 587. [CrossRef] [PubMed]
75. Messaoudi, K.; Clavreul, A.; Lagarce, F. Toward an effective strategy in glioblastoma treatment. Part I: Resistance mechanisms and strategies to overcome resistance of glioblastoma to temozolomide. *Drug Discov. Today* **2015**, *20*, 899–905. [CrossRef] [PubMed]
76. Ohgaki, H.; Kleihues, P. Genetic alterations and signaling pathways in the evolution of gliomas. *Cancer Sci.* **2009**, *100*, 2235–2241. [CrossRef]
77. An, Z.; Aksoy, O.; Zheng, T.; Fan, Q.W.; Weiss, W.A. Epidermal growth factor receptor and EGFRvIII in glioblastoma: Signaling pathways and targeted therapies. *Oncogene* **2018**, *37*, 1561–1575. [CrossRef]
78. Takano, S.; Ishikawa, E.; Sakamoto, N.; Matsuda, M.; Akutsu, H.; Noguchi, M.; Kato, Y.; Yamamoto, T.; Matsumura, A. Immunohistochemistry on IDH 1/2, ATRX, p53 and Ki-67 substitute molecular genetic testing and predict patient prognosis in grade III adult diffuse gliomas. *Brain Tumor Pathol.* **2016**, *33*, 107–116. [CrossRef]
79. Armocida, D.; Frati, A.; Salvati, M.; Santoro, A.; Pesce, A. Is Ki-67 index overexpression in IDH wild type glioblastoma a predictor of shorter Progression Free survival? A clinical and Molecular analytic investigation. *Clin. Neurol. Neurosurg.* **2020**, *198*, 106126. [CrossRef]
80. Nakod, P.S.; Kim, Y.; Rao, S.S. The Impact of Astrocytes and Endothelial Cells on Glioblastoma Stemness Marker Expression in Multicellular Spheroids. *Cell. Mol. Bioeng.* **2021**, *14*, 639–651. [CrossRef]
81. Kim, S.J.; Lee, H.J.; Kim, M.S.; Choi, H.J.; He, J.; Wu, Q.; Aldape, K.; Weinberg, J.S.; Yung, W.K.; Conrad, C.A.; et al. Macitentan, a Dual Endothelin Receptor Antagonist, in Combination with Temozolomide Leads to Glioblastoma Regression and Long-term Survival in Mice. *Clin. Cancer Res.* **2015**, *21*, 4630–4641. [CrossRef]
82. Horng, S.; Therattil, A.; Moyon, S.; Gordon, A.; Kim, K.; Argaw, A.T.; Hara, Y.; Mariani, J.N.; Sawai, S.; Flodby, P.; et al. Astrocytic tight junctions control inflammatory CNS lesion pathogenesis. *J. Clin. Investig.* **2017**, *127*, 3136–3151. [CrossRef] [PubMed]
83. Lai, C.P.; Kim, E.Y.; Badr, C.E.; Weissleder, R.; Mempel, T.R.; Tannous, B.A. Visualization and tracking of tumour extracellular vesicle delivery and RNA translation using multiplexed reporters. *Nat. Commun.* **2015**, *6*, 7029. [CrossRef] [PubMed]
84. Skog, J.; Würdinger, T.; van Rijn, S.; Meijer, D.H.; Gainche, L.; Sena-Esteves, M.; Breakefield, X.O. Glioblastoma microvesicles transport RNA and proteins that promote tumour growth and provide diagnostic biomarkers. *Nat. Cell Biol.* **2008**, *10*, 1470–1476. [CrossRef]
85. Al-Nedawi, K.; Meehan, B.; Kerbel, R.S.; Allison, A.C.; Rak, J. Endothelial expression of autocrine VEGF upon the uptake of tumor-derived microvesicles containing oncogenic EGFR. *Proc. Natl. Acad. Sci. USA* **2009**, *106*, 3794–3799. [CrossRef] [PubMed]

86. Jahan, S.; Mukherjee, S.; Ali, S.; Bhardwaj, U.; Choudhary, R.K.; Balakrishnan, S.; Naseem, A.; Mir, S.A.; Banawas, S.; Alaidarous, M.; et al. Pioneer Role of Extracellular Vesicles as Modulators of Cancer Initiation in Progression, Drug Therapy, and Vaccine Prospects. *Cells* **2022**, *11*, 490. [CrossRef]
87. Murgoci, A.N.; Duhamel, M.; Raffo-Romero, A.; Mallah, K.; Aboulouard, S.; Lefebvre, C.; Salzet, M. Location of neonatal microglia drives small extracellular vesicles content and biological functions in vitro. *J. Extracell. Vesicles* **2020**, *9*, 1727637. [CrossRef]
88. Muñoz Herrera, O.M.; Zivkovic, A.M. Microglia and Cholesterol Handling: Implications for Alzheimer's Disease. *Biomedicines* **2022**, *10*, 3105. [CrossRef]
89. Marschallinger, J.; Iram, T.; Zardeneta, M.; Lee, S.E.; Lehallier, B.; Haney, M.S.; Pluvinage, J.V.; Mathur, V.; Hahn, O.; Morgens, D.W.; et al. Lipid-Droplet-Accumulating Microglia Represent a Dysfunctional and Proinflammatory State in the Aging Brain. *Nat. Neurosci.* **2020**, *23*, 194–208. [CrossRef]
90. Colombo, A.; Dinkel, L.; Müller, S.A.; Sebastian Monasor, L.; Schifferer, M.; Cantuti-Castelvetri, L.; König, J.; Vidatic, L.; Bremova-Ertl, T.; Lieberman, A.P.; et al. Loss of NPC1 Enhances Phagocytic Uptake and Impairs Lipid Trafficking in Microglia. *Nat. Commun.* **2021**, *12*, 1158. [CrossRef]
91. Hanisch, U.K.; Kettenmann, H. Microglia: Active Sensor and Versatile Effector Cells in the Normal and Pathologic Brain. *Nat. Neurosci.* **2007**, *10*, 1387–1394. [CrossRef]
92. Bernardini, J.P.; Lazarou, M.; Dewson, G. Parkin and mitophagy in cancer. *Oncogene* **2017**, *36*, 1315–1327. [CrossRef] [PubMed]
93. Maiti, P.; Scott, J.; Sengupta, D.; Al-Gharaibeh, A.; Dunbar, G.L. Curcumin and solid lipid curcumin particles induce autophagy, but inhibit mitophagy and the PI3K-Akt/mTOR pathway in cultured glioblastoma cells. *Int. J. Mol. Sci.* **2019**, *20*, 399. [CrossRef] [PubMed]
94. Liu, C.; Peng, Z.; Li, P.; Fu, H.; Feng, J.; Zhang, Y.; Liu, T.; Liu, Y.; Liu, Q.; Liu, Q.; et al. lncRNA RMST suppressed glioblastoma cell mitophagy through enhancing FUS SUMOylation. *Mol. Ther. Nucleic Acids* **2020**, *19*, 1198–1208. [CrossRef] [PubMed]
95. Huang, T.; Xu, T.; Wang, Y.; Zhou, Y.; Yu, D.; Wang, Z.; He, L.; Chen, Z.; Zhang, Y.; Davidson, D.; et al. Cannabidiol inhibits human glioma by induction of lethal mitophagy through activating TRPV4. *Autophagy* **2021**, *17*, 3592–3606. [CrossRef] [PubMed]
96. He, C.; Lu, S.; Wang, X.Z.; Wang, C.C.; Wang, L.; Liang, S.P.; Luo, T.-F.; Wang, Z.-C.; Piao, M.-H.; Chi, G.-F.; et al. FOXO3a protects glioma cells against temozolomide-induced DNA double strand breaks via promotion of BNIP3-mediated mitophagy. *Acta Pharmacol. Sin.* **2021**, *2*, 1324–1337. [CrossRef]
97. Zhang, J.; Liu, L.; Xue, Y.; Ma, Y.; Liu, X.; Li, Z.; Li, Z.; Liu, Y. Endothelial monocyte-activating polypeptide-II induces BNIP3-mediated mitophagy to enhance temozolomide cytotoxicity of glioma stem cells via down-regulating MiR-24-3p. *Front. Mol. Neurosci.* **2018**, *11*, 92. [CrossRef]

biomedicines

MDPI

*Article*

# Biomarkers in Adult-Type Diffuse Gliomas: Elevated Levels of Circulating Vesicular Heat Shock Protein 70 Serve as a Biomarker in Grade 4 Glioblastoma and Increase NK Cell Frequencies in Grade 3 Glioma

**Philipp Lennartz [1,2,†], Dennis Thölke [1,2,†], Ali Bashiri Dezfouli [1,3], Mathias Pilz [1,2], Dominik Lobinger [4], Verena Messner [1,2], Hannah Zanth [1,2], Karen Ainslie [1,2], Morteza Hasanzadeh Kafshgari [1,5], Gerhard Rammes [6], Markus Ballmann [6], Martin Schlegel [6], Gemma Ann Foulds [7], Alan Graham Pockley [7], Friederike Schmidt-Graf [8] and Gabriele Multhoff [1,2,*]**

1 Central Institute for Translational Cancer Research Technische Universität München (TranslaTUM), Klinikum rechts der Isar, TUM School of Medicine and Health, 81675 Munich, Germany; philipp.lennartz@tum.de (P.L.); dennis.thoelke@hotmail.de (D.T.); ali.bashiri@tum.de (A.B.D.); ge54may@mytum.de (V.M.); hannah.zanth@tum.de (H.Z.)

2 Department of Radiation Oncology, Klinikum rechts der Isar, TUM School of Medicine and Health, 81675 Munich, Germany

3 Department of Otolaryngology, Head and Neck Surgery, Klinikum rechts der Isar, TUM School of Medicine and Health, 81675 Munich, Germany

4 Department of Thoracic Surgery, München Klinik Bogenhausen, Lehrkrankenhaus der TUM, 81925 Munich, Germany; dominiklobinger@web.de

5 Department of Biomedical Electronics, Central Instititute for Translational Cancer Research, Technische Universität München (TranslaTUM), Klinikum rechts der Isar, TUM School of Medicine and Health, 81675 Munich, Germany

6 Department of Anaesthesiology and Intensive Care Medicine, Klinikum rechts der Isar, TUM School of Medicine and Health, 81675 Munich, Germany; martin.schlegel@tum.de (M.S.)

7 John van Geest Cancer Research Centre, School of Science and Technology, Nottingham Trent University, Nottingham NG11 8NS, UK; gemma.foulds@ntu.ac.uk (G.A.F.); graham.pockley@ntu.ac.uk (A.G.P.)

8 Department of Neurology, Klinikum rechts der Isar, TUM School of Medicine and Health, 81675 Munich, Germany; f.schmidt-graf@tum.de

* Correspondence: gabriele.multhoff@tum.de; Tel.: +49-89-4140-4514

† These authors contributed equally to this work.

**Abstract:** The presence of circulating Hsp70 levels and their influence on the immunophenotype of circulating lymphocyte subsets were examined as diagnostic/prognostic biomarkers for the overall survival (OS) in patients with IDH-mutant WHO grade 3 oligodendroglioma, astrocytoma, and IDH-wildtype grade 4 glioblastoma (GBM). Vesicular and free Hsp70 in the plasma/serum was measured using the Hsp70-exo and R&D Systems DuoSet® Hsp70 ELISAs. The immunophenotype and membrane Hsp70 status was determined by multiparameter flow cytometry on peripheral blood lymphocytes and single-cell suspensions of tumor specimens and cultured cells. Compared to healthy controls, circulating vesicular Hsp70 levels were significantly increased in patients with GBM, concomitant with a significant decrease in the proportion of CD3+/CD4+ helper T cells, whereas the frequency of NK cells was most prominently increased in patients with grade 3 gliomas. Elevated circulating Hsp70 levels and a higher prevalence of activated CD3−/CD56+/CD94+/CD69+ NK cells were associated with an improved OS in grade 3 gliomas, whereas high Hsp70 levels and low CD3+/CD4+ frequencies were associated with an adverse OS in GBM. It is assumed that a reduced membrane Hsp70 density on grade 4 versus grade 3 primary glioma cells and reduced CD3+/CD4+ T cell counts in GBM might drive an immunosuppressive tumor microenvironment.

**Keywords:** biomarkers; liquid biopsy; extracellular free and vesicular Hsp70; plasma membrane bound Hsp70; glioblastoma; oligodendroglioma; astrocytoma; isocitrate dehydrogenase 1/2; NK cells; immunophenotyping

## 1. Introduction

Malignant brain tumors comprise approximately 30% of all primary brain tumors, of which 80% are adult-type diffuse gliomas—a group of highly infiltrative and largely incurable malignancies [1]. The 2021 update of the WHO classification of central nervous system tumors (WHO CNS5) identifies tissue biopsy-based molecular biomarkers for defining neoplastic entities. The two most important molecular biomarkers, wildtype or mutant isocitrate dehydrogenase 1/2 (IDH) and the 1p19q codeletion, differentiate adult-type diffuse gliomas into oligodendroglioma (IDH mutant and 1p19q codeleted), astrocytoma (IDH mutant), and GBM (IDH wildtype) [2]. In general, patients with IDH-mutated glioma have a better outcome compared to patients with IDH-wildtype GBM. IDH-mutant gliomas can be further differentiated by the *ATRX*, *CDKN2A/B*, and *TP53* status, whereas the *TERT* mutation, *EGFR* amplification status, *MGMT* promotor methylation status, and whole gain of chromosome 7 and whole loss of chromosome 10 status can further differentiate IDH-wildtype GBM [2–4].

Grade 4 GBM is the most common and aggressive primary malignant brain tumor and accounts for approximately 15% of all intracranial neoplasms and 45–50% of all primary malignant brain tumors [1]. The disease course is typically rapid, and despite intensive research and multimodal therapeutic approaches, outcomes remain very poor with a typical median overall survival (OS) of 11–20 months [5–7]. It is well known that important factors that contribute to the poor prognosis in GBM are inter- and intratumor molecular and cellular heterogeneity, inadequate drug or agent delivery across the blood brain barrier, redundant and anti-apoptotic signaling pathways, and an immunosuppressive tumor microenvironment (TME) [8,9].

Currently, the diagnosis and classification of diffuse gliomas is mainly based on neuroimaging combined with a histopathological and molecular analysis of tumor biopsies and resected tumor samples [2,10,11]. In addition to tissue biopsies, liquid biopsies have gained importance as a minimally invasive method which is generally well accepted by patients and can be used to identify different molecular biomarkers [12]. With respect to the significant level of heterogeneity in gliomas, especially GBM [8,9], the diagnostic value of tissue samples from singular locations is limited. As a consequence, blood-based biomarkers have particular value for the molecular characterization of diffuse gliomas and the potential to improve patient care in terms of diagnostics, prognostics, and the delivery of adaptive therapies. The major goal of this study was to investigate the potential of molecular and immunological biomarkers in the circulation of glioma patients at initial diagnosis to assess the presence and clinical course of different grade 3 and 4 gliomas. In this regard, levels of the highly conserved, major stress-inducible 70 kD heat shock protein 70 (Hsp70) in the circulation of tumor patients represents a promising candidate.

It is known that Hsp70 is overexpressed in the cytosol of a large variety of highly aggressive tumor entities, for example, non-small cell lung carcinoma (NSCLC) [13,14], colorectal carcinoma [15,16], and prostate carcinoma [17,18], as well as GBM [19–21]. Physiologically, Hsp70 is expressed in nearly all subcellular compartments where it controls protein homeostasis by regulating the maturation and degradation of proteins during cell growth and differentiation, preventing protein aggregation, and transporting other proteins across membranes [22,23]. Cellular stress, such as ischemia, hypoxia, or nutrient deprivation due to increased cell proliferation, which is typical for tumor cells, highly upregulates the synthesis of Hsp70 [23,24]. As cytosolic Hsp70 affects signaling pathways that regulate the apoptosis, proliferation, and differentiation of cells, elevated intracellular Hsp70 levels in tumor cells contribute to stress resistance by preventing apoptotic cell death [22–26].

In addition to its intracellular chaperoning functions, we have previously shown that Hsp70 plays a crucial role as a tumor-specific biomarker when it is presented on the plasma membrane of tumor cells ('membrane Hsp70', mHsp70) or when tumor-derived microvesicles expressing mHsp70 are released into the extracellular milieu [20,27,28]. Membrane Hsp70 (mHsp70) expression, which is enabled by a tumor-specific remodeling of the lipid composition [29], stabilizes tumor cell membranes and thereby mediates therapy resistance and increases tumor aggressiveness [30,31] and also serves as a tumor-specific target for activated natural killer (NK) cells [19,32]. With regards to the latter, it has been previously demonstrated that IL-2 and Hsp70-primed CD3−/CD56+ NK cells, which exhibit an increased cell surface density of the C-type lectin receptor complex CD94/NKG2A/C and the activatory receptor CD69, can recognize and kill tumor cells expressing mHsp70 via granzyme B-mediated apoptosis [33]. A randomized phase II clinical trial in patients with advanced NSCLC expressing mHsp70 demonstrated good tolerability and clinical responses after the i.v. injection of autologous NK cells that had been ex vivo activated with low-dose IL-2 and an Hsp70-derived 14mer peptide ('TKD') [34].

In the circulation of patients with cancer, Hsp70 exists either as a free protein or is expressed on the surface and in the lumen of extracellular lipid microvesicles [20,27,28]. Free Hsp70 predominantly originates from dying or inflamed cells, whereas vesicular Hsp70, which represents the majority of the circulating Hsp70 in patients with cancer, is actively released from viable tumor cells expressing mHsp70 [20,27,35] and has been shown to serve as a biomarker of viable tumor masses in human patients [36]. The newly developed Hsp70-exo enzyme-linked immunoassay (ELISA) can measure both free and vesicular Hsp70 in the peripheral blood of tumor patients [20], whereas the R&D Systems DuoSet® Hsp70 ELISA only detects free Hsp70 [20].

The expression of activating receptors (e.g., C-type lectin receptors and natural cytotoxicity receptors NCRs) enables NK cells to control tumor growth [37], and stress-induced danger-associated molecular patterns (DAMPs) in the circulation can trigger NK cell activity [38,39]. However, in highly aggressive tumor entities, the ability of NK cells to control tumor growth is often impaired by the immunosuppressive nature of the tumor microenvironment (TME) or by antigen loss [40]. The immunosuppressive TME in GBM is considered as one of the key features associated with treatment failure [41–43].

Because circulating levels of vesicular Hsp70, as measured using the Hsp70-exo ELISA, are known to correlate with viable tumor masses and to have an impact on the composition of circulating NK cell subpopulations [36,44], the aim of this study was to investigate circulating Hsp70 levels and the immunophenotype of lymphocyte subpopulations, with a focus on NK cells, in the peripheral blood at the initial diagnosis of patients with grade 3 and 4 gliomas. Of particular interest was the identification of a novel chaperone- and immune-related biomarker signature to improve the differentiation of glioma subtypes and to better predict disease prognosis and progression.

## 2. Materials and Methods

### 2.1. Patients

This study was approved by the local ethics committee of the Medical Faculty of the Klinikum rechts der Isar, Technische Universität München (TUM, 324/15s), and written informed consent was obtained from all patients with different types of brain tumors and healthy donors before the start of this study. Blood sampling (EDTA anticoagulation) was conducted in accordance with the Declaration of Helsinki in the Department of Neurology at the Klinikum rechts der Isar, TUM.

### 2.2. Immunohistochemical Analysis of IDH1 R132H Point Mutation

The *IDH1 R132H* point mutation was determined on deparaffinized and rehydrated formalin-fixed paraffin-embedded brain tumor sections by immunohistochemistry, as recommended by the guidelines. Sections with a high tumor burden were stained with the prediluted IDH1 antibody (DIA-H09-L RTU, Dianova, Hamburg, Germany) using the automated Ventana Benchmark XT stainer. A granular cytosolic staining pattern was considered as IDH-mutant positive. No sequencing was performed in the IDH-mutant-negative cases, but in unclear cases, a positive finding was confirmed by IDH1 and IDH2 sequencing. The *CDKN2A/B* gene status (2021 WHO CNS5) was determined in our hospital from the year 2020 onwards. Since the patients of this study were recruited between 2015 and 2019, the *CDKN2A/B* status was not assessed.

### 2.3. *Measurement of Intracellular Hsp70 by Immunohistochemistry on Tumor Sections*

Formalin-fixed paraffin-embedded (FFPE) sections (3 μm) of grade 3 and grade 4 brain tumors were stained with the cmHsp70.1 mAb, as described previously [18]. Briefly, after dewaxing and rehydrating in xylol (Karl Roth GmbH, Karlsruhe, Germany) and descending concentrations of ethanol (96% $v/v$ Otto Fischar GmbH, Saarbrücken, Germany), sections were heated at 60 °C for 30 min for antigen retrieval. After a washing and blocking step (5% $v/v$ rabbit serum in antibody diluent, S20022, Agilent Dako, St. Clara, CA, USA), sections were incubated with cmHsp70.1 mAb (1:500, multimmune GmbH, Munich, Germany) at 4 °C overnight. After incubation with an HRP-conjugated secondary antibody and staining with diaminobenzidine (DAB) chromogen, the nucleus was counterstained with hematoxylin and eosin (T865.2, Karl Roth GmbH).

### 2.4. *Measurement of Free and Vesicular Hsp70 in the Blood Using the Hsp70-exo ELISA*

Plasma prepared from EDTA anticoagulated blood by centrifugation (15 min at 1500× $g$) at 4 °C was aliquoted (300 μL) and stored at −80 °C. Levels of free and vesicular Hsp70 were determined in freshly prepared or frozen plasma samples using the Hsp70-exo ELISA protocol, as described recently [20]. Briefly, 96-well MaxiSorp Nunc Immuno plates (Thermo Fisher, Darmstadt, Germany) were coated overnight with the cmHsp70.2 mAb (1 μg/mL; multimmune GmbH, Munich, Germany) to which, after a washing and blocking step, plasma samples diluted in StabilZyme Select Stabilizer (1:5; Diarect GmbH, Freiburg i. Breisgau, Germany) in duplicate were added alongside an 8-point Hsp70 protein standard (1–100 ng/mL), also diluted in StabilZyme Select (1:5). After washing, plates were incubated with a biotinylated cmHsp70.1 mAb diluted in HRP Protector (200 ng/mL; Candor Biosciences, Candor Biosciences GmbH, Wangen i. Allgäu, Germany) for 30 min. After washing, BioFX-TMB substrate (Surmodics Inc., Eden Prairie, MN, USA) was added and the resulting absorbance read at 450 nm (reference absorbance at 570 nm) using a VICTOR™ X4 Multilabel Plate Reader (PerkinElmer, Waltham, MA, USA).

### 2.5. *Measurement of Free Hsp70 in the Blood Using the R&D Systems DuoSet® ELISA*

Levels of free Hsp70 in freshly prepared or thawed serum/plasma samples were determined using an R&D Systems DuoSet® Total Hsp70 ELISA (R&D Systems/Bio-Techne GmbH, Wiesbaden-Nordenstadt, Germany) in duplicates, following the manufacturer's protocol. Colorimetric measurements were performed on a VICTOR™ X4 Multilabel Plate Reader (PerkinElmer, Waltham, MA, USA), as described above. We have previously shown that there are no significant differences in the measured levels of Hsp70 between serum and plasma samples [20].

### 2.6. *Exosome Characterization*

Extracellular vesicles were isolated from the plasma (10 mL diluted in PBS, 1:1) by sequential centrifugation steps (Optima L-100XP, Beckman Coulter GmbH, Krefeld, Germany) at 300× $g$ for 10 min, 2000× $g$ for 20 min to remove cellular debris, 12,000× $g$ for 45 min to remove other microvesicles, and twice 100,000× $g$ for 70 min at room temperature. Freshly isolated extracellular vesicles were resuspended in PBS (pH 7.4) for size and affinity measurements and a flow cytometric analysis. The diameter of the vesicles was determined on a Zetasizer Nano SZ instrument (Malvern, Worcestershire, UK) by dynamic light scattering with a refractive index of 1.38 and an adsorption of 0.01, and the data were analyzed using the FLEX Analysis 3.4 software. The binding affinity ($K_D$) of the surface Hsp70-positive extracellular vesicles to cmHsp70.1 mAb (multimmune GmbH, Munich, Germany) was determined by Microscale Thermopheresis (MST) on the Monolith NT.115

(NanoTemper Technologies GmbH, Munich Germany), as described recently [20]. Typical exosomal cell surface markers such as CD9, CD81, and Hsp70 were measured by flow cytometry on a MACSQuant® instrument (Miltenyi Biotec, Bergisch Gladbach, Germany) using fluorescence-labeled mAbs.

### 2.7. Peripheral Blood Immunophenotyping by Multiparameter Flow Cytometry

EDTA anticoagulated blood was immunophenotyped using a BD FACSCalibur™ flow cytometer (BD Biosciences, Heidelberg, Germany). Briefly, 100 µL of EDTA blood was incubated with different panels of fluorescence-labeled antibody combinations. Combining four different fluorescence-labeled antibodies (FITC/PE/PerCP/APC) allowed for the following lymphocyte subpopulations to be determined: CD3−/CD19+ B cells, CD3+/CD45+ T cells, CD3+/CD4+ helper T cells, CD3+/CD8+ cytotoxic T cells, CD3+/CD69+ activated T cells, CD3+/NKG2D+ T cells, CD3+/CD94+ T cells, CD3+/CD4+/CD25+/FoxP3+ regulatory T cells, CD3+/CD8+/CD25+/FoxP3+ regulatory T cells, CD3+/CD56+ NK-like T cells, CD3−/CD56+, CD3−/CD16+, CD3−/CD94+, CD3−/NKG2D+, CD3−/NKp30+, CD3−/NKp46+ NK cells, and CD3−/CD69+ activated NK cells. The antibody combinations used in this study have been described previously [19]. Isotype-matched control antibodies were used as negative controls. The respective percentage of a certain lymphocyte subpopulation is defined as the proportion of cells stained positively for a specific antibody within a defined viable lymphocyte gate.

### 2.8. Flow Cytometry of Membrane-Bound Hsp70 on Viable Primary Brain Tumor Cells, Astrocytes, and Glioblastoma Cell Cultures

The flow cytometric analysis of mHsp70 on viable brain tumor cells and astrocytes has been described previously [19]. In short, tissue biopsies of human brain tumors were digested in a digestion medium (EDTA, 2 mM; L-cysteine 5 mM; papain, 5 U/mL) for 8 min at 37 °C and then forced through a sterile cell strainer (70 µm). Single-cell suspensions of human brain tumor biopsies and cultured primary human astrocytes (InnoProt, Valencia, Spain) and U87MG (ACTT HTB-14) glioblastoma cells were cultured in an astrocyte growth medium (P60101; InnoProt, Valencia, Spain, astrocyte basal medium supplemented with 10% *v*/*v* fetal bovine serum (FBS), astrocyte growth supplement, and penicillin/streptomycin) on poly-L-lysine coated culture flasks. After treatment with trypsin/EDTA for 2 min at 37 °C and a washing step, single cells were incubated with IgG1-FITC/APC/PE (BD Biosciences, Heidelberg, Germany), CD45-APC (FAB114A, R&D Systems, Minneapolis, MN, USA), cmHsp70.1-FITC (multimmune GmbH, Munich, Germany), pan-HLA class I-FITC (F5662, Sigma-Aldrich/Merck, Darmstadt, Germany) mAbs for 30 min at 4 °C in the dark. Cells were incubated with a 7-AAD viability stain (BD Biosciences) before the cytometric analysis to enable the specific analysis of viable cells. Only viable (7-AAD negative) and CD45-negative brain cells were analyzed using a BD FACSCalibur™ flow cytometer (BD Biosciences). The pan-HLA class I antibody staining provides a positive control and the isotype-matched control antibodies a negative control.

### 2.9. Statistical Analysis

Multiple comparisons among all groups were conducted using one-way ANOVA if the data were normally distributed or the Kruskal–Wallis test if the data were not normally distributed. Normal distribution was tested using the Shapiro–Wilk test. The comparison between the percentage of mHsp70-positive tumor cells in grade 3 vs. grade 4 glioma was conducted using the Mann–Whitney test. Comparisons of the Kaplan Meier curves were conducted using the log-rank test. Differences are considered to be statistically significant as follows: not significant, * $p < 0.05$, ** $p < 0.01$, *** $p < 0.001$, **** $p < 0.0001$.

## 3. Results

### 3.1. Patient Characteristics and Overall Survival

Between 2015 and 2019, samples from a total of 130 patients with different brain tumors were included in this clinical study, the aim of which was to identify novel biomarkers for diagnosis and predicting outcomes. As shown in the CONSORT diagram (Figure 1), 31 patients were diagnosed as having WHO grade 3 glioma ($n$ = 31), of which 15 were classified as oligodendroglioma ($n$ = 15) and 16 as astrocytoma ($n$ = 16). Another 99 patients were diagnosed as having WHO grade 4 GBM ($n$ = 99). Unexpectedly, none of the grade 4 GBM patients were *IDH R132H* mutant. An explanation for this finding could be the fact that potential cases with non-canonical IDH mutations (i.e., *IDH1 R132S/C/G* or *IDH2 R172* mutations) might have been missed. The male–female ratio of the patient cohort (76 to 23) was higher than the expected ratio of 1.6 to 1. The *CDKN2A/B* status was determined only from the year 2020 onwards in our hospital and, therefore, was not assessed in this patient cohort. Patients with WHO grade 1 or 2 glioma or central nervous system (CNS) metastasis were excluded from this study. EDTA anticoagulated blood was collected from therapy-naïve patients at the initial tumor diagnosis (t = 0) for measuring circulating free Hsp70 levels using the R&D Systems DuoSet® IC Human/Mouse/Rat Total Hsp70 ELISA and free and vesicular Hsp70 levels using the Hsp70-exo ELISA and for the immunophenotyping of major lymphocyte subpopulations. Circulating Hsp70 levels were assessed in the plasma samples of 118 healthy donors ($n$ = 118; age range 21–77 years), 10 patients with oligodendroglioma ($n$ = 10), 13 patients with astrocytoma grade 3 ($n$ = 13), and 76 patients with GBM ($n$ = 76). Multiparameter flow cytometric immunophenotyping was conducted in samples from 16 healthy donors ($n$ = 16; age range 21–85 years), 13 patients with oligodendroglioma grade 3 ($n$ = 13), 15 patients with grade 3 astrocytoma ($n$ = 15), and 81 patients with GBM ($n$ = 81). Patient characteristics, IDH, 1p19q status, and therapy regimens are summarized in Table 1.

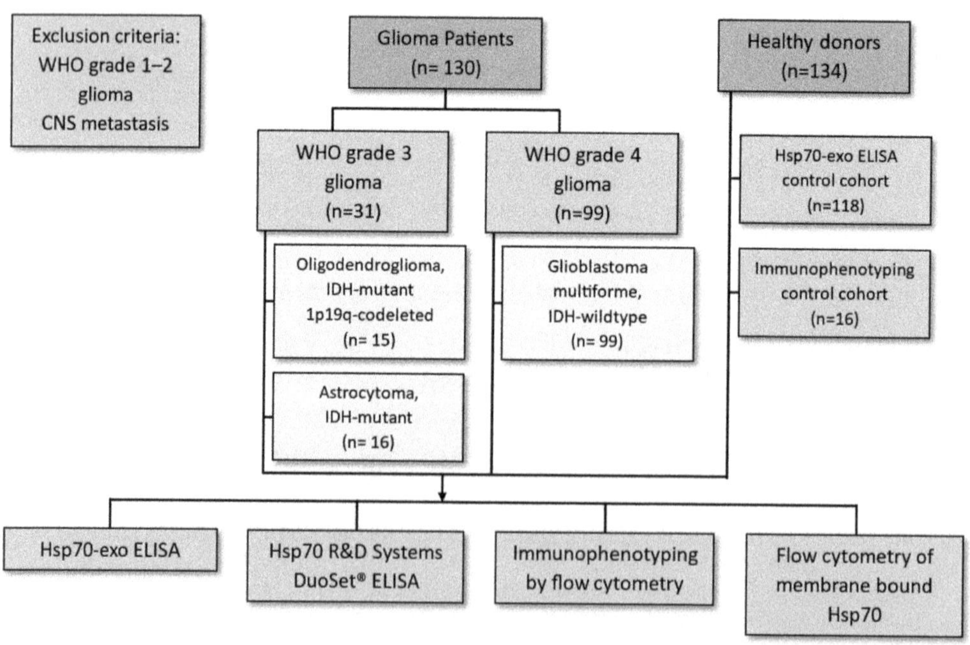


**Figure 1.** CONSORT diagram of the brain tumor patients and healthy controls included in this study.

Kaplan–Meier curves of the analyzed patient cohorts show a median overall survival (OS) for patients with oligodendroglioma and astrocytoma of 88 and 32 months, respectively

(Figure 2A, $p = 0.10$), a difference which does not reach statistical significance but is in accordance with data derived from other clinical studies [45–47]. The median OS of the GBM patients was 11 months, as determined by a Kaplan–Meier analysis, a finding which is also in accordance with other patient cohorts treated with multimodal therapies [5–7].

**Table 1.** Characteristics, IDH, and 1p19q status and therapy of patients with grade 3 glioma (oligodendroglioma and astrocytoma) and grade 4 glioblastoma (GBM).

| Parameters | Grade 3 Glioma | | Grade 4 Glioma |
|---|---|---|---|
| Tumor histology | Oligodendroglioma | Astrocytoma | GBM |
| Number of patients (*n*) | 15 | 16 | 99 |
| Gender (f/m) | 6/9 | 7/9 | 26/73 |
| Age range | 29–75 | 29–81 | 25–94 |
| Median age | 54 | 49 | 65 |
| Isocitrate dehydrogenase status (IDH) | IDH mutant | IDH mutant | IDH wildtype |
| 1p19q status | Codeleted | Non-codeleted | - |
| Therapy regimen | Maximal resection + RT + sequential PC(V) | Maximal resection + RT + sequential or simultaneous TMZ | Maximal resection + RT + simultaneous and sequential TMZ |

Abbreviations: RT, Radiotherapy (59.4 Gy/1.8 Gy or 60 Gy/2 Gy per day, 5× per week for both grade 3 and grade 4 glioma); PCV, Procarbazine, Lomustine, Vincristine; TMZ, Temozolomide.

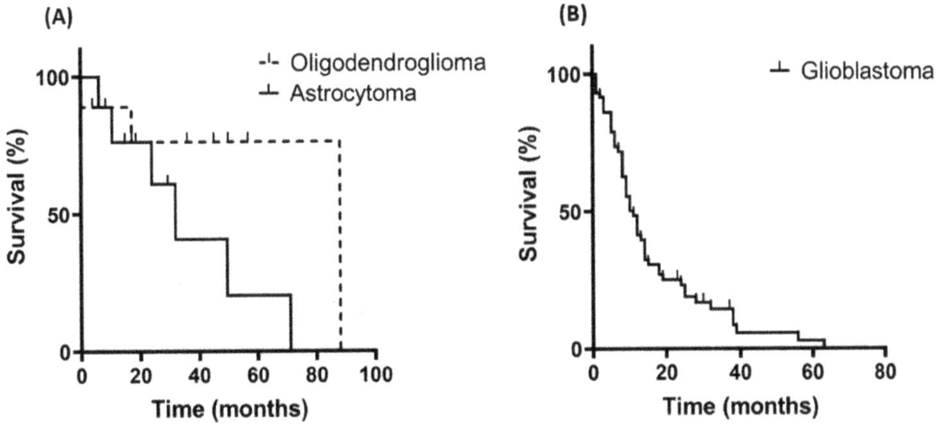


**Figure 2.** Kaplan–Meier curves of the overall survival (OS) of patients with grade 3 and grade 4 glioma. (**A**) OS of patients with oligodendroglioma ($n$ = 15; median survival 88 months) and astrocytoma ($n$ = 16; median survival 32 months). (**B**) OS of patients with glioblastoma (GBM; $n$ = 78; median survival 11 months).

### *3.2. Comparison of Intracellular and Circulating Hsp70 Levels in Patients with Oligodendroglioma, Astrocytoma, and Glioblastoma (GBM) and Characterization of Extracellular Vesicles*

Representative views of intracellular Hsp70 levels in tumor areas of FFPE sections (3 µm) derived from grade 3 and grade 4 gliomas determined by immunohistochemical staining with the cmHsp70.1 mAb revealed a higher Hsp70 staining intensity in grade 4 GBM (Figure 3). The surrounding normal brain tissue in grade 4 GBM showed a very low Hsp70 staining intensity. In line with previous data [18], grade 3 gliomas show predominantly a nuclear staining pattern, whereas grade 4 GBM shows a positive staining in both the nucleus and cytosol.

The concentration of free and vesicular Hsp70 in the plasma, as measured using the Hsp70-exo ELISA, is significantly elevated in patients with GBM (**** $p < 0.0001$) compared to the healthy donors (median Hsp70 levels: 45.4 ng/mL (GBM) versus 18.9 ng/mL (healthy) (Figure 4A). The median circulating Hsp70 levels in patients with astrocytoma (27.5 ng/mL) and oligodendroglioma (14.8 ng/mL) were not significantly different to those in the healthy individuals.

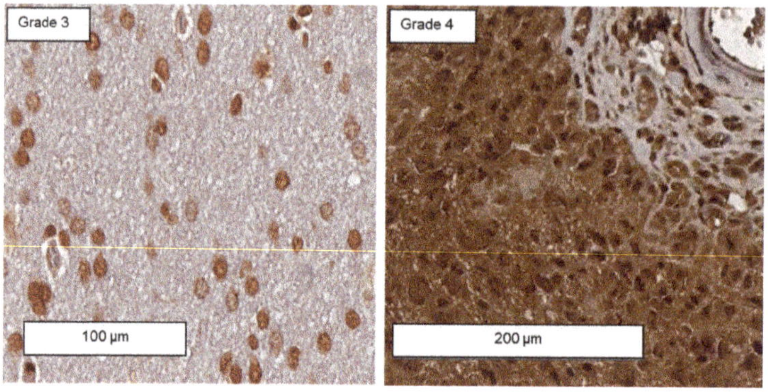


**Figure 3.** Representative immunohistochemical staining of intracellular Hsp70 levels in tumor sections of grade 3 and grade 4 gliomas using cmHsp70.1 mAb. Scale bars, 100 µm and 200 µm.

The concentration of free Hsp70 in the serum samples, as measured with the R&D Systems DuoSet® Total Hsp70 ELISA, was also significantly elevated in patients with GBM (** $p < 0.01$) compared to the healthy individuals (Figure 4B), with median Hsp70 levels of 3.5 ng/mL (GBM) and 2.6 ng/mL (healthy), respectively. The median levels of free Hsp70 in patients with oligodendroglioma and astrocytoma were 2.4 ng/mL and 3.4 ng/mL, respectively, and these levels were not significantly different to those in the healthy control cohort. Overall, the level of circulating vesicular Hsp70 (Figure 4A) in tumor patients is significantly higher than the level of free Hsp70 (Figure 4B).

The characteristics of the extracellular vesicles derived from a patient with brain tumor were determined by measuring the size by dynamic light scattering using the Zeta sizer (Figure 4C), the binding affinity (Figure 4D) of Hsp70 carrying exosomes to cmHsp70.1 mAb, and typical exosomal surface markers (Figure 4E) such as the tetraspanins CD9, CD81, and mHsp70 by flow cytometry. A representative example of the average diameter (100 nm), $K_D$ (9.3 nM), and the surface expression of CD9 (27.3%), CD81 (29.4%), and Hsp70 (98.2%) are illustrated in Figure 4C–E. The Hsp70 value determined with the Hsp70-exo ELISA in the exosomal fraction of this patient was 129.5 ng/mL and that of the exosome-depleted fraction of the same patient was 14.8 ng/mL.

### 3.3. Immunophenotyping of Major Lymphocyte Subpopulations in the Peripheral Blood of Patients with Oligodendroglioma, Astrocytoma, and Glioblastoma (GBM)

A multiparameter flow cytometric analysis of whole blood samples from 109 patients with glioma using the fluorescence-labeled mAbs described in the Materials and Methods section was undertaken to measure the frequencies of the following major lymphocyte subpopulations: CD3−/CD19+ B cells, CD3+/CD45+ T cells, CD3+/CD4+ helper T cells, CD3+/CD8+ cytotoxic T cells, CD3+/CD69+ activated T cells, CD3+/NKG2D+ T cells, CD3+/CD94+ T cells, CD3+/CD4+/CD25+/FoxP3+ regulatory T cells, CD3+/CD8+/CD25+/FoxP3+ regulatory T cells, CD3+/CD56+ NK-like T cells, CD3−/CD56+, CD3−/CD16+, CD3−/CD94+, CD3−/NKG2D+, CD3−/NKp30+, CD3−/NKp46+ NK cells, and CD3−/CD69+ activated NK cells. Differences in the composition of the lymphocyte subpopulations were seen only in CD3+/CD4+ helper T cells, CD3+/CD8+ cytotoxic T cells, CD3−/CD56+ NK cells, CD56+/CD94+ NK cells, and CD56+/CD69+ NK cells when compared to the healthy control cohort (Figure 5).

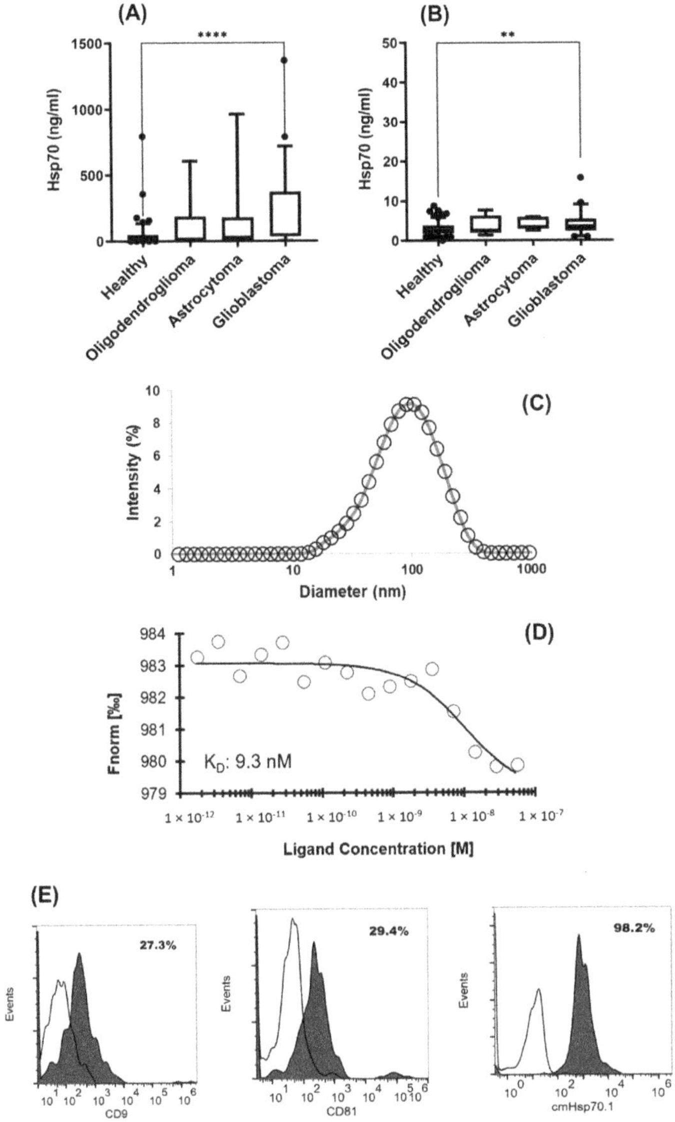


**Figure 4.** (**A**) Free and vesicular Hsp70 (ng/mL) levels measured in the plasma of healthy donors ($n$ = 118; Hsp70 median, 18.9 ng/mL) and patients with oligodendroglioma ($n$ = 10; Hsp70 median, 14.8 ng/mL), astrocytoma ($n$ = 13; Hsp70 median, 27.5 ng/mL), and GBM ($n$ = 76; Hsp70 median, 45.4 ng/mL), as measured using the Hsp70-exo ELISA. (**B**) Free Hsp70 levels measured in the serum of healthy donors ($n$ = 150; Hsp70 median, 2.6 ng/mL) and patients with oligodendroglioma ($n$ = 7; Hsp70 median, 2.4 ng/mL), astrocytoma ($n$ = 5, Hsp70 median, 3.4 ng/mL), and GBM ($n$ = 51; Hsp70 median, 3.5 ng/mL) using the DuoSet® IC Human/Mouse/Rat Total Hsp70 ELISA. Statistically significant differences ** $p < 0.01$, **** $p < 0.0001$. (**C**) Size distribution of extracellular vesicles, (**D**) binding affinity ($K_D$) of extracellular vesicles to cmHsp70.1 mAb, and (**E**) typical exosomal surface markers such as CD9, CD81, and Hsp70 on extracellular vesicles derived from a patient with astrocytoma.

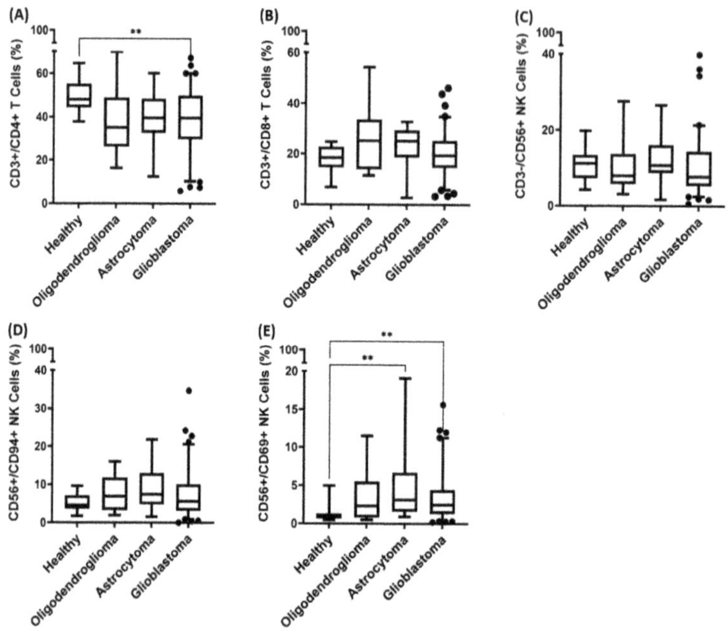


**Figure 5.** Proportions of lymphocyte subsets in the peripheral blood of healthy donors ($n$ = 16) and patients with oligodendroglioma ($n$ = 13), astrocytoma ($n$ = 15), and glioblastoma (GBM) ($n$ = 81). (**A**) CD3+/CD4+ helper T cells, as determined by multiparameter flow cytometry. (**B**) CD3+/CD8+ cytotoxic T cells, (**C**) CD3−/CD56+ NK cells, (**D**) CD56+/CD94+ NK cells, and (**E**) CD56+/CD69+ NK cells. Statistically significant differences ** $p < 0.01$.

The frequency of CD3+/CD4+ helper T cells in the peripheral blood of all patients with different malignant brain tumor entities was lower than that in the healthy control cohort (Figure 5A). Patients with GBM exhibited a significantly lower frequency of CD3+/CD4+ T cells than that in the healthy control (median values of 39.5% (GBM) versus 47.9% (healthy donors), ** $p < 0.001$). Although patients with oligodendroglioma (median 35.0%; $p = 0.066$) and astrocytoma (median 39.5%; $p = 0.085$) also exhibited reduced proportions of CD3+/CD4+ helper T cells compared to the healthy control cohort, these differences did not reach statistical significance.

Although the median frequency of CD3+/CD8+ cytotoxic T cells (Figure 5B) was marginally increased in the blood of patients with oligodendroglioma, astrocytoma, and GBM (median frequencies of 25.2%, 25.1%, and 19.4%, respectively), no significant differences were observed when compared to the healthy controls (median frequency of 18.4%).

The prevalence of CD3−/CD56+ NK cells (Figure 5C) is similar in patients with oligodendroglioma, astrocytoma, and GBM (median frequencies of 7.9%, 10.7%, and 7.7%, respectively), with no significant differences to the frequency of this cell type in the healthy controls (median frequency of 11.1%). In contrast, the frequency of activated CD56+/CD69+ NK cells is significantly increased in patients with astrocytoma (** $p < 0.01$) and GBM (** $p < 0.01$) compared to the healthy controls (median frequencies of 3.1% (astrocytoma) and 2.5% (GBM) versus 0.9% (healthy donors)). Patients with oligodendroglioma also showed a slightly elevated frequency of CD56+/CD69+ NK cells (median 2.3%); however, this difference did not reach statistical significance.

Patients with oligodendroglioma, astrocytoma, and GBM exhibited elevated median frequencies of CD56+/CD94+ NK cells compared to the healthy controls (6.9%, 7.4%, and 5.6%, respectively, versus 4.5% in the healthy control cohort, Figure 5D). Although this trend was similar to that observed for CD56+/CD69+ NK cells, no statistically significant differences were reached.

### 3.4. Correlation of Hsp70$^{low}$/Hsp70$^{high}$ Levels and Frequencies of NK Cell Subpopulations in Patients with Grade 3 and Grade 4 Gliomas

To investigate the potential relationship between and influence of circulating Hsp70 levels on the composition of NK cell subpopulations in the peripheral blood, we segregated patients with grade 3 and grade 4 glioma into groups with low (Hsp70$^{low}$) and high (Hsp70$^{high}$) circulating Hsp70 levels according to the median circulating Hsp70 levels of each cohort. Patients with grade 3 glioma with Hsp70 plasma levels below 29.8 ng/mL were classified as Hsp70$^{low}$, and patients above this value were classified as Hsp70$^{high}$. For the grade 3 glioma patients, the Hsp70$^{low}$ group comprises 4 patients with oligodendroglioma and 7 patients with astrocytoma ($n = 11$), whereas the Hsp70$^{high}$ group comprises 5 patients with oligodendroglioma and 6 patients with astrocytoma ($n = 11$). Patients with grade 4 glioma, which were all diagnosed as GBM, were considered as Hsp70$^{low}$ ($n = 30$) when their plasma Hsp70 levels were below 45.4 ng/mL (Figure 4A), and patients with plasma levels above this value were considered as Hsp70$^{high}$ ($n = 32$). The Hsp70$^{low}$ and Hsp70$^{high}$ groups were subsequently compared with respect to the frequency of CD3+/CD4+ helper T cells, CD3+/CD8+ cytotoxic T cells, CD3−/CD56+, CD56+/CD94+, and CD56+/CD69+ NK cells. Significant differences in the lymphocyte frequencies between the Hsp70$^{low}$ and Hsp70$^{high}$ glioma patients were observed for CD3−/CD56+, CD56+/CD94+, and CD56+/CD69+ NK cells (Figure 6). None of the other studied major lymphocyte populations showed significant differences between the Hsp70$^{low}$ and Hsp70$^{high}$ groups of patients.

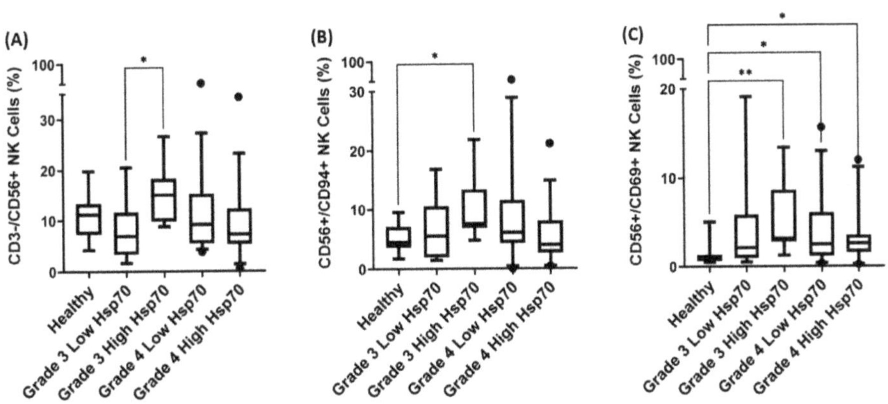


**Figure 6.** Correlation of low and high Hsp70 values in the circulation and the proportion of different NK cell subpopulations in patients with grade 3 glioma (Hsp70$^{low}$ $n = 11$; Hsp70$^{high}$ $n = 11$) and grade 4 GBM (Hsp70$^{low}$ $n = 30$; Hsp70$^{high}$ $n = 32$) compared to a healthy control cohort ($n = 16$). (**A**) CD3−/CD56+ NK cells in grade 3 and grade 4 GBM patients with high (Hsp70$^{high}$) and low (Hsp70$^{low}$) Hsp70 plasma levels. (**B**) CD56+/CD94+ NK cells in grade 3 and grade 4 GBM patients with high (Hsp70$^{high}$) and low (Hsp70$^{low}$) Hsp70 plasma levels. (**C**) CD56+/CD69+ NK cells in grade 3 and grade 4 GBM patients with high (Hsp70$^{high}$) and low (Hsp70$^{low}$) Hsp70 plasma levels. Statistically significant differences * $p < 0.05$, ** $p < 0.01$.

Patients with Hsp70$^{high}$ grade 3 glioma exhibited a significantly elevated frequency of CD3−/CD56+ NK cells (Figure 6A) compared to the Hsp70$^{low}$ patient group (* $p < 0.05$), with median CD3−/CD56+ NK cell frequencies of 15.1% versus 6.9%, respectively. The frequency of CD3−/CD56+ NK cells in patients with Hsp70$^{high}$ and Hsp70$^{low}$ grade 4 GBM did not show any significant differences (Figure 6A).

There were no statistically significant differences in the frequency of CD56+/CD94+ NK cells in the peripheral blood of patients with Hsp70$^{low}$ and Hsp70$^{high}$ grade 3 glioma

(Figure 6B). Nevertheless, the frequency of CD56+/CD94+ NK cells in patients with $Hsp70^{high}$ grade 3 glioma were significantly higher than those in the healthy controls (* $p < 0.05$), whereas there was no significant difference between patients with $Hsp70^{low}$ grade 3 glioma and the controls. Similarly, the frequency of CD56+/CD69+ NK cells in patients with $Hsp70^{high}$ grade 3 glioma was significantly higher than that in the controls (** $p < 0.01$, Figure 6B), whereas there was no difference between patients with $Hsp70^{low}$ grade 3 glioma and the controls. $Hsp70^{high}$ and $Hsp70^{low}$ patients with grade 4 GBM did not exhibit any significant differences in the frequencies of CD56+/CD94+ or CD56+/CD69+ NK cells (Figure 6A–C).

A comparison of the frequencies of the NK cell subpopulations indicated above in patients with oligodendroglioma and astrocytoma with low and high circulating Hsp70 levels revealed significantly higher frequencies in CD3−/CD56+, CD56+/CD94+, and CD56+/CD69+ NK cells in patients with oligodendroglioma and high Hsp70 levels compared to those with low Hsp70 levels and/or with healthy individuals (Supplementary Figure S1).

### 3.5. Correlation of $Hsp70^{low}$/$Hsp70^{high}$ Levels, Lymphocyte Frequencies, and Overall Survival in Patients with Oligodendroglioma, Astrocytoma, and Glioblastoma (GBM)

In addition to its potential use as a diagnostic marker for glioma (Figure 4), circulating Hsp70 levels may also provide a prognostic value, particularly for patients with GBM who exhibit significantly elevated levels of circulating Hsp70 (Figure 4) compared to healthy individuals. To investigate the association of circulating Hsp70 levels on the Overall Survival (OS) of patients with GBM, we segregated the GBM patients into groups exhibiting low and high levels of circulating Hsp70 ($Hsp70^{low}$ and $Hsp70^{high}$) according to the median circulating Hsp70 levels in each cohort. GBM patients with circulating Hsp70 levels below a concentration of 45.4 ng/mL were considered $Hsp70^{low}$ ($n = 32$), whereas patients above this concentration were considered as being $Hsp70^{high}$ ($n = 27$). Although not statistically different, the $Hsp70^{low}$ patients had a median survival of 11 months, whereas the $Hsp70^{high}$ patients had a median survival of 10 months (Figure 7, $p > 0.05$). Patients with grade 3 glioma, who generally had lower circulating Hsp70 levels than the patients with GBM, did not show any differences in the OS when classified into $Hsp70^{low}$ and $Hsp70^{high}$ groups.

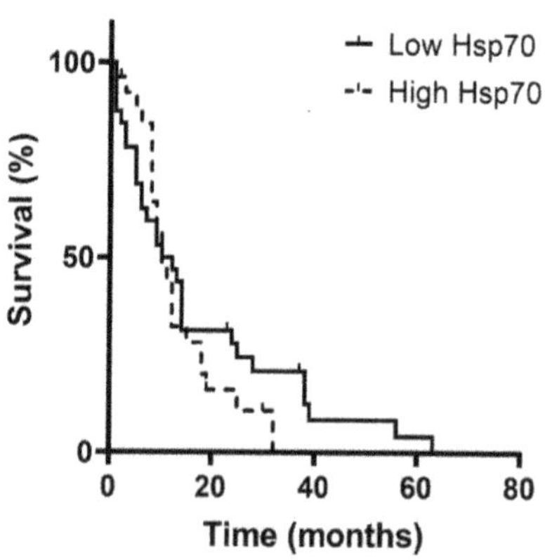


**Figure 7.** Correlation of circulating $Hsp70^{low}$ and $Hsp70^{high}$ values and overall survival (OS) in patients with grade 4 GBM (low, $n = 32$; high, $n = 27$), as determined by the Kaplan–Meier analysis.

To determine possible associations between the frequencies of certain lymphocyte subpopulations and the OS of patients with glioma, patients with grade 3 and grade 4 glioma were segregated into groups having low or high frequencies of different lymphocyte subpopulations such as CD3+/CD4+ helper T cells, CD3+/CD8+ cytotoxic T cells,

CD3−/CD56+, CD56+/CD94+, and CD56+/CD69+ NK cells. The cut-off value for the low and high groups corresponds to the median lymphocyte frequency of each group. In the grade 3 glioma patients, significant differences in the OS were only observed for low and high CD3−/CD56+ and CD56+/CD94+ NK cells (Figure 8), but not for other lymphocyte subpopulations.

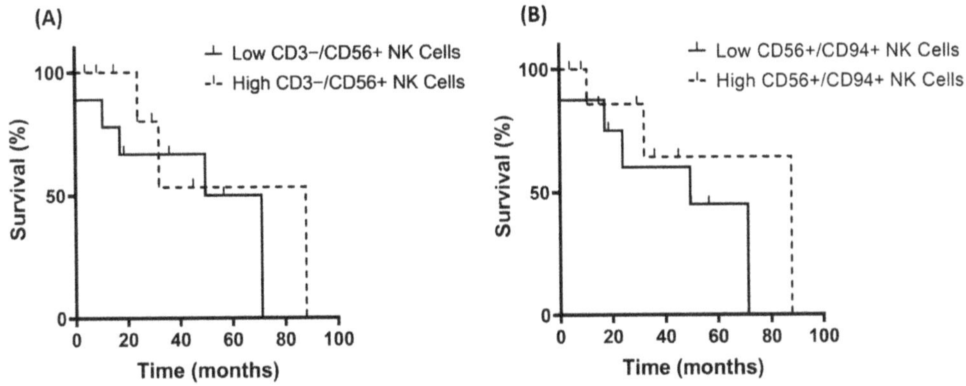


**Figure 8.** Correlation of the frequency of CD3−/CD56+ and CD56+/CD94+ NK cell subpopulations and the overall survival (OS) in patients with grade 3 glioma (low NK cells $n$ = 14; high NK cells $n$ = 14) as determined by the Kaplan–Meier analysis. (**A**) OS in grade 3 glioma patients with low and high frequencies of CD3−/CD56+ NK cells. (**B**) OS in grade 3 glioma patients with low and high frequencies of CD56+/CD94+ NK cells.

Patients with grade 3 glioma having a frequency of CD3−/CD56+ NK cells below a cut-off value of 10% were considered as part of the low CD3−/CD56+ NK cell group ($n$ = 14; comprising 8 oligodendroglioma and 6 astrocytoma patients), whereas patients with a frequency of these NK cells above 10% were considered as being in the high CD3−/CD56+ NK cell group ($n$ = 14; 5 oligodendroglioma, 9 astrocytoma). Patients with grade 3 glioma exhibiting a low frequency of CD3−/CD56+ NK cells had a median OS of 60 months, whereas patients with a high frequency of CD3−/CD56+ NK cells had a median OS of 88 months (Figure 8A, $p > 0.05$). Due to the relatively small patient cohorts, the differences in the OS did not reach statistical significance. In contrast, the patients with grade 4 GBM showed no differences in OS, irrespective of the frequency of CD3−/CD56+ NK cells in the peripheral blood.

Similar results were observed for CD56+/CD94+ NK cells. Patients with grade 3 glioma exhibiting a frequency of CD56+/CD94+ NK cells below a cut-off value of 7.2% were considered as the low CD56+/CD94+ NK cell group ($n$ = 14; 7 oligodendroglioma, 7 astrocytoma), whereas patients with a frequency of CD56+/CD94+ NK cells above 7.2% were considered as the high CD56+/CD94+ NK cell group ($n$ = 14; 6 oligodendroglioma, 8 astrocytoma). Patients with a low frequency of CD56+/CD94+ NK cells had a median OS of 50 months, whereas patients with a high frequency of CD56+/CD94+ NK cells had a median OS of 88 months (Figure 8B, $p > 0.05$). Patients with grade 4 GBM showed no differences in OS, irrespective of their CD56+/CD94+ NK cell frequency.

### *3.6. Differences in the Expression of Membrane-Bound Hsp70 in Grade 3 and Grade 4 Gliomas, Astrocytes, and Cultured Glioblastoma Cells*

The percentage of viable tumor cells expressing mHsp70, as measured by flow cytometry using the cmHsp70.1 mAb, is significantly higher in grade 3 gliomas (94.3%) compared to grade 4 GBM (64.7%, ** $p < 0.01$) (Figure 9). All grade 3 gliomas included show an IDH mutation, whereas all grade 4 GBM are IDH-wildtype. A comparison of the mHsp70 expression on grade 3 oligodendroglioma and astrocytoma revealed no statistically significant differences (Supplementary Figure S2).

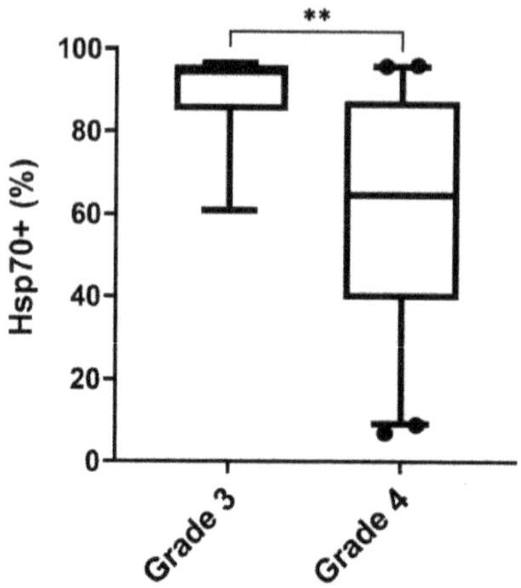


**Figure 9.** Differences between the percentages of grade 3 glioma ($n$ = 6) and grade 4 GBM ($n$ = 45) expressing mHsp70. Statistically significant differences ** $p < 0.01$.

However, a comparison of the mHsp70 expression as determined by flow cytometry on cultured U87MG glioblastoma cells and primary human astrocytes revealed an mHsp70 positivity on viable U87MG cells (61.5%) but not on primary astrocytes (1.7%) (Supplementary Figure S3). Moreover, circulating Hsp70 was detectable only in the supernatant of cultured U87MG cells but not in that of primary human astrocytes, as determined by ELISA. These data indicate that soluble Hsp70 originates from membrane Hsp70-positive glioma cells but not from primary human astrocytes.

## 4. Discussion

Despite multimodal therapeutic approaches such as maximal resection and adjuvant radiochemotherapy, the mortality of intracranial gliomas remains disproportionately high [48]. The lack of well-established biomarkers for diffuse adult-type gliomas highlights the need for a tumor-specific, universally applicable biomarker which can be measured in liquid biopsies. The expression of the major stress-inducible heat shock protein 70 (Hsp70) is uniquely elevated in a variety of tumor types, including GBM [19–21]. Al-though damaged or necrotic tumor cells release free Hsp70 from subcellular compartments into the surrounding tumor environment, a significant amount of Hsp70 is actively released by viable tumor cells in the form of vesicular Hsp70 [20,27,28].

The Hsp70-exo ELISA used in this study utilizes the unique ability of the cmHsp70.1 and cmHsp70.2 mAbs to detect circulating levels of both free and vesicular Hsp70 derived from mHsp70-positive tumor cells in the form of a liquid biopsy [20]. We show that patients with GBM exhibit significantly higher levels of free and vesicular circulating Hsp70 when compared to a healthy cohort, a result which is in line with previous studies [19–21]. Crucially, we show that vesicular Hsp70 represents the vast majority of the total circulating Hsp70 in all patient groups studied, including oligodendroglioma, astrocytoma, and GBM, thereby reinforcing the potential value of biomarkers expressed on the surface of extracellular vesicles such as exosomes for gliomas. This finding suggests that the blood–brain barrier does not pose a large obstacle to the measurement of vesicular Hsp70 in liquid biopsies for gliomas that are localized within the central nervous system. As free Hsp70 can be elevated in a variety of systemic inflammatory diseases such as COPD or liver cirrhosis [14,49], membrane-bound Hsp70, selectively expressed on the surface of tumor cells and extracellular vesicles, may be key in differentiating between inflammation and malignant cancer diseases.

As known from other studies, elevated levels of circulating Hsp70 are not only indicative of the presence of a tumor but also correlate with the gross tumor volume, aggressiveness, and cancer stage [20,36]. For tumors of the central nervous system, the extent of damage to the blood–brain barrier may be another, tumor-specific event which results in elevated levels of circulating Hsp70 in different glioma types and stages of disease [50]. In line with this, we show that the elevation of Hsp70 is greater in grade 4 GBM when compared to grade 3 glioma, and this might be reflective of the more aggressive nature of this tumor grade. No differences were observed in circulating Hsp70 levels in grade 3 oligodendroglioma and astrocytoma. Additionally, patients with grade 4 GBM and elevated levels of circulating Hsp70 at the time of initial tumor diagnosis show a trend towards a lower overall survival, compared to patients with initially low Hsp70 levels. This reinforces the proposition that Hsp70 levels in patients with grade 4 GBM might also have prognostic implications, which has previously been shown for patients with colon carcinoma [51,52]. In this regard, a measurement of Hsp70 in regular intervals after tumor diagnosis, which is possible when using a liquid biopsy-based approach, could provide more insight into the development of the disease over time and might offer interesting and potentially important prospects for earlier therapy adaptations.

The overexpression of Hsp70 imparts several, in part conflicting, functions to tumor cells [28]. As a molecular chaperone, Hsp70 is integral to the protein folding, correction, or removal of misfolded proteins and the prevention of protein aggregation, all of which are vital functions for tumor cells experiencing high levels of cellular stress [22–24]. Additionally, cytosolic Hsp70 is involved in the inhibition of apoptotic pathways in aberrant tumor cells [22–26]. Membrane-bound Hsp70 stabilizes membranes and mediates therapy resistance [30,31]. On the other hand, membrane-bound Hsp70 contributes to the activation of a patient's immune system and thereby tumor control by stimulating lymphocytes via antigen cross-presentation [44,53]. In contrast to T cells, NK cells play a pivotal role in the ability of a patient's immune system to provide early tumor defense without prior antigen stimulation [54]. The targeted response of NK cells to membrane-bound Hsp70 on viable tumor cells has been shown previously [32–34]. One identified pathway is the interaction between mHsp70 and the heterodimer of the CD94 lectin receptor and NKG2A or NKG2C on NK cells primed with the TKD 14mer peptide (TKDNNLLGRFELSG, aa 450–463), a sequence derived from the Hsp70 protein expressed on the membrane of tumor cells [32–34]. Ultimately, Hsp70-primed NK cells in the presence of IL-2 induce cell death via granzyme B-mediated apoptosis [33,40]; however, associations with NKG2A, NKG2C, NKG2D, and the natural cytotoxicity receptors (NCRs) NKp30, NKp44, and NKp46 [55] suggest that a variety of both activatory and inhibitory pathways can be initiated via membrane-bound Hsp70. The response ultimately depends on the immunocompetence of each individual patient and the TME, as an adequate proinflammatory environment, including factors such as the interleukins IL-2 and IL-15, has been shown to amplify the stimulatory effect of membrane-bound Hsp70 on NK cells [56–58]. Interleukin 1 is upregulated in GBM and exhibits suppressing effects mediated by IL-2 [59–62]. Other factors such as IL-10 and TGF-β and cell surface changes such as the upregulation of PD-L1 may further inhibit tumor control by NK cells in GBM [63–65]. A significant reduction in CD4+ helper T cells, as shown in this study, contributes to the immunosuppressive effects induced by GBM.

Although both, grade 3 glioma and grade 4 GBM, show membrane positivity for Hsp70 [19], the flow cytometry of tumor cell suspensions using the cmHsp70.1 mAb allows us to quantify the percentage of tumor cells expressing Hsp70 on the cell membrane. We show that the percentage of tumor cells expressing mHsp70 is greater in grade 3 glioma compared to grade 4 GBM. This finding is in line with data showing that mutations of the tumor suppressor protein p53, which are more frequently found in grade 4 GBM compared to grade 3 glioma, are associated with a reduced Hsp70 expression [66,67]. Antigen presenting MHC class I and II molecules can be downregulated in invasive GBM to escape the T cell-mediated immune responses [68]. We show that mHsp70 follows this trend and may induce similar immunosuppressive effects towards NK cells together with decreased CD3+/CD4+ T cell counts.

Patients with grade 3 glioma retain a greater ability to stimulate NK cells, possibly due to a higher mHsp70 expression density on their tumor cells in a less immunosuppressive TME. Patients with grade 3 glioma expressing high levels of Hsp70 exhibit a significantly increased frequency of CD3−/CD56+ NK cells, thereby supporting the potency of mHsp70 and vesicular Hsp70 in stimulating NK cells in these patients. Additionally, an improved OS in grade 3 glioma patients was found to be associated with a higher frequency of CD3−/CD56+/CD94+ NK cells and an upregulated expression of the activatory marker CD69 on NK cells. These data suggest that activated NK cells can effectively target mHsp70-positive tumor cells and thereby slow the disease progression. In patients with grade 4 GBM with significantly decreased proportions of CD4+ T cells and a highly immunosuppressed TME, elevated Hsp70 levels in the circulation are unable to stimulate an Hsp70-mediated NK cell activity.

The heterogeneous group of brain tumors exhibit very different outcomes in terms of survival and therapeutic response that can be estimated by WHO grading and molecular biomarkers [2–4]. An increased expression of mHsp70 on the cell surface of tumor cells may be beneficial for the recognition by activated NK cells to control tumor growth in grade 3 oligodendroglioma and astrocytoma and thereby increasing OS, whereas lower levels of mHsp70 expression and an immunosuppressive TME [9,69–71] may inhibit this effect in patients with grade 4 GBM. In a randomized phase II clinical trial, we have shown previously that the adoptive transfer of ex vivo Hsp70-peptide TKD and IL-2-activated autologous (patient-derived) NK cells provides a survival benefit in patients with advanced non-small-cell lung carcinoma [34]. Since these data have been confirmed in a rat model of GBM [57], we assume that Hsp70-activated NK cells may also provide similarly beneficial effects in human glioma patients.

## 5. Conclusions

In summary, we show that Hsp70 in the circulation and on the membrane of tumor cells, in the presence of IL-2, stimulates NK cells in patients with grade 3 glioma and possibly confers a survival benefit to these patients, whereas in patients with grade 4 GBM with low CD4+ helper T cells, the NK cell stimulation is hindered. However, in the latter patients, circulating levels of Hsp70 are significantly elevated from the onset of the tumor disease and may serve as a diagnostic and prognostic tumor marker.

## 6. Patents

The Hsp70-exo ELISA is patented by multimmune GmbH, Munich, Germany (US 11,460,472—other applications pending).

**Supplementary Materials:** The following supporting information can be downloaded at: https://www.mdpi.com/article/10.3390/biomedicines11123235/s1, Figure S1: Correlation of low and high circulating Hsp70 values and different NK cell subpopulations CD3−/CD56+ (A), CD56+/CD94+ (B), CD56+/CD69+ (C) in patients with oligodendroglioma and astrocytoma; Figure S2: Percentage of mHsp70 expressing cells in grade 3 oligodendroglioma and astrocytoma; Figure S3: Gating strategy and representative histograms of cultured mHsp70 positive U87 cells (A) and primary human astrocytes (B).

**Author Contributions:** Writing—original draft preparation—methodology, investigation, and formal analysis, P.L. and D.T.; methodology and investigation, A.B.D., M.P., V.M., H.Z., K.A., M.H.K., G.R., M.B., M.S. and G.A.F.; clinical material and data analysis, D.L. and F.S.-G.; proof-reading and data curation, A.G.P.; conceptualization, data curation, writing and editing, validation, funding acquisition, and project administration, G.M. All authors have read and agreed to the published version of the manuscript.

**Funding:** This research was funded by grants of BMBF 02NUK064A, BMWi, ZF4320102CS7, Bayerische Forschungsstiftung BAYCELLator AZ-1568-22, Dr med Erwin Braun Stiftung.

**Institutional Review Board Statement:** This study was conducted in accordance with the Declaration of Helsinki and approved by the Institutional Review Board (Ethics Committee, 324/15s, 28 June 2015) of the Medical Faculty of the University Hospital Klinikum rechts der Isar, Technische Universität München (TUM).

**Informed Consent Statement:** Informed consent was obtained from all subjects involved in this study.

**Data Availability Statement:** The data presented in this study are available upon request from the corresponding author.

**Acknowledgments:** The authors want to thank Anett Lange for editorial work, as well as Nicholas Taylor and Erika Roberts for help with the ELISA measurements.

**Conflicts of Interest:** The authors declare no conflict of interest. G.M. is Founder and CSO of multimmune GmbH, Munich, Germany. A.G.P. is the CEO of multimmune GmbH. The funders have no role in the design of this study, in the collection of material, analysis, or interpretation of the data, in the writing of the manuscript, or in the decision to publish the results.

## References

1. Ostrom, Q.T.; Patil, N.; Cioffi, G.; Waite, K.; Kruchko, C.; Barnholtz-Sloan, J.S. CBTRUS Statistical Report: Primary Brain and Other Central Nervous System Tumors Diagnosed in the United States in 2013–2017. *Neuro-Oncology* **2020**, *22*, iv1–iv96. [CrossRef] [PubMed]
2. Louis, D.N.; Perry, A.; Wesseling, P.; Brat, D.J.; Cree, I.A.; Figarella-Branger, D.; Hawkins, C.; Ng, H.K.; Pfister, S.M.; Reifenberger, G.; et al. The 2021 WHO Classification of Tumors of the Central Nervous System: A Summary. *Neuro-Oncology* **2021**, *23*, 1231–1251. [CrossRef] [PubMed]
3. Yan, H.; Parsons, D.W.; Jin, G.; McLendon, R.; Rasheed, B.A.; Yuan, W.; Kos, I.; Batinic-Haberle, I.; Jones, S.; Riggins, G.J.; et al. IDH1 and IDH2 Mutations in Gliomas. *N. Engl. J. Med.* **2009**, *360*, 765–773. [CrossRef] [PubMed]
4. Whitfield, B.T.; Huse, J.T. Classification of Adult-Type Diffuse Gliomas: Impact of the World Health Organization 2021 Update. *Brain Pathol.* **2022**, *32*, e13062. [CrossRef] [PubMed]
5. Stupp, R.; Mason, W.P.; van den Bent, M.J.; Weller, M.; Fisher, B.; Taphoorn, M.J.B.; Belanger, K.; Brandes, A.A.; Marosi, C.; Bogdahn, U.; et al. Radiotherapy plus Concomitant and Adjuvant Temozolomide for Glioblastoma. *N. Engl. J. Med.* **2005**, *352*, 987–996. [CrossRef] [PubMed]
6. Stupp, R.; Taillibert, S.; Kanner, A.; Read, W.; Steinberg, D.M.; Lhermitte, B.; Toms, S.; Idbaih, A.; Ahluwalia, M.S.; Fink, K.; et al. Effect of Tumor-Treating Fields Plus Maintenance Temozolomide vs Maintenance Temozolomide Alone on Survival in Patients With Glioblastoma: A Randomized Clinical Trial. *JAMA* **2017**, *318*, 2306–2316. [CrossRef]
7. Witthayanuwat, S.; Pesee, M.; Supaadirek, C.; Supakalin, N.; Thamronganantasakul, K.; Krusun, S. Survival Analysis of Glioblastoma Multiforme. *Asian Pac. J. Cancer Prev.* **2018**, *19*, 2613–2617. [CrossRef]
8. Aum, D.J.; Kim, D.H.; Beaumont, T.L.; Leuthardt, E.C.; Dunn, G.P.; Kim, A.H. Molecular and Cellular Heterogeneity: The Hallmark of Glioblastoma. *Neurosurg. Focus FOC* **2014**, *37*, E11. [CrossRef]
9. DeCordova, S.; Shastri, A.; Tsolaki, A.G.; Yasmin, H.; Klein, L.; Singh, S.K.; Kishore, U. Molecular Heterogeneity and Immunosuppressive Microenvironment in Glioblastoma. *Front. Immunol.* **2020**, *11*, 1402. [CrossRef]
10. Silantyev, A.S.; Falzone, L.; Libra, M.; Gurina, O.I.; Kardashova, K.S.; Nikolouzakis, T.K.; Nosyrev, A.E.; Sutton, C.W.; Mitsias, P.D.; Tsatsakis, A. Current and Future Trends on Diagnosis and Prognosis of Glioblastoma: From Molecular Biology to Proteomics. *Cells* **2019**, *8*, 863. [CrossRef]
11. Delgado-Martín, B.; Medina, M.Á. Advances in the Knowledge of the Molecular Biology of Glioblastoma and Its Impact in Patient Diagnosis, Stratification, and Treatment. *Adv. Sci.* **2020**, *7*, 1902971. [CrossRef] [PubMed]
12. Tamai, S.; Ichinose, T.; Nakada, M. Liquid Biomarkers in Glioma. *Brain Tumor Pathol.* **2023**, *40*, 66–77. [CrossRef] [PubMed]
13. Małusecka, E.; Zborek, A.; Krzyzowska-Gruca, S.; Krawczyk, Z. Expression of Heat Shock Proteins HSP70 and HSP27 in Primary Non-Small Cell Lung Carcinomas. An Immunohistochemical Study. *Anticancer Res.* **2001**, *21*, 1015–1021. [PubMed]
14. Zimmermann, M.; Nickl, S.; Lambers, C.; Hacker, S.; Mitterbauer, A.; Hoetzenecker, K.; Rozsas, A.; Ostoros, G.; Laszlo, V.; Hofbauer, H.; et al. Discrimination of Clinical Stages in Non-Small Cell Lung Cancer Patients by Serum HSP27 and HSP70: A Multi-Institutional Case–Control Study. *Clin. Chim. Acta* **2012**, *413*, 1115–1120. [CrossRef] [PubMed]
15. Hwang, T.S.; Han, H.S.; Choi, H.K.; Lee, Y.J.; KIM, Y.-J.; HAN, M.-Y.; PARK, Y.-M. Differential, Stage-dependent Expression of Hsp70, Hsp110 and Bcl-2 in Colorectal Cancer. *J. Gastroenterol. Hepatol.* **2003**, *18*, 690–700. [CrossRef] [PubMed]
16. Gráf, L.; Barabás, L.; Madaras, B.; Garam, N.; Maláti, É.; Horváth, L.; Prohászka, Z.; Horváth, Z.; Kocsis, J. High Serum Hsp70 Level Predicts Poor Survival in Colorectal Cancer: Results Obtained in an Independent Validation Cohort. *Cancer Biomark.* **2018**, *23*, 539–547. [CrossRef]

17. Abe, M.; Manola, J.B.; Oh, W.K.; Parslow, D.L.; George, D.J.; Austin, C.L.; Kantoff, P.W. Plasma Levels of Heat Shock Protein 70 in Patients with Prostate Cancer: A Potential Biomarker for Prostate Cancer. *Clin. Prostate Cancer* **2004**, *3*, 49–53. [CrossRef] [PubMed]
18. Kumar, S.; Gurshaney, S.; Adagunodo, Y.; Gage, E.; Qadri, S.; Sharma, M.; Malik, S.; Manne, U.; Singh, U.P.; Singh, R. Hsp70 and Gama-Semino Protein as Possible Prognostic Marker of Prostate Cancer. *Front. Biosci. Landmark Ed.* **2018**, *23*, 1987.
19. Lobinger, D.; Gempt, J.; Sievert, W.; Barz, M.; Schmitt, S.; Nguyen, H.T.; Stangl, S.; Werner, C.; Wang, F.; Wu, Z.; et al. Potential Role of Hsp70 and Activated NK Cells for Prediction of Prognosis in Glioblastoma Patients. *Front. Mol. Biosci.* **2021**, *8*, 669366. [CrossRef]
20. Werner, C.; Stangl, S.; Salvermoser, L.; Schwab, M.; Shevtsov, M.; Xanthopoulos, A.; Wang, F.; Dezfouli, A.B.; Thölke, D.; Ostheimer, C. Hsp70 in Liquid Biopsies—A Tumor-Specific Biomarker for Detection and Response Monitoring in Cancer. *Cancers* **2021**, *13*, 3706. [CrossRef]
21. Elstner, A.; Stockhammer, F.; Nguyen-Dobinsky, T.-N.; Nguyen, Q.L.; Pilgermann, I.; Gill, A.; Guhr, A.; Zhang, T.; von Eckardstein, K.; Picht, T.; et al. Identification of Diagnostic Serum Protein Profiles of Glioblastoma Patients. *J. Neuro-Oncol.* **2011**, *102*, 71–80. [CrossRef] [PubMed]
22. Mayer, M.P.; Bukau, B. Hsp70 Chaperones: Cellular Functions and Molecular Mechanism. *Cell. Mol. Life Sci.* **2005**, *62*, 670–684. [CrossRef] [PubMed]
23. Radons, J. The Human HSP70 Family of Chaperones: Where Do We Stand? *Cell Stress Chaperones* **2016**, *21*, 379–404. [CrossRef] [PubMed]
24. Sharp, F.R.; Zhan, X.; Liu, D.-Z. Heat Shock Proteins in the Brain: Role of Hsp70, Hsp 27, and HO-1 (Hsp32) and Their Therapeutic Potential. *Transl. Stroke Res.* **2013**, *4*, 685–692. [CrossRef] [PubMed]
25. Mosser, D.D.; Caron, A.W.; Bourget, L.; Denis-Larose, C.; Massie, B. Role of the Human Heat Shock Protein Hsp70 in Protection against Stress-Induced Apoptosis. *Mol. Cell. Biol.* **1997**, *17*, 5317–5327. [CrossRef]
26. Roufayel, R.; Kadry, S. Molecular Chaperone HSP70 and Key Regulators of Apoptosis-a Review. *Curr. Mol. Med.* **2019**, *19*, 315–325. [CrossRef]
27. Chanteloup, G.; Cordonnier, M.; Isambert, N.; Bertaut, A.; Hervieu, A.; Hennequin, A.; Luu, M.; Zanetta, S.; Coudert, B.; Bengrine, L.; et al. Monitoring HSP70 Exosomes in Cancer Patients' Follow up: A Clinical Prospective Pilot Study. *J. Extracell. Vesicles* **2020**, *9*, 1766192. [CrossRef]
28. Vostakolaei, M.A.; Hatami-Baroogh, L.; Babaei, G.; Molavi, O.; Kordi, S.; Abdolalizadeh, J. Hsp70 in Cancer: A Double Agent in the Battle between Survival and Death. *J. Cell. Physiol.* **2021**, *236*, 3420–3444. [CrossRef]
29. Gehrmann, M.; Liebisch, G.; Schmitz, G.; Anderson, R.; Steinem, C.; De Maio, A.; Pockley, G.; Multhoff, G. Tumor-Specific Hsp70 Plasma Membrane Localization Is Enabled by the Glycosphingolipid Gb3. *PLoS ONE* **2008**, *3*, e1925. [CrossRef]
30. Nylandsted, J.; Gyrd-Hansen, M.; Danielewicz, A.; Fehrenbacher, N.; Lademann, U.; Høyer-Hansen, M.; Weber, E.; Multhoff, G.; Rohde, M.; Jäättelä, M. Heat Shock Protein 70 Promotes Cell Survival by Inhibiting Lysosomal Membrane Permeabilization. *J. Exp. Med.* **2004**, *200*, 425–435. [CrossRef]
31. Elmallah, M.I.; Cordonnier, M.; Vautrot, V.; Chanteloup, G.; Garrido, C.; Gobbo, J. Membrane-Anchored Heat-Shock Protein 70 (Hsp70) in Cancer. *Cancer Lett.* **2020**, *469*, 134–141. [CrossRef]
32. Multhoff, G. Heat Shock Protein 70 (Hsp70): Membrane Location, Export and Immunological Relevance. *Methods* **2007**, *43*, 229–237. [CrossRef]
33. Zhang, H.; Liu, R.; Huang, W. A 14-Mer Peptide from HSP70 Protein Is the Critical Epitope Which Enhances NK Activity against Tumor Cells in Vivo. *Immunol. Investig.* **2007**, *36*, 233–246. [CrossRef]
34. Multhoff, G.; Seier, S.; Stangl, S.; Sievert, W.; Shevtsov, M.; Werner, C.; Pockley, A.G.; Blankenstein, C.; Hildebrandt, M.; Offner, R.; et al. Targeted Natural Killer Cell–Based Adoptive Immunotherapy for the Treatment of Patients with NSCLC after Radiochemotherapy: A Randomized Phase II Clinical Trial. *Clin. Cancer Res.* **2020**, *26*, 5368–5379. [CrossRef]
35. Albakova, Z.; Armeev, G.A.; Kanevskiy, L.M.; Kovalenko, E.I.; Sapozhnikov, A.M. HSP70 Multi-Functionality in Cancer. *Cells* **2020**, *9*, 587. [CrossRef]
36. Gunther, S.; Ostheimer, C.; Stangl, S.; Specht, H.M.; Mozes, P.; Jesinghaus, M.; Vordermark, D.; Combs, S.E.; Peltz, F.; Jung, M.P.; et al. Correlation of Hsp70 Serum Levels with Gross Tumor Volume and Composition of Lymphocyte Subpopulations in Patients with Squamous Cell and Adeno Non-Small Cell Lung Cancer. *Front. Immunol.* **2015**, *6*, 556. [CrossRef]
37. Cooper, M.A.; Fehniger, T.A.; Caligiuri, M.A. The Biology of Human Natural Killer-Cell Subsets. *Trends Immunol.* **2001**, *22*, 633–640. [CrossRef]
38. Vulpis, E.; Soriani, A.; Cerboni, C.; Santoni, A.; Zingoni, A. Cancer Exosomes as Conveyors of Stress-Induced Molecules: New Players in the Modulation of NK Cell Response. *Int. J. Mol. Sci.* **2019**, *20*, 611. [CrossRef]
39. Zingoni, A.; Fionda, C.; Borrelli, C.; Cippitelli, M.; Santoni, A.; Soriani, A. Natural Killer Cell Response to Chemotherapy-Stressed Cancer Cells: Role in Tumor Immunosurveillance. *Front. Immunol.* **2017**, *8*, 1194. [CrossRef]
40. Vuletić, A.; Mirjačić Martinović, K.; Tišma Miletić, N.; Zoidakis, J.; Castellvi-Bel, S.; Čavić, M. Cross-Talk between Tumor Cells Undergoing Epithelial to Mesenchymal Transition and Natural Killer Cells in Tumor Microenvironment in Colorectal Cancer. *Front. Cell Dev. Biol.* **2021**, *9*, 750022. [CrossRef]
41. Wang, M.; Zhao, J.; Zhang, L.; Wei, F.; Lian, Y.; Wu, Y.; Gong, Z.; Zhang, S.; Zhou, J.; Cao, K. Role of Tumor Microenvironment in Tumorigenesis. *J. Cancer* **2017**, *8*, 761. [CrossRef]

42. Erices, J.I.; Bizama, C.; Niechi, I.; Uribe, D.; Rosales, A.; Fabres, K.; Navarro-Martínez, G.; Torres, Á.; San Martín, R.; Roa, J.C. Glioblastoma Microenvironment and Invasiveness: New Insights and Therapeutic Targets. *Int. J. Mol. Sci.* **2023**, *24*, 7047. [CrossRef]
43. Bikfalvi, A.; da Costa, C.A.; Avril, T.; Barnier, J.-V.; Bauchet, L.; Brisson, L.; Cartron, P.F.; Castel, H.; Chevet, E.; Chneiweiss, H. Challenges in Glioblastoma Research: Focus on the Tumor Microenvironment. *Trends Cancer* **2023**, *9*, 9–27. [CrossRef]
44. Komarova, E.Y.; Suezov, R.V.; Nikotina, A.D.; Aksenov, N.D.; Garaeva, L.A.; Shtam, T.A.; Zhakhov, A.V.; Martynova, M.G.; Bystrova, O.A.; Istomina, M.S.; et al. Hsp70-Containing Extracellular Vesicles Are Capable of Activating of Adaptive Immunity in Models of Mouse Melanoma and Colon Carcinoma. *Sci. Rep.* **2021**, *11*, 21314. [CrossRef]
45. Shin, D.-W.; Lee, S.; Song, S.W.; Cho, Y.H.; Hong, S.H.; Kim, J.H.; Kim, H.S.; Park, J.E.; Nam, S.J.; Kim, Y.-H. Survival Outcome and Prognostic Factors in Anaplastic Oligodendroglioma: A Single-Institution Study of 95 Cases. *Sci. Rep.* **2020**, *10*, 20162. [CrossRef]
46. Reuss, D.E.; Mamatjan, Y.; Schrimpf, D.; Capper, D.; Hovestadt, V.; Kratz, A.; Sahm, F.; Koelsche, C.; Korshunov, A.; Olar, A.; et al. IDH Mutant Diffuse and Anaplastic Astrocytomas Have Similar Age at Presentation and Little Difference in Survival: A Grading Problem for WHO. *Acta Neuropathol.* **2015**, *129*, 867–873. [CrossRef]
47. Dong, X.; Noorbakhsh, A.; Hirshman, B.R.; Zhou, T.; Tang, J.A.; Chang, D.C.; Carter, B.S.; Chen, C.C. Survival Trends of Grade I, II, and III Astrocytoma Patients and Associated Clinical Practice Patterns between 1999 and 2010: A SEER-Based Analysis. *Neuro-Oncol. Pract.* **2016**, *3*, 29–38. [CrossRef]
48. Ostrom, Q.T.; Gittleman, H.; Stetson, L.; Virk, S.M.; Barnholtz-Sloan, J.S. Epidemiology of Gliomas. In *Current Understanding and Treatment of Gliomas*; Raizer, J., Parsa, A., Eds.; Springer International Publishing: Cham, Switzerland, 2015; pp. 1–14. ISBN 978-3-319-12048-5.
49. Wang, Z.; Gou, W.; Liu, M.; Sang, W.; Chu, H.; Zhang, W. Expression of P53 and HSP70 in Chronic Hepatitis, Liver Cirrhosis, and Early and Advanced Hepatocellular Carcinoma Tissues and Their Diagnostic Value in Hepatocellular Carcinoma: An Immunohistochemical Study. *Med. Sci. Monit. Int. Med. J. Exp. Clin. Res.* **2015**, *21*, 3209. [CrossRef]
50. Sarkaria, J.N.; Hu, L.S.; Parney, I.F.; Pafundi, D.H.; Brinkmann, D.H.; Laack, N.N.; Giannini, C.; Burns, T.C.; Kizilbash, S.H.; Laramy, J.K.; et al. Is the Blood–Brain Barrier Really Disrupted in All Glioblastomas? A Critical Assessment of Existing Clinical Data. *Neuro-Oncology* **2018**, *20*, 184–191. [CrossRef]
51. Gao, G.; Liu, S.; Yao, Z.; Zhan, Y.; Chen, W.; Liu, Y. The Prognostic Significance of Hsp70 in Patients with Colorectal Cancer Patients: A PRISMA-Compliant Meta-Analysis. *BioMed Res. Int.* **2021**, *2021*, 5526327. [CrossRef]
52. Bauer, K.; Nitsche, U.; Slotta-Huspenina, J.; Drecoll, E.; von Weyhern, C.H.; Rosenberg, R.; Höfler, H.; Langer, R. High HSP27 and HSP70 Expression Levels Are Independent Adverse Prognostic Factors in Primary Resected Colon Cancer. *Cell. Oncol.* **2012**, *35*, 197–205. [CrossRef]
53. Binder, R.J. Heat-Shock Protein-Based Vaccines for Cancer and Infectious Disease. *Expert Rev. Vaccines* **2008**, *7*, 383–393. [CrossRef]
54. Bald, T.; Krummel, M.F.; Smyth, M.J.; Barry, K.C. The NK Cell–Cancer Cycle: Advances and New Challenges in NK Cell–Based Immunotherapies. *Nat. Immunol.* **2020**, *21*, 835–847. [CrossRef]
55. Hudspeth, K.; Silva-Santos, B.; Mavilio, D. Natural Cytotoxicity Receptors: Broader Expression Patterns and Functions in Innate and Adaptive Immune Cells. *Front. Immunol.* **2013**, *4*, 69. [CrossRef]
56. Firouzi, J.; Hajifathali, A.; Azimi, M.; Parvini, N.; Ghaemi, F.; Asl, N.S.; Asl, A.A.H.; Safa, M.; Ebrahimi, M. Hsp70, in Combination with IL-15 and PD-1 Blocker, Interferes with the Induction of Cytotoxic NK Cells in Relapsed Acute Myeloid Leukemia Patients. *Cell J.* **2023**, *25*, 92.
57. Sharifzad, F.; Mardpour, S.; Mardpour, S.; Fakharian, E.; Taghikhani, A.; Sharifzad, A.; Kiani, S.; Heydarian, Y.; Łos, M.J.; Azizi, Z.; et al. HSP70/IL-2 Treated NK Cells Effectively Cross the Blood Brain Barrier and Target Tumor Cells in a Rat Model of Induced Glioblastoma Multiforme (GBM). *Int. J. Mol. Sci.* **2020**, *21*, 2263. [CrossRef]
58. Hromadnikova, I.; Li, S.; Kotlabova, K.; Dickinson, A.M. Influence of in Vitro IL-2 or IL-15 Alone or in Combination with Hsp 70 Derived 14-Mer Peptide (TKD) on the Expression of NK Cell Activatory and Inhibitory Receptors on Peripheral Blood T Cells, B Cells and NKT Cells. *PLoS ONE* **2016**, *11*, e0151535. [CrossRef]
59. Fontana, A.; Hengartner, H.; de Tribolet, N.; Weber, E. Glioblastoma Cells Release Interleukin 1 and Factors Inhibiting Interleukin 2-Mediated Effects. *J. Immunol.* **1984**, *132*, 1837–1844. [CrossRef]
60. Masood, A.; Kayani, M.; Batool, S. Targetting Interleukins Involved in Glioblastoma—A New Pharmacological Approach. *JLBSR* **2020**, *1*, 82–88. [CrossRef]
61. Zhu, V.F.; Yang, J.; LeBrun, D.G.; Li, M. Understanding the Role of Cytokines in Glioblastoma Multiforme Pathogenesis. *Cancer Lett.* **2012**, *316*, 139–150. [CrossRef] [PubMed]
62. Tarassishin, L.; Lim, J.; Weatherly, D.B.; Angeletti, R.H.; Lee, S.C. Interleukin-1-Induced Changes in the Glioblastoma Secretome Suggest Its Role in Tumor Progression. *J. Proteom.* **2014**, *99*, 152–168. [CrossRef] [PubMed]
63. Himes, B.T.; Geiger, P.A.; Ayasoufi, K.; Bhargav, A.G.; Brown, D.A.; Parney, I.F. Immunosuppression in Glioblastoma: Current Understanding and Therapeutic Implications. *Front. Oncol.* **2021**, *11*, 770561. [CrossRef] [PubMed]
64. Nduom, E.K.; Weller, M.; Heimberger, A.B. Immunosuppressive Mechanisms in Glioblastoma. *Neuro-Oncology* **2015**, *17*, vii9–vii14. [CrossRef] [PubMed]
65. Berghoff, A.S.; Kiesel, B.; Widhalm, G.; Rajky, O.; Ricken, G.; Wöhrer, A.; Dieckmann, K.; Filipits, M.; Brandstetter, A.; Weller, M.; et al. Programmed Death Ligand 1 Expression and Tumor-Infiltrating Lymphocytes in Glioblastoma. *Neuro-Oncology* **2015**, *17*, 1064–1075. [CrossRef] [PubMed]

66. Agoff, S.N.; Hou, J.; Linzer, D.I.; Wu, B. Regulation of the Human Hsp70 Promoter by P53. *Science* **1993**, *259*, 84–87. [CrossRef] [PubMed]
67. Zylicz, M.; King, F.W.; Wawrzynow, A. Hsp70 Interactions with the P53 Tumour Suppressor Protein. *EMBO J.* **2001**, *20*, 4634–4638. [CrossRef] [PubMed]
68. Zagzag, D.; Salnikow, K.; Chiriboga, L.; Yee, H.; Lan, L.; Ali, M.A.; Garcia, R.; Demaria, S.; Newcomb, E.W. Downregulation of Major Histocompatibility Complex Antigens in Invading Glioma Cells: Stealth Invasion of the Brain. *Lab. Investig.* **2005**, *85*, 328–341. [CrossRef] [PubMed]
69. Li, L.; Yu, R.; Cai, T.; Chen, Z.; Lan, M.; Zou, T.; Wang, B.; Wang, Q.; Zhao, Y.; Cai, Y. Effects of Immune Cells and Cytokines on Inflammation and Immunosuppression in the Tumor Microenvironment. *Int. Immunopharmacol.* **2020**, *88*, 106939. [CrossRef]
70. Martinez-Lage, M.; Lynch, T.M.; Bi, Y.; Cocito, C.; Way, G.P.; Pal, S.; Haller, J.; Yan, R.E.; Ziober, A.; Nguyen, A.; et al. Immune Landscapes Associated with Different Glioblastoma Molecular Subtypes. *Acta Neuropathol. Commun.* **2019**, *7*, 203. [CrossRef]
71. Kmiecik, J.; Poli, A.; Brons, N.H.C.; Waha, A.; Eide, G.E.; Enger, P.Ø.; Zimmer, J.; Chekenya, M. Elevated CD3+ and CD8+ Tumor-Infiltrating Immune Cells Correlate with Prolonged Survival in Glioblastoma Patients despite Integrated Immunosuppressive Mechanisms in the Tumor Microenvironment and at the Systemic Level. *J. Neuroimmunol.* **2013**, *264*, 71–83. [CrossRef]

biomedicines

Article

# The Roles of AGTRAP, ALKBH3, DIVERSIN, NEDD8 and RRM1 in Glioblastoma Pathophysiology and Prognosis

Claudia Alexandra Dumitru [1,*], Nikolas Walter [1], Carl Ludwig Raven Siebert [1], Frederik Till Alexander Schäfer [1], Ali Rashidi [1], Belal Neyazi [1], Klaus-Peter Stein [1], Christian Mawrin [2] and Ibrahim Erol Sandalcioglu [1]

1 Department of Neurosurgery, Otto-von-Guericke University, 39120 Magdeburg, Germany; erol.sandalcioglu@med.ovgu.de (I.E.S.)
2 Department of Neuropathology, Otto-von-Guericke University, 39120 Magdeburg, Germany
* Correspondence: claudia.dumitru@med.ovgu.de

**Abstract:** This study determined the expression of five novel biomarker candidates in IDH wild-type glioblastoma (GBM) tissues compared to non-malign brain parenchyma, as well as their prognostic relevance for the GBM patients' outcomes. The markers were analysed by immunohistochemistry in tumour tissues (n = 186) and healthy brain tissues (n = 54). The association with the patients' overall survival (OS) and progression-free survival (PFS) was assessed by Kaplan–Meier and log-rank test. The prognostic value of the markers was determined using multivariate Cox proportional hazard models. AGTRAP, DIVERSIN, cytoplasmic NEDD8 (NEDD8c) and RRM1 were significantly overexpressed in tumour tissues compared to the healthy brain, while the opposite was observed for ALKBH3. AGTRAP, ALKBH3, NEDD8c and RRM1 were significantly associated with OS in univariate analysis. AGTRAP and RRM1 were also independent prognostic factors for OS in multivariate analysis. For PFS, only AGTRAP and NEDD8c reached significance in univariate analysis. Additionally, AGTRAP was an independent prognostic factor for PFS in multivariate models. Finally, combined analysis of the markers enhanced their prognostic accuracy. The combination AGTRAP/ALKBH3 had the strongest prognostic value for the OS of GBM patients. These findings contribute to a better understanding of the GBM pathophysiology and may help identify novel therapeutic targets in this type of cancer.

**Keywords:** IDH wild-type glioblastoma; biomarkers; overall survival; progression-free survival; prognostic accuracy; marker combinations

## 1. Introduction

Glioblastoma (GBM) is the most prevalent malign brain tumour in adult patients and one of the most deadly types of cancer overall, with 5-year survival rates of less than 7% [1,2]. Since the newest classification from 2021, only the WHO grade 4 diffuse gliomas that occur in adults and have an IDH wild-type phenotype are considered as 'true' glioblastomas [3]. Because of its highly infiltrative growth, this tumour cannot be completely removed by surgery and necessitates a multimodal therapeutic approach. However, after the development of the 'Stupp protocol' almost 20 years ago [4], there have been no other major breakthroughs in the therapy of GBM. The focus of recent research has therefore shifted towards understanding the pathophysiology of GBM, in order to develop specific targeted therapies. To reach this goal, it is necessary to identify novel cellular and molecular factors that control GBM progression and could serve as therapeutic targets in this type of cancer. An important first step in this process is the identification of biomarkers which are differentially expressed in GBM compared to the healthy brain tissue and/or associate with the patients' outcome.

In this study, we investigated five potential biomarker candidates that have been linked to tumour progression and prognosis in several types of cancer. AGTRAP (Angiotensin II Type I Receptor-associated Protein) is a member of the renin–angiotensin system, which specifically interacts with the AT1 receptor to regulate cardiovascular and fluid homeostasis [5]. In cancer, AGTRAP can promote tumour progression through activation of the MAPK and AKT/mTOR signaling pathways [6,7], but potentially also through regulation of immunological processes, such as T-cell exhaustion or recruitment of pro-inflammatory and immunosuppressive immune cells [6,8]. ALKBH3 (AlkB Homolog 3)/PCA-1 (Prostate Cancer Antigen-1) is a DNA/RNA repair enzyme, which acts as a demethylase to regulate cellular metabolism (reviewed in [9]). Currently, there is no consensus regarding the role of ALKBH3 in cancer. While some studies found that ALKBH3 was overexpressed in tumour tissues and exhibited tumour-promoting functions [10–12], others showed that ALKBH3 was epigenetically inactivated in certain malignancies. The latter phenomenon associated with a poor outcome of the respective cancer patients [13–15]. DIVERSIN, also known as ANKRD6 (Ankyrin Repeat Domain-containing Protein 6), is a modulator of the Wnt and Wnt/JNK signaling pathways [16]. Thus far, most studies consider DIVERSIN to be a pro-tumour factor, as it enhances tumour proliferation and invasion in a variety of malignancies, such as non-small cell lung cancer, colorectal carcinoma, breast cancer and glioma [17–20]. Additionally, recent studies linked this marker to immunological processes in colon cancer, where DIVERSIN expression correlated positively with the numbers of tumour-promoting M2 macrophages and negatively with the numbers of anti-tumour effector cells [21]. NEDD8 (Neuronal precursor cell-Expressed Developmentally Down-regulated protein 8) is a ubiquitin-like molecule, which conjugates to a lysine residue of the substrate protein [22]. This mechanism called 'neddylation' can promote—when overactivated—the degradation of tumour suppressor proteins, such as p21 and p27, thereby facilitating carcinogenesis and tumour progression (reviewed in [23]). Importantly, neddylation can be inhibited by the FDA-approved drug MLN4924/pevonedistat, which makes it a promising target for cancer therapy (reviewed in [24]). RRM1 (Ribonucleotide Reductase large subunit M1) is the catalytic subunit of the ribonucleotide reductase [25]. Similar to ALKBH3, there have been contradictory reports in the literature regarding the role of RRM1 in cancer [26–29], reviewed in [30]. In glioma however, RRM1 appears to be a tumour-promoting factor, as RRM1 silencing induced cell cycle arrest and inhibited the proliferation of the tumour cells [31]. Furthermore, very recent studies linked high levels of RRM1 to the poor outcome of GBM patients and identified RRM1 as a promising target for drug combination therapy in this type of cancer [32].

The role of these markers in GBM pathophysiology and prognosis is insufficiently characterized thus far. In particular, studies on homogenous cohorts of IDH wild-type GBM patients employing multivariate proportional hazard models are largely missing. The purpose of our study was to assess (1) the expression of the markers in IDH wild-type GBM tissues compared to non-malign adjacent brain parenchyma, (2) the association between marker expression and the outcome of the patients, (3) the prognostic value of the markers for the patients' outcome using multivariate models and (4) the prognostic accuracy of the markers analysed individually or in combination.

## 2. Materials and Methods

### 2.1. Study Subjects

This study was conducted on adult patients with newly diagnosed glioblastoma (GBM). Only the patients with confirmed IDH wild-type status were included in the analysis (n = 186). The median age of the patients was 66 years. All patients were treated at the Department of Neurosurgery, University Hospital Magdeburg between 2008 and 2018. The study was conducted in accordance with the Declaration of Helsinki issued in 1975 and revised in 2013. The ethics committee of the Otto-von-Guericke University Magdeburg approved the study (No. 146/2019) and waivered the need for informed written consent. The main clinical characteristics of the GBM patients are summarized in Table 1.

**Table 1.** Clinical characteristics of the GBM patients included in this study. RTX: radiotherapy; CTX: chemotherapy; RCTX: radiochemotherapy; n.d.: not determinable.

| Clinical Parameter | Number | Percentage |
|---|---|---|
| **Sex** | | |
| female | 85 | 45.7 |
| male | 101 | 54.3 |
| **Karnofsky performance scale** | | |
| 10 | 3 | 1.6 |
| 20 | 2 | 1.1 |
| 30 | 11 | 5.9 |
| 40 | 8 | 4.3 |
| 50 | 29 | 15.6 |
| 60 | 33 | 17.7 |
| 70 | 60 | 32.3 |
| 80 | 19 | 10.2 |
| 90 | 19 | 10.2 |
| 100 | 0 | 0 |
| n.d. | 2 | 1.1 |
| **Therapy** | | |
| surgery | 35 | 18.8 |
| surgery+RTX | 21 | 11.3 |
| surgery+CTX | 1 | 0.5 |
| surgery+RCTX | 123 | 66.1 |
| n.d. | 6 | 3.2 |
| **Extent of surgical resection** | | |
| total | 65 | 34.9 |
| subtotal | 111 | 59.7 |
| n.d. | 10 | 5.4 |
| **MGMT methylation status** | | |
| unmethylated | 70 | 37.6 |
| methylated | 116 | 62.4 |
| n.d. | 0 | 0 |
| **IDH mutation status** | | |
| wild-type | 186 | 100 |
| mutated | 0 | 0 |
| n.d. | 0 | 0 |

### 2.2. Marker Expression and Scoring

The expression of the biomarker candidates was assessed in FFPE (formalin-fixed paraffin-embedded) tissues, which were processed as tissue microarrays (TMAs) as in previous studies [33,34]. Both tumour tissues (n = 186) and non-malign brain parenchymas (n = 54) were included in the analysis. The latter consisted of brain tissue distant from the solid tumour area in order to minimize the effect of the infiltrating/diffuse-growing tumour cells on the subsequent analysis. The samples were stained with the following primary antibodies: 193 ng/mL polyclonal rabbit anti-AGTRAP, 293 ng/mL polyclonal rabbit anti-ALKBH3, 333 ng/mL polyclonal rabbit anti-ANKRD6/DIVERSIN, 114 ng/mL monoclonal rabbit anti-NEDD8 (19E3) and 248 ng/mL monoclonal rabbit anti-RRM1 (D12F12). The AGTRAP, ALKBH3 and DIVERSIN antibodies were purchased from Proteintech Europe (Manchester, UK), while NEDD8 and RRM1 were from Cell Signaling Technology (Frankfurt am Main, Germany). After staining, the samples were digitalized with an Aperio VERSA high-resolution whole slide scanner and analysed with the Aperio ImageScope V12.1.0.5029 software (both from Leica Biosystems, Nussloch, Germany). Authors C.A.D., N.W., C.L.R.S. and F.T.A.S. performed blinded histological scoring independently.

AGTRAP, ALKBH3, DIVERSIN and RRM1 displayed a cytoplasmic subcellular localization. NEDD8 displayed both a cytoplasmic (NEDD8c) and a nuclear (NEDD8n) subcellular localization. According to the intensity, we categorized the cytoplasmic staining as 'weak' (1 point), 'medium' (2 points) and 'strong' (3 points) (Figure 1A–D,F). To account for potential heterogeneity in the staining pattern among samples, we assessed the final expression levels of the cytoplasmic markers using the H-Score: $(1 \times X) + (2 \times Y) + (3 \times Z)$, where $X + Y + Z = 100\%$ of the total tumour area. For NEDD8n, we assessed the percentage of positive cells using a 5-tier scoring system: <20%, 20–40%, 41–60%, 61–80% and >80%, and we assigned 0, 1, 2, 3 and 4 points, respectively (Figure 1E). A high-magnification micrograph showing both positive and negative NEDD8n tumour cells is presented in Supplementary Figure S1. At least three different fields per TMA spot were analysed at 200-fold magnification, and the values were subsequently averaged.

### 2.3. *Statistical Analysis*

The data were analysed with the SPSS version 29.0.1.0 (171) software (IBM Corporation, Armonk, NY, USA). The difference in marker expression between tumour and non-malign brain parenchyma was analysed using box-whisker plots. For this set of studies, the statistical significance was assessed with the Mann–Whitney U-test. Survival curves (36 months for overall survival and 12 months for progression-free survival) were generated with the Kaplan–Meier method. The significance was tested both by univariate analysis using the log-rank test and by multivariate Cox proportional hazard regression models adjusted for age, Karnofsky performance scale, extent of surgical resection and MGMT methylation status.

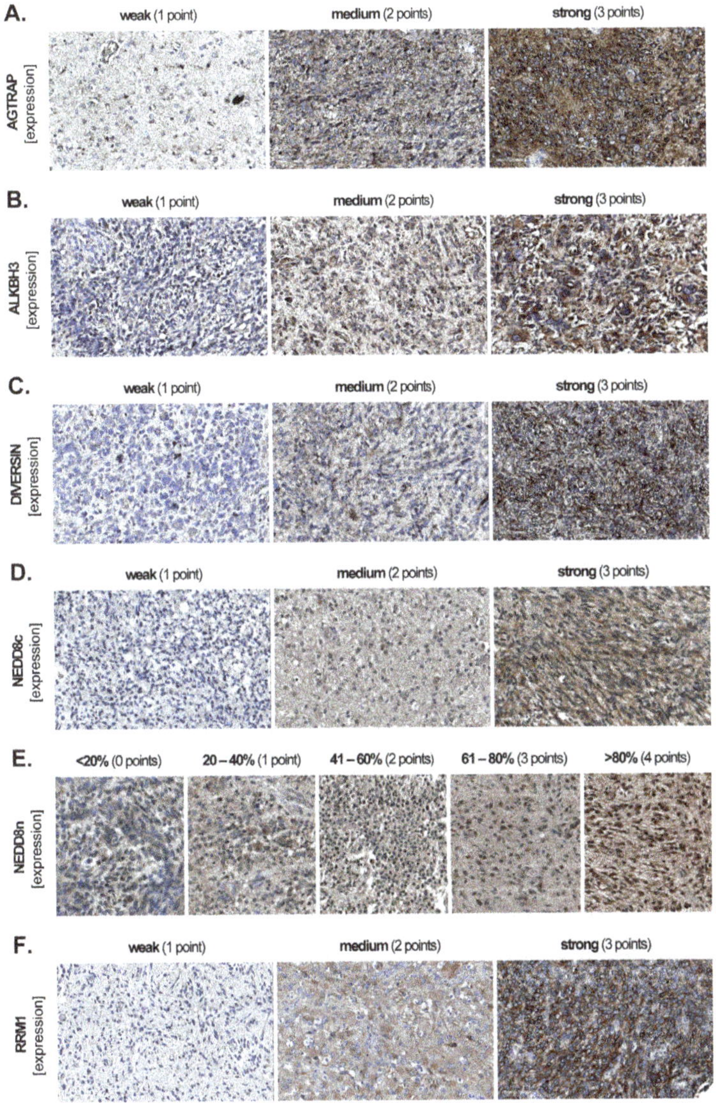


**Figure 1. Marker expression in GBM tissues.** Representative micrographs showing weak (1 point), medium (2 points) and strong (3 points) cytoplasmic expression of (**A**) AGTRAP, (**B**) ALKBH3, (**C**) DIVERSIN, (**D**) NEDD8c and (**F**) RRM1. The H-score was subsequently calculated according to the formula $(1 \times X) + (2 \times Y) + (3 \times Z)$, where $X + Y + Z = 100\%$ of the total tumour area. (**E**) The 5-tier score for nuclear NEDD8 (NEDD8n) according to the percentage of positive cells.

## 3. Results

### 3.1. Marker Expression in Tumour versus Healthy Brain Tissues

In the first set of studies, we sought to determine whether our biomarker candidates were differentially expressed in glioblastoma (GBM) compared to the adjacent non-malign brain tissue. The results showed that the levels of AGTRAP (Figure 2A,B), DIVERSIN (Figure 2E,F), NEDD8c (Figure 2G,H) and RRM1 (Figure 2I,J) were significantly higher in the tumour tissues compared to the healthy brain parenchyma. In contrast, ALKBH3 was significantly downregulated in GBM compared to the healthy brain (Figure 2C,D). The expression of NEDD8n was not significantly different between the two groups of samples ($p = 0.737$, Mann–Whitney U). These results were confirmed when the group sizes were adjusted to include only the cases where both healthy and tumour tissues were available (n = 54) (Supplementary Figure S2A–E).

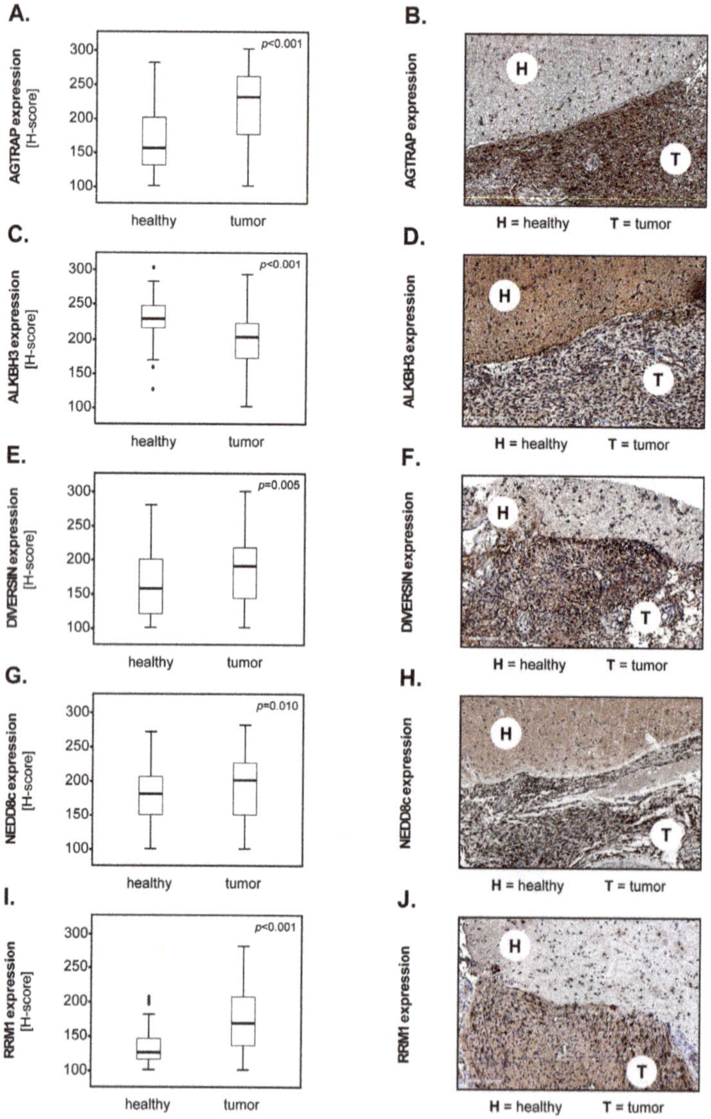


**Figure 2. Marker expression in healthy versus GBM tissues.** Expression of (**A**) AGTRAP, (**C**) ALKBH3, (**E**) DIVERSIN, (**G**) NEDD8c and (**I**) RRM1 in GBM (n = 186) and tumour-free adjacent brain tissues (n = 54). The medians are shown as black lines and the percentiles (25th and 75th) as vertical boxes with error bars. The outliers are indicated by circles. Statistical analysis was performed with the Mann–Whitney U test, and the *p*-values are indicated in the upper-right corner of each plot. (**B**,**D**,**F**,**H**,**J**) Representative micrographs showing the expression of the markers in the solid tumour area (T) versus the adjacent, tumour-free tissue area (H).

### 3.2. Marker Expression and the Overall Survival of GBM Patients

Next, we tested whether our biomarker candidates might associate with the overall survival (OS) of the GBM patients. To this end, we first dichotomised the expression levels of the markers into 'low' and 'high' according to the median-split method. We subsequently plotted 36-month Kaplan–Meier survival curves for each marker and performed univariate survival analyses using the log-rank test. The results showed that patients with high levels of AGTRAP had a significantly shorter OS compared to patients with low AGTRAP levels ($p < 0.001$, log-rank; Figure 3A). Similarly, high levels of NEDD8c were significantly associated with a shorter OS of the GBM patients ($p = 0.03$, log-rank; Figure 3D). ALKBH3 was significantly associated with the OS of the GBM patients as well, but in this case, the patients with high levels of ALKBH3 survived significantly better than the ALKBH3$^{low}$ patients ($p = 0.039$, log-rank; Figure 3B).

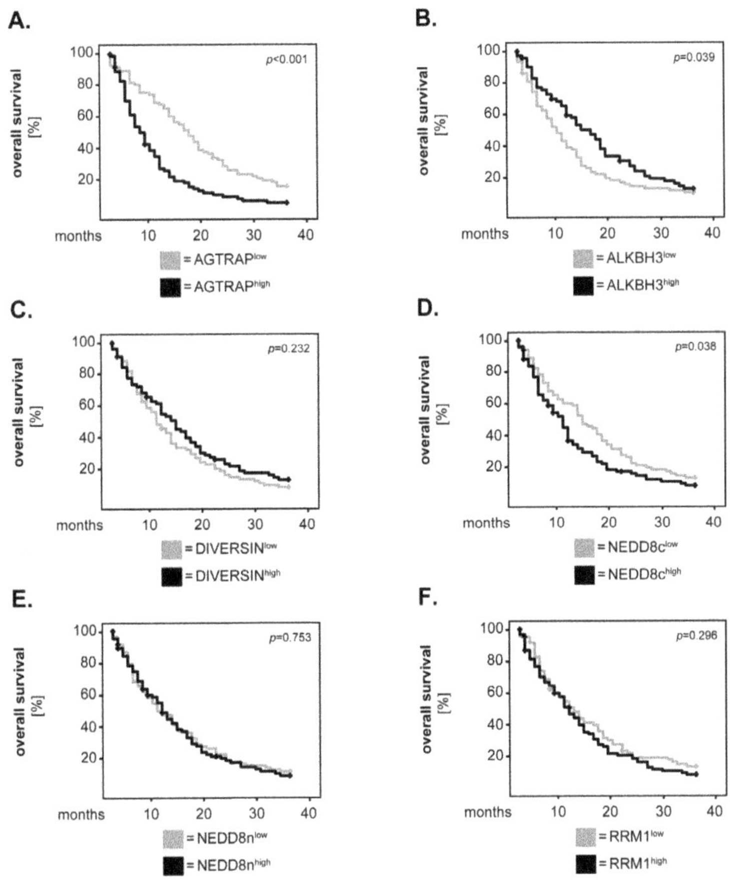


**Figure 3. Marker expression and the overall survival of GBM patients—univariate analysis.** (**A–F**) The expression levels of the markers were dichotomised into 'low' and 'high' according to the median-split method. Kaplan–Meier curves were generated for the 36-month overall survival, and statistical analysis was performed with the log-rank test. The $p$-values are indicated in the upper-right corner of each plot.

For DIVERSIN (Figure 3C), NEDD8n (Figure 3E) and RRM1 (Figure 3F), there was no significant difference regarding OS between patients with high and those with low expression levels of these markers ($p = 0.232$, $p = 0.753$ and $p = 0.296$, log-rank, respectively).

We consequently sought to determine whether the dichotomisation of these markers into 'low' and 'high' using a new cut-off might yield other results. Since DIVERSIN and RRM1 were differentially expressed in healthy brain compared to the GBM tissues, we selected the median values of the healthy samples as the new cut-off. Even under these conditions, there was no significant difference in OS between DIVERSIN$^{high}$ and DIVERSIN$^{low}$ patients ($p$ = 0.929, log-rank; Figure 4A). However, we now found a significant association between high levels of RRM1 and a shorter survival of the GBM patients ($p$ = 0.031, log-rank; Figure 4B). NEDD8n could not be analysed with the new cut-off, as its expression levels were similar in both healthy and tumour tissues.

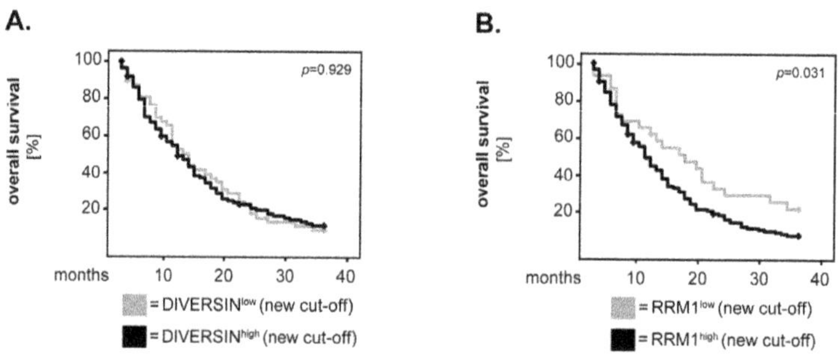


**Figure 4. New univariate analysis of DIVERSIN and RRM1 in relation to the overall survival of GBM patients.** The expression levels of (**A**) DIVERSIN and (**B**) RRM1 were dichotomised into 'low' and 'high' according to the median values of the healthy tissues. Kaplan–Meier curves were generated for the 36-month overall survival, and statistical analysis was performed with the log-rank test. The $p$-values are indicated in the upper-right corner of each plot.

In further studies, we performed multivariate survival analyses using Cox proportional hazard models adjusted for potential confounders, such as age, Karnofsky Performance Scale (KPS), extent of surgical resection and MGMT methylation status [35–38]. We only carried out these tests for the markers that were significantly associated with OS in the univariate analysis. The multivariate models showed that high levels of AGTRAP and of RRM1 significantly predicted the shorter OS of GBM patients (AGTRAP: HR = 1.814, 95% CI = 1.260–2.612, $p$ = 0.001; RRM1: HR = 1.746, 95% CI = 1.099–2.775, $p$ = 0.018). In contrast, ALKBH3 and NEDD8c did not reach significant prognostic values regarding the OS of these patients. The results of the multivariate analysis are summarized in Table 2.

**Table 2.** Multivariate Cox regression analysis of individual markers in relation to overall survival. The reference variable (dummy) is indicated by HR = 1. The model was adjusted for age, Karnofsky performance scale, surgical resection and MGMT methylation status. HR: Hazard Ratio; CI [95%]: 95% confidence interval.

| Cox Regression: OS | HR | CI [95%] | $p$-Value |
|---|---|---|---|
| AGTRAP$^{low}$ | 1 | | |
| AGTRAP$^{high}$ | 1.814 | 1.260–2.612 | 0.001 |
| ALKBH3$^{high}$ | 1 | | |
| ALKBH3$^{low}$ | 1.345 | 0.938–1.929 | 0.107 |
| NEDD8c$^{low}$ | 1 | | |
| NEDD8c$^{high}$ | 1.370 | 0.980–1.914 | 0.065 |
| RRM1$^{low}$ (new cut-off) | 1 | | |
| RRM1$^{high}$ (new cut-off) | 1.746 | 1.099–2.775 | 0.018 |

### 3.3. Marker Expression and the Progression-Free Survival of GBM Patients

Next, we tested whether our biomarker candidates might associate with the 12-month progression-free survival (PFS) of the GBM patients. The survival analysis was performed as described above. From all markers tested, only two showed significant associations with the PFS of GBM patients. Specifically, patients with high levels of AGTRAP (Figure 5A) and NEDD8c (Figure 5D) had a significantly shorter PFS than patients with low levels of these markers (AGTRAP: $p < 0.001$, log-rank; NEDD8c: $p = 0.028$, log-rank). ALKBH3, DIVERSIN, NEDD8n and RRM1 did not associate with the PFS of GBM patients, regardless if the cut-off was set according to the median-split method (Figure 5B,C,E,F) or based on the median values of the healthy tissues.

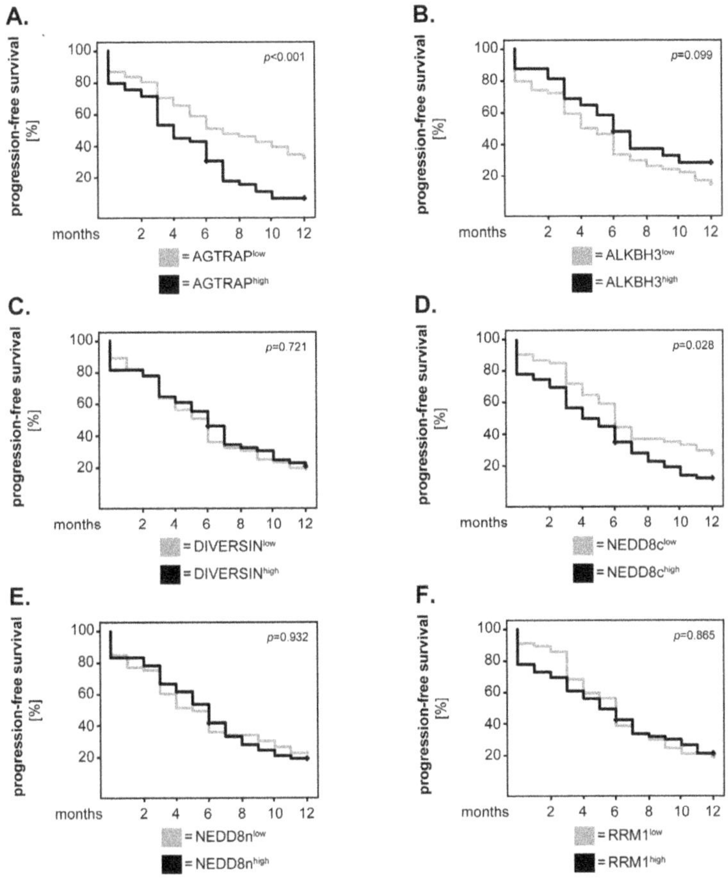


**Figure 5. Marker expression and the progression-free survival of GBM patients—univariate analysis.** (**A–F**) The expression levels of the markers were dichotomised into 'low' and 'high' according to the median-split method. Kaplan–Meier curves were generated for the 12-month progression-free survival and statistical analysis was performed with the log-rank test. The *p*-values are indicated in the upper-right corner of each plot.

In multivariate analysis, high levels of AGTRAP significantly predicted the shorter PFS of GBM patients (HR = 1.921, [95% CI] = 1.200–3.075, $p = 0.007$). Patients with high levels of NEDD8c had an increased hazard ratio compared to NEDD8c$^{low}$ patients, but this marker did not reach statistical significance in our cohort. The results of the multivariate analysis are summarized in Table 3.

**Table 3.** Multivariate Cox regression analysis of individual markers in relation to progression-free survival. The reference variable (dummy) is indicated by HR = 1. The model was adjusted for age, Karnofsky performance scale, surgical resection and MGMT methylation status. HR: Hazard Ratio; CI [95%]: 95% confidence interval.

| Cox Regression: PFS | HR | CI [95%] | *p*-Value |
|---|---|---|---|
| AGTRAP$^{low}$ | 1 | | |
| AGTRAP$^{high}$ | 1.921 | 1.200–3.075 | 0.007 |
| NEDD8c$^{low}$ | 1 | | |
| NEDD8c$^{high}$ | 1.498 | 0.979–2.293 | 0.063 |

### 3.4. Marker Combinations and the Survival of GBM Patients

In the final set of studies, we sought to determine whether our biomarker candidates would have stronger prognostic values when analysed in combination rather than individually. We only carried out these analyses for the markers that were significantly associated with OS and PFS in the univariate analysis of survival. For the analysis of OS, we selected the following test groups of patients: AGTRAP$^{high}$ALKBH3$^{low}$, AGTRAP$^{high}$NEDD8c$^{high}$, AGTRAP$^{high}$RRM1$^{high}$, ALKBH3$^{low}$NEDD8c$^{high}$, ALKBH3$^{low}$RRM1$^{high}$ and NEDD8c$^{high}$RRM1$^{high}$. Within each combination of markers, the reference group consisted of all remaining GBM patients taken together (i.e., AGTRAP$^{high}$ALKBH3$^{low}$ versus Rest, where 'Rest' included patients with AGTRAP$^{low}$ALKBH3$^{low}$, AGTRAP$^{low}$ALKBH3$^{high}$ and AGTRAP$^{high}$ALKBH3$^{high}$ phenotypes). The results of the multivariate analysis showed that all marker combinations were significant predictors of poor OS (Table 4). The combination of AGTRAP and ALKBH3 had the strongest prognostic value for the OS of GBM patients (HR = 2.059, 95% CI = 1.343–3.157, $p < 0.001$; Table 4). The combination of ALKBH3 and NEDD8c significantly improved the prognostic value of the individual markers (HR = 1.893, 95% CI = 1.263–2.836, $p = 0.002$; Table 4).

**Table 4.** Multivariate Cox regression analysis of combined markers in relation to overall and progression-free survival. The reference variable (dummy) is indicated by HR = 1. The model was adjusted for age, Karnofsky performance scale, surgical resection and MGMT methylation status. HR: Hazard Ratio; CI [95%]: 95% confidence interval.

| Cox Regression: OS | HR | CI [95%] | *p*-Value |
|---|---|---|---|
| Rest | 1 | | |
| AGTRAP$^{high}$ALKBH3$^{low}$ | 2.059 | 1.343–3.157 | <0.001 |
| AGTRAP$^{high}$NEDD8c$^{high}$ | 1.516 | 1.025–2.243 | 0.037 |
| AGTRAP$^{high}$RRM1$^{high}$ | 1.742 | 1.223–2.480 | 0.002 |
| ALKBH3$^{low}$NEDD8c$^{high}$ | 1.893 | 1.263–2.836 | 0.002 |
| ALKBH3$^{low}$RRM1$^{high}$ | 1.576 | 1.096–2.266 | 0.014 |
| NEDD8c$^{high}$RRM1$^{high}$ | 1.416 | 1.017–1.971 | 0.039 |
| **Cox regression: PFS** | **HR** | **CI [95%]** | ***p*-value** |
| Rest | 1 | | |
| AGTRAP$^{high}$NEDD8c$^{high}$ | 1.834 | 1.123–2.995 | 0.015 |

PFS was analysed in a similar manner. However, since only two markers were significant in the univariate analysis, there was only one possible combination in this case: AGTRAP with NEDD8c. The results showed that the AGTRAP$^{high}$NEDD8c$^{high}$ phenotype was a significant predictor of poor PFS in GBM patients (HR = 1.834, 95% CI = 1.123–2.995, $p = 0.015$; Table 4).

## 4. Discussion

This study assessed the expression levels of five biomarker candidates in IDH wild-type glioblastoma (GBM) compared to healthy brain tissues and tested their prognostic values regarding overall (OS) and progression-free survival (PFS) of the patients with this type of cancer.

For AGTRAP, we found a significantly higher expression of this marker in GBM compared to the adjacent non-malign brain parenchyma. These results are in line with a recent study by Hong et al., demonstrating that AGTRAP was overexpressed at both mRNA levels (TCGA+GTEx database) and protein levels (CPTAC database) in many tumours, including GBM [8]. Our study further showed that GBM patients with high levels of AGTRAP had a significantly shorter OS and PFS than patients with low levels of this marker. In the multivariate analysis, AGTRAP proved to be an independent prognostic factor for the OS and PFS of the GBM patients. While no detailed biomarker studies in IDH wild-type GBM were available for comparison, studies in other types of solid cancer, such as hepatocellular carcinoma [6,39], cervix carcinoma [40] and tongue squamous cell carcinoma [41], showed that AGTRAP associated with the poor outcome of those patients. Taken together, these data indicate that AGTRAP may be a valuable prognostic marker in GBM. Additionally, AGTRAP may be involved in the tumourigenesis and progression of GBM. Indeed, previous studies linked AGTRAP with the activation of well-established tumourigenic pathways such as MAPK and AKT/mTOR [6,7]. Furthermore, very recent studies found a direct correlation between AGTRAP levels and the numbers of tumour-infiltrating M2 macrophages in GBM [8]. Since M2 macrophages can drive tumour progression through immunosuppression in gliomas [42], it would be interesting for future studies to elucidate the exact involvement of AGTRAP in this context.

The role of ALKBH3 in cancer is characterised by a striking dichotomy. Previous studies in pancreatic cancer and in head and neck squamous cell carcinoma found that ALKBH3 was overexpressed in tumour compared to the healthy tissues [11,12]. In contrast, ALKBH3 levels were lower in tumour compared to healthy tissues in other types of cancer, such as breast and lung carcinomas [14,15]. In a similar manner, high levels of ALKBH3 associated with the poor outcome of patients with pancreatic cancer and renal cell carcinoma [10,12], while the opposite was observed in breast cancer, lung adenocarcinoma and Hodgkin lymphoma [13–15]. Recent studies by Feng et al. assessed the expression of AlkB family members, including ALKBH3, at both mRNA and protein levels in GBM. Interestingly, while the mRNA analysis using the GEPIA2 database found increased ALKBH3 levels in GBM compared to the healthy brain tissues, protein analysis using the Human Protein Atlas showed that this marker was only weakly expressed in GBM [43]. The latter data are in line with our study, which found significantly lower protein levels of ALKBH3 in GBM compared to the healthy brain parenchyma, as well as an association of the ALKBH3$^{low}$ phenotype with a short OS of the GBM patients. A possible explanation for the discrepancy between mRNA and protein data is the relatively high instability of the ALKBH3 protein due to proteasome targeting. This is supported by comprehensive biochemical studies on ALKBH3 ubiquitination, which found that ALKBH3 mRNA levels did not correlate well with ALKBH3 protein expression in prostate, breast or lung carcinoma cell lines [44]. While the mechanisms of ALKBH3 ubiquitination in GBM still require characterisation, it is not excluded that a similar phenomenon occurs in this type of cancer as well.

DIVERSIN/ANKRD6 is a tumour-promoting factor, which has been proposed as prognostic biomarker in several types of cancer. Specifically, DIVERSIN was found to be overexpressed in tumours compared to healthy adjacent tissues in colorectal carcinoma, non-small cell lung cancer and breast cancer [17,18,20]. For the same tumour entities, patients with high levels of DIVERSIN had a significantly worse outcome than patients with low levels of this marker [17,18,20,21]. Our study did not find an association between DIVERSIN expression and the OS and/or PFS of GBM patients. However, we did find significantly higher levels of DIVERSIN in GBM compared to non-malign brain tissues. These results are supported by previous studies by Wang et al., showing that DIVERSIN was overexpressed in glioma tissues and positively correlated with the WHO grade of the gliomas [19]. Using siRNA knockdown, the authors additionally showed that DIVERSIN was involved in the proliferation and invasion of glioma cell lines. Together, these data indicate that DIVERSIN—while not suitable as a prognostic marker—may be nevertheless a tumourigenic factor and, thus, a potential therapeutic target in GBM.

Accumulating evidence indicates that neddylation plays important roles in the pathophysiology and therapy of cancer (reviewed in [23,24]). This mechanism received particular attention in recent years, since the neddylation inhibitor MLN4924/pevonedistat showed promising anti-tumour effects in phase I–III clinical trials [24]. It is therefore surprising that only few studies addressed the prognostic relevance of NEDD8 expression for the outcome of cancer patients thus far. Specifically, Tian et al. showed that bladder carcinoma patients with high NEDD8 tumour levels had a significantly shorter OS and PFS compared to their NEDD8$^{low}$ counterparts [45]. Similar results were obtained by Xian et al. in esophageal squamous cell carcinoma [46]. In gliomas, Hua et al. found that NEDD8 expression correlated with the WHO grade, and that high NEDD8 levels associated with a shorter OS in WHO grade 4 patients [47]. Our own results are in line with these findings. However, it should be pointed out that the scoring system used in previous studies did not take into consideration the subcellular localisation of NEDD8. In the present study, we observed that NEDD8 had both a cytoplasmic and a nuclear distribution. We therefore assessed the expression levels of NEDD8 in the two cellular compartments separately (NEDD8c versus NEDD8n). Interestingly, we found that only NEDD8c was overexpressed in GBM tissues compared to the healthy brain. Furthermore, high levels of NEDD8c significantly associated with a shorter OS and PFS of the GBM patients. In contrast, NEDD8n was equally expressed in tumour and healthy tissues and did not associate with the patients' outcome. These findings suggest that the tumour-promoting functions of NEDD8 may also be a consequence of its cytoplasmic localisation. Although no direct evidence is currently available to support this hypothesis, previous studies found that NEDD8 could translocate from the nucleus to the cytoplasm in response to Interleukin-1β [48]. Since this cytokine is known to promote GBM proliferation, invasion and angiogenesis (reviewed in [49]), it would be tempting to speculate that an Interleukin-1β/NEDD8 signaling axis is also involved in the progression of GBM.

The last marker tested in our study was RRM1. This protein was proposed as a prognostic factor for OS in gastric cancer [28], pancreatic cancer [27] and biliary tract cancer [26]. A meta-analysis of 23 studies on non-small cell lung cancer found that high levels of RRM1 significantly associated with a shorter OS of the patients, though the prognostic significance was not reached in Cox regression analysis [29]. Using the CGGA and Rembrandt databases, Jiang et al. found that the mRNA levels of RRM1 were increased in high-grade gliomas (WHO grade 3 and 4) compared to healthy brain tissues. Within the WHO grade 4 group, there was a significant association between high RRM1 mRNA expression and poor OS of these patients [31]. Very recently, Ariey-Bonnet et al. elegantly demonstrated that RRM1 was a top target for drug combination therapy in GBM. The authors additionally evaluated RRM1 expression by immunohistochemistry in a cohort of 97 GBM patients and showed that high levels of RRM1 predicted poor OS of the patients in a Cox regression model adjusted by type of surgery [32]. Our study found that GBM tissues had significantly higher levels of RRM1 compared to the healthy brain tissues. When the cut-off was set according to the baseline expression of the healthy tissues, the RRM1$^{high}$ phenotype was significantly associated with the poor OS of the GBM patients. Furthermore, we identified RRM1 as a significant prognostic factor for the OS of GBM patients, which was independent of age, Karnofsky performance scale, extent of surgical resection and MGMT methylation status. Taken together, these data indicate that RRM1 may be a valuable biomarker and, importantly, a promising therapeutic target in GBM.

Previous studies showed that combined analysis of multiple markers had superior prognostic value compared to the analysis of individual markers [50–53]. Here, we performed two-by-two combinations of the markers that significantly associated with the OS

(AGTRAP, ALKBH3, NEDD8c and RRM1) or the PFS (AGTRAP and NEDD8c) of GBM patients in univariate survival analysis. We found that all combinations of markers were significant predictors for the OS or PFS in these patients. However, only two combinations (AGTRAP/ALKBH3 and ALKBH3/NEDD8c) had superior prognostic value compared to the individual markers. The other four combinations did not have significantly increased prognostic values compared to AGTRAP or RRM1 analysed individually. An overview of the prognostic accuracy of the markers evaluated in this study is provided in Figure 6.

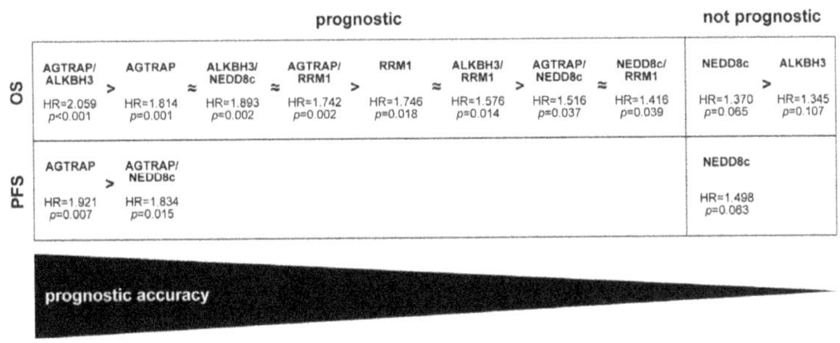


**Figure 6. Prognostic accuracy of individual and combination markers regarding OS and PFS of GBM patients.** Only the markers that reached significance in the univariate survival analysis are included in this overview. Markers with stronger prognostic value are indicated by '>'. Markers with approximately equal prognostic value are indicated by '≈'.

## 5. Conclusions

This study identifies novel individual and combination markers with prognostic relevance for the outcome of IDH wild-type GBM patients. Our results indicate that AGTRAP and RRM1 are prognostic markers for the OS, while AGTRAP is also a prognostic marker for the PFS of these patients. Combining AGTRAP with ALKBH3 may improve the prognostic accuracy for the OS of the GBM patients. Since ALKBH3 and NEDD8c did not reach prognostic significance when analysed individually, their combination could be useful to predict the patients' OS more accurately. Together with our studies on tumour versus non-malign brain tissues, these findings contribute to a better understanding of GBM pathophysiology and may help identify future targets for improved therapeutic strategies in this type of cancer.

**Supplementary Materials:** The following supporting information can be downloaded at: https://www.mdpi.com/article/10.3390/biomedicines12040926/s1, Figure S1: NEDD8n expression in GBM tissues—high magnification; Figure S2: Marker expression in healthy versus GBM tissues (n = 54).

**Author Contributions:** Conceptualization, C.A.D. and I.E.S.; methodology, C.A.D., N.W. and C.L.R.S.; validation, A.R., B.N., K.-P.S. and C.M.; formal analysis, C.A.D., N.W., C.L.R.S. and F.T.A.S.; investigation, C.A.D., N.W., C.L.R.S., F.T.A.S., A.R., B.N. and K.-P.S.; resources, C.M.; writing—original draft preparation, C.A.D.; writing—review and editing, A.R., B.N., K.-P.S., C.M. and I.E.S.; visualization, C.A.D.; supervision, C.A.D. and I.E.S.; project administration, I.E.S.; funding acquisition, I.E.S. All authors have read and agreed to the published version of the manuscript.

**Funding:** This study was partially supported by the Hartmann Stiftung (grant to I.E.S.). The funding body had no role in the design of the study, data collection, analysis and interpretation, or writing of the manuscript.

**Institutional Review Board Statement:** All studies were carried out in accordance to the declaration of Helsinki from 1975, revised in 2013, and were approved by the ethics committee of the Otto-von-Guericke University Magdeburg (Study No. 146/2019).

**Informed Consent Statement:** All studies involving GBM patients were retrospectively performed on material collected during routine surgical and diagnostic procedures. The ethics committee provided a waiver of the need for informed consent.

**Data Availability Statement:** The datasets on patient material are not publicly available as they also contain individual participants' details. These data can be obtained from the corresponding author in anonymised form upon reasonable request.

**Acknowledgments:** We thank Max Pätzold and Josephine Böhme (Department of Neurosurgery, Otto-von-Guericke University Magdeburg, Germany) for excellent technical assistance. We are grateful to Dirk Schomburg (Department of Biometry and Medical Informatics, Otto-von-Guericke University Magdeburg, Germany) for his help with the statistical analysis of the data.

**Conflicts of Interest:** The authors declare no conflict of interest.

## References

1. Ostrom, Q.T.; Cioffi, G.; Gittleman, H.; Patil, N.; Waite, K.; Kruchko, C.; Barnholtz-Sloan, J.S. CBTRUS Statistical Report: Primary Brain and Other Central Nervous System Tumors Diagnosed in the United States in 2012–2016. *Neuro Oncol.* **2019**, *21*, v1–v100. [CrossRef] [PubMed]
2. Tan, A.C.; Ashley, D.M.; Lopez, G.Y.; Malinzak, M.; Friedman, H.S.; Khasraw, M. Management of glioblastoma: State of the art and future directions. *CA Cancer J. Clin.* **2020**, *70*, 299–312. [CrossRef] [PubMed]
3. Louis, D.N.; Perry, A.; Wesseling, P.; Brat, D.J.; Cree, I.A.; Figarella-Branger, D.; Hawkins, C.; Ng, H.K.; Pfister, S.M.; Reifenberger, G.; et al. The 2021 WHO Classification of Tumors of the Central Nervous System: A summary. *Neuro Oncol.* **2021**, *23*, 1231–1251. [CrossRef] [PubMed]
4. Hegi, M.E.; Diserens, A.C.; Gorlia, T.; Hamou, M.F.; de Tribolet, N.; Weller, M.; Kros, J.M.; Hainfellner, J.A.; Mason, W.; Mariani, L.; et al. MGMT gene silencing and benefit from temozolomide in glioblastoma. *N. Engl. J. Med.* **2005**, *352*, 997–1003. [CrossRef] [PubMed]
5. Lopez-Ilasaca, M.; Liu, X.; Tamura, K.; Dzau, V.J. The angiotensin II type I receptor-associated protein, ATRAP, is a transmembrane protein and a modulator of angiotensin II signaling. *Mol. Biol. Cell* **2003**, *14*, 5038–5050. [CrossRef] [PubMed]
6. Liu, S.; Zhao, W.; Li, X.; Zhang, L.; Gao, Y.; Peng, Q.; Du, C.; Jiang, N. AGTRAP Is a Prognostic Biomarker Correlated with Immune Infiltration in Hepatocellular Carcinoma. *Front. Oncol.* **2021**, *11*, 713017. [CrossRef] [PubMed]
7. Wang, D.; Jin, X.; Lei, M.; Jiang, Y.; Liu, Y.; Yu, F.; Guo, Y.; Han, B.; Yang, Y.; Sun, W.; et al. USF1-ATRAP-PBX3 Axis Promote Breast Cancer Glycolysis and Malignant Phenotype by Activating AKT/mTOR Signaling. *Int. J. Biol. Sci.* **2022**, *18*, 2452–2471. [CrossRef]
8. Hong, K.; Zhang, Y.; Yao, L.; Zhang, J.; Sheng, X.; Song, L.; Guo, Y.; Guo, Y. Pan-cancer analysis of the angiotensin II receptor-associated protein as a prognostic and immunological gene predicting immunotherapy responses in pan-cancer. *Front. Cell Dev. Biol.* **2022**, *10*, 913684. [CrossRef]
9. Rajecka, V.; Skalicky, T.; Vanacova, S. The role of RNA adenosine demethylases in the control of gene expression. *Biochim. Biophys. Acta Gene Regul. Mech.* **2019**, *1862*, 343–355. [CrossRef]
10. Hotta, K.; Sho, M.; Fujimoto, K.; Shimada, K.; Yamato, I.; Anai, S.; Harada, H.; Tsujikawa, K.; Konishi, N.; Shinohara, N.; et al. Clinical significance and therapeutic potential of prostate cancer antigen-1/ALKBH3 in human renal cell carcinoma. *Oncol. Rep.* **2015**, *34*, 648–654. [CrossRef]
11. Pilzys, T.; Marcinkowski, M.; Kukwa, W.; Garbicz, D.; Dylewska, M.; Ferenc, K.; Mieczkowski, A.; Kukwa, A.; Migacz, E.; Wolosz, D.; et al. ALKBH overexpression in head and neck cancer: Potential target for novel anticancer therapy. *Sci. Rep.* **2019**, *9*, 13249. [CrossRef] [PubMed]
12. Yamato, I.; Sho, M.; Shimada, K.; Hotta, K.; Ueda, Y.; Yasuda, S.; Shigi, N.; Konishi, N.; Tsujikawa, K.; Nakajima, Y. PCA-1/ALKBH3 contributes to pancreatic cancer by supporting apoptotic resistance and angiogenesis. *Cancer Res.* **2012**, *72*, 4829–4839. [CrossRef] [PubMed]
13. Esteve-Puig, R.; Climent, F.; Piñeyro, D.; Domingo-Domènech, E.; Davalos, V.; Encuentra, M.; Rea, A.; Espejo-Herrera, N.; Soler, M.; Lopez, M.; et al. Epigenetic loss of m1A RNA demethylase ALKBH3 in Hodgkin lymphoma targets collagen, conferring poor clinical outcome. *Blood* **2021**, *137*, 994–999. [CrossRef] [PubMed]
14. Stefansson, O.A.; Hermanowicz, S.; van der Horst, J.; Hilmarsdottir, H.; Staszczak, Z.; Jonasson, J.G.; Tryggvadottir, L.; Gudjonsson, T.; Sigurdsson, S. CpG promoter methylation of the ALKBH3 alkylation repair gene in breast cancer. *BMC Cancer* **2017**, *17*, 469. [CrossRef] [PubMed]
15. Wu, G.T.; Yan, Y.L.; Cai, Y.; Peng, B.; Li, J.N.; Huang, J.Z.; Xu, Z.J.; Zhou, J.H. ALKBH1-8 and FTO: Potential Therapeutic Targets and Prognostic Biomarkers in Lung Adenocarcinoma Pathogenesis. *Front. Cell Dev. Biol.* **2021**, *9*, 633927. [CrossRef]
16. Schwarz-Romond, T.; Asbrand, C.; Bakkers, J.; Kuhl, M.; Schaeffer, H.J.; Huelsken, J.; Behrens, J.; Hammerschmidt, M.; Birchmeier, W. The ankyrin repeat protein Diversin recruits Casein kinase Iepsilon to the beta-catenin degradation complex and acts in both canonical Wnt and Wnt/JNK signaling. *Genes. Dev.* **2002**, *16*, 2073–2084. [CrossRef]
17. Luan, L.; Li, N.Y.; Zhang, K.Y.; Wang, X.J.; Pan, H. Diversin upregulates the proliferative ability of colorectal cancer by inducing cell cycle proteins. *Exp. Mol. Pathol.* **2023**, *129*, 104850. [CrossRef] [PubMed]

18. Luan, L.; Zhao, Y.; Xu, Z.; Jiang, G.; Zhang, X.; Fan, C.; Liu, D.; Zhao, H.; Xu, K.; Wang, M.; et al. Diversin increases the proliferation and invasion ability of non-small-cell lung cancer cells via JNK pathway. *Cancer Lett.* **2014**, *344*, 232–238. [CrossRef] [PubMed]
19. Wang, M.H.; Yu, X.M.; Dong, Q.Z.; Wang, Y.J. Diversin is overexpressed in human gliomas and its depletion inhibits proliferation and invasion. *Tumor Biol.* **2014**, *35*, 7905–7909. [CrossRef]
20. Yu, X.; Wang, M.; Dong, Q.; Jin, F. Diversin is overexpressed in breast cancer and accelerates cell proliferation and invasion. *PLoS ONE* **2014**, *9*, e98591. [CrossRef]
21. Bai, R.; Wu, D.H.; Shi, Z.; Hu, W.X.; Li, J.; Chen, Y.Y.; Ge, W.T.; Yuan, Y.; Zheng, S. Pan-cancer analyses demonstrate that ANKRD6 is associated with a poor prognosis and correlates with M2 macrophage infiltration in colon cancer. *Chin. J. Cancer Res.* **2021**, *33*, 93–102. [CrossRef] [PubMed]
22. Kamitani, T.; Kito, K.; Nguyen, H.P.; Yeh, E.T. Characterization of NEDD8, a developmentally down-regulated ubiquitin-like protein. *J. Biol. Chem.* **1997**, *272*, 28557–28562. [CrossRef] [PubMed]
23. Zhou, L.S.; Jiang, Y.Y.; Luo, Q.; Li, L.H.; Jia, L.J. Neddylation: A novel modulator of the tumor microenvironment. *Mol. Cancer* **2019**, *18*, 77. [CrossRef] [PubMed]
24. Zhou, L.; Jia, L. Targeting Protein Neddylation for Cancer Therapy. *Adv. Exp. Med. Biol.* **2020**, *1217*, 297–315. [CrossRef] [PubMed]
25. Nordlund, P.; Reichard, P. Ribonucleotide reductases. *Annu. Rev. Biochem.* **2006**, *75*, 681–706. [CrossRef] [PubMed]
26. Chun, J.W.; Lee, B.; Park, W.S.; Han, N.; Hong, E.K.; Park, E.Y.; Han, S.S.; Park, S.J.; Kim, T.H.; Lee, W.J.; et al. RRM1 Expression as a Prognostic Biomarker for Unresectable or Recurrent Biliary Tract Cancer Treated with Gemcitabine plus Cisplatin. *J. Clin. Med.* **2021**, *10*, 4652. [CrossRef] [PubMed]
27. Lin, X.Y.; Tan, Y.; Pan, L.L.; Tian, Z.F.; Lin, L.J.; Su, M.X.; Ou, G.S.; Chen, Y.T. Prognostic value of RRM1 and its effect on chemoresistance in pancreatic cancer. *Cancer Chemother. Pharm.* **2023**, *93*, 237–251. [CrossRef] [PubMed]
28. Wang, Q.C.; Liu, X.Y.; Zhou, J.C.; Huang, Y.S.; Zhang, S.J.; Shen, J.G.; Loera, S.; Yuan, X.M.; Chen, W.J.; Jin, M.; et al. Ribonucleotide Reductase Large Subunit M1 Predicts Poor Survival Due to Modulation of Proliferative and Invasive Ability of Gastric Cancer. *PLoS ONE* **2013**, *8*, e70191. [CrossRef] [PubMed]
29. Zhu, C.M.; Lian, X.Y.; Bi, Y.H.; Hu, C.C.; Liang, Y.W.; Li, Q.S. Prognostic value of ribonucleotide reductase subunit M1 (RRM1) in non-small cell lung cancer: A meta-analysis. *Clin. Chim. Acta* **2018**, *485*, 67–73. [CrossRef]
30. Jordheim, L.P.; Seve, P.; Tredan, O.; Dumontet, C. The ribonucleotide reductase large subunit (RRM1) as a predictive factor in patients with cancer. *Lancet Oncol.* **2011**, *12*, 693–702. [CrossRef]
31. Jiang, K.; Zhi, T.L.; Xu, W.H.; Xu, X.P.; Wu, W.N.; Yu, T.F.; Nie, E.; Zhou, X.; Bao, Z.Y.; Jin, X.; et al. MicroRNA-1468-5p inhibits glioma cell proliferation and induces cell cycle arrest by targeting RRM1. *Am. J. Cancer Res.* **2017**, *7*, 784–800. [PubMed]
32. Ariey-Bonnet, J.; Berges, R.; Montero, M.P.; Mouysset, B.; Piris, P.; Muller, K.; Pinna, G.; Failes, T.W.; Arndt, G.M.; Morando, P.; et al. Combination drug screen targeting glioblastoma core vulnerabilities reveals pharmacological synergisms. *Ebiomedicine* **2023**, *95*, 104752. [CrossRef] [PubMed]
33. Dumitru, C.A.; Schroder, H.; Schafer, F.T.A.; Aust, J.F.; Kresse, N.; Siebert, C.L.R.; Stein, K.P.; Haghikia, A.; Wilkens, L.; Mawrin, C.; et al. Progesterone Receptor Membrane Component 1 (PGRMC1) Modulates Tumour Progression, the Immune Microenvironment and the Response to Therapy in Glioblastoma. *Cells* **2023**, *12*, 2498. [CrossRef] [PubMed]
34. Kresse, N.; Schroder, H.; Stein, K.P.; Wilkens, L.; Mawrin, C.; Sandalcioglu, I.E.; Dumitru, C.A. PLOD2 Is a Prognostic Marker in Glioblastoma That Modulates the Immune Microenvironment and Tumor Progression. *Int. J. Mol. Sci.* **2022**, *23*, 6037. [CrossRef]
35. Chambless, L.B.; Kistka, H.M.; Parker, S.L.; Hassam-Malani, L.; McGirt, M.J.; Thompson, R.C. The relative value of postoperative versus preoperative Karnofsky Performance Scale scores as a predictor of survival after surgical resection of glioblastoma multiforme. *J. Neuro-Oncol.* **2015**, *121*, 359–364. [CrossRef] [PubMed]
36. Han, Q.; Liang, H.; Cheng, P.; Yang, H.; Zhao, P. Gross Total vs. Subtotal Resection on Survival Outcomes in Elderly Patients with High-Grade Glioma: A Systematic Review and Meta-Analysis. *Front. Oncol.* **2020**, *10*, 151. [CrossRef] [PubMed]
37. Schiffgens, S.; Wilkens, L.; Brandes, A.A.; Meier, T.; Franceschi, E.; Ermani, M.; Hartmann, C.; Sandalcioglu, I.E.; Dumitru, C.A. Sex-specific clinicopathological significance of novel (Frizzled-7) and established (MGMT, IDH1) biomarkers in glioblastoma. *Oncotarget* **2016**, *7*, 55169–55180. [CrossRef] [PubMed]
38. Siker, M.L.; Wang, M.; Porter, K.; Nelson, D.F.; Curran, W.J.; Michalski, J.M.; Souhami, L.; Chakravarti, A.; Yung, W.K.; Delrowe, J.; et al. Age as an independent prognostic factor in patients with glioblastoma: A Radiation Therapy Oncology Group and American College of Surgeons National Cancer Data Base comparison. *J. Neuro-Oncol.* **2011**, *104*, 351–356. [CrossRef]
39. Wang, L.; Zhang, W.; Yang, T.; He, L.; Liao, Y.; Lu, J. Construction and Comprehensive Analysis of a Stratification System Based on AGTRAP in Patients with Hepatocellular Carcinoma. *Dis. Markers* **2021**, *2021*, 6144476. [CrossRef]
40. Mendaza, S.; Fernandez-Irigoyen, J.; Santamaria, E.; Zudaire, T.; Guarch, R.; Guerrero-Setas, D.; Vidal, A.; Santos-Salas, J.; Matias-Guiu, X.; Ausin, K.; et al. Absence of Nuclear p16 Is a Diagnostic and Independent Prognostic Biomarker in Squamous Cell Carcinoma of the Cervix. *Int. J. Mol. Sci.* **2020**, *21*, 2125. [CrossRef]
41. Zeng, H.; Li, H.; Zhao, Y.; Chen, L.; Ma, X. Transcripto-based network analysis reveals a model of gene activation in tongue squamous cell carcinomas. *Head. Neck* **2019**, *41*, 4098–4110. [CrossRef] [PubMed]
42. Vidyarthi, A.; Agnihotri, T.; Khan, N.; Singh, S.; Tewari, M.K.; Radotra, B.D.; Chatterjee, D.; Agrewala, J.N. Predominance of M2 macrophages in gliomas leads to the suppression of local and systemic immunity. *Cancer Immunol. Immunother.* **2019**, *68*, 1995–2004. [CrossRef] [PubMed]

43. Feng, S.S.; Xu, Z.J.; Peng, J.W.; Zhang, M.Y. The AlkB Family: Potential Prognostic Biomarkers and Therapeutic Targets in Glioblastoma. *Front. Oncol.* **2022**, *12*, 847821. [CrossRef] [PubMed]
44. Zhao, Y.; Majid, M.C.; Soll, J.M.; Brickner, J.R.; Dango, S.; Mosammaparast, N. Noncanonical regulation of alkylation damage resistance by the OTUD4 deubiquitinase. *EMBO J.* **2015**, *34*, 1687–1703. [CrossRef] [PubMed]
45. Tian, D.W.; Wu, Z.L.; Jiang, L.M.; Gao, J.; Wu, C.L.; Hu, H.L. Neural precursor cell expressed, developmentally downregulated 8 promotes tumor progression and predicts poor prognosis of patients with bladder cancer. *Cancer Sci.* **2019**, *110*, 458–467. [CrossRef] [PubMed]
46. Xian, J.R.; Wang, S.W.; Jiang, Y.Y.; Li, L.H.; Cai, L.L.; Chen, P.; Liu, Y.; Zeng, X.F.; Chen, G.A.; Ding, C.; et al. Overexpressed NEDD8 as a potential therapeutic target in esophageal squamous cell carcinoma. *Cancer Biol. Med.* **2022**, *19*, 504–517. [CrossRef] [PubMed]
47. Hua, W.; Li, C.J.; Yang, Z.X.; Li, L.H.; Jiang, Y.A.; Yu, G.Y.; Zhu, W.; Liu, Z.Y.; Duan, S.Z.; Chu, Y.W.; et al. Suppression of glioblastoma by targeting the overactivated protein neddylation pathway. *Neuro-Oncol.* **2015**, *17*, 1333–1343. [CrossRef] [PubMed]
48. Balasubramaniam, M.; Parcon, P.A.; Bose, C.; Liu, L.; Jones, R.A.; Farlow, M.R.; Mrak, R.E.; Barger, S.W.; Griffin, W.S.T. Interleukin-1β drives NEDD8 nuclear-to-cytoplasmic translocation, fostering parkin activation via NEDD8 binding to the P-ubiquitin activating site. *J. Neuroinflamm.* **2019**, *16*, 275. [CrossRef] [PubMed]
49. Yeung, Y.T.; McDonald, K.L.; Grewal, T.; Munoz, L. Interleukins in glioblastoma pathophysiology: Implications for therapy. *Br. J. Pharmacol.* **2013**, *168*, 591–606. [CrossRef]
50. Molenaar, R.J.; Verbaan, D.; Lamba, S.; Zanon, C.; Jeuken, J.W.; Boots-Sprenger, S.H.; Wesseling, P.; Hulsebos, T.J.; Troost, D.; van Tilborg, A.A.; et al. The combination of IDH1 mutations and MGMT methylation status predicts survival in glioblastoma better than either IDH1 or MGMT alone. *Neuro-Oncol.* **2014**, *16*, 1263–1273. [CrossRef]
51. Nystrom, H.; Tavelin, B.; Bjorklund, M.; Naredi, P.; Sund, M. Improved tumour marker sensitivity in detecting colorectal liver metastases by combined type IV collagen and CEA measurement. *Tumour Biol.* **2015**, *36*, 9839–9847. [CrossRef] [PubMed]
52. Weller, P.; Bankfalvi, A.; Gu, X.; Dominas, N.; Lehnerdt, G.F.; Zeidler, R.; Lang, S.; Brandau, S.; Dumitru, C.A. The role of tumour FoxP3 as prognostic marker in different subtypes of head and neck cancer. *Eur. J. Cancer* **2014**, *50*, 1291–1300. [CrossRef] [PubMed]
53. Zemans, R.L.; Jacobson, S.; Keene, J.; Kechris, K.; Miller, B.E.; Tal-Singer, R.; Bowler, R.P. Multiple biomarkers predict disease severity, progression and mortality in COPD. *Respir. Res.* **2017**, *18*, 117. [CrossRef] [PubMed]

biomedicines

Article

# Prognostic Factors for Survival in Glioblastoma: A Retrospective Analysis Focused on the Role of Hemoglobin

Zuzana Pleskacova [1,*], Michael Bartos [2], Hana Vosmikova [3], Rafael Dolezal [4], Petr Krupa [2], Barbora Vitovcova [5], Petra Kasparova [3], Emil Rudolf [5], Veronika Skarkova [5], Denisa Pohankova [1], Veronika Novotna [1] and Jiri Petera [1]

1 Department of Oncology and Radiotherapy, University Hospital and Medical Faculty, 500 05 Hradec Kralove, Czech Republic; denisa.pohankova@fnhk.cz (D.P.); veronika.bockayova@fnhk.cz (V.N.); jiri.petera@fnhk.cz (J.P.)

2 Department of Neurosurgery, University Hospital and Medical Faculty, 500 05 Hradec Kralove, Czech Republic; michael.bartos2@fnhk.cz (M.B.); petr.krupa@fnhk.cz (P.K.)

3 The Fingerland Department of Pathology, University Hospital and Medical Faculty, 500 05 Hradec Kralove, Czech Republic; hana.vosmikova@fnhk.cz (H.V.); petra.kasparova@fnhk.cz (P.K.)

4 Biomedical Research Centre, University hospital and Medical Faculty, 500 05 Hradec Kralove, Czech Republic; rafael.dolezal@fnhk.cz

5 Department of Medical Biology and Genetics, Medical Faculty, 500 05 Hradec Kralove, Czech Republic; vitovcob@lfhk.cuni.cz (B.V.); rudolf.emil@lfhk.cuni.cz (E.R.); hanusovav@lfhk.cuni.cz (V.S.)

* Correspondence: pleskacova@multiscan.cz or zuzana.pleskacova@fnhk.cz

**Abstract:** *Background*: Although several prognostic factors for survival have been identified in glioblastoma, there are numerous other potential markers (such as hemoglobin) whose role has not yet been confirmed. The aim of this study was to evaluate a wide range of potential prognostic factors, including HIF-1α and hemoglobin levels, for survival in glioblastoma. A secondary aim was to determine whether hemoglobin levels were associated with HIF-1α expression. *Methods:* A retrospective study of 136 patients treated for glioblastoma at our institution between 2012 and 2021 was performed. Cox univariate and multivariate analyses were carried out. Kaplan–Meier survival curves were generated. In addition, bivariate non-parametric correlation analyses were performed for key variables. *Results:* Median survival was 11.9 months (range: 0–119.4). According to the univariate analysis, 13 variables were significantly associated with survival: age, performance status, extent of surgery, tumor depth, tumor size, epilepsy, postoperative chemoradiotherapy, IDH mutations, CD44, HIF-1α, HIF-1β, vimentin, and PDFGR. According to the multivariate regression analysis, only four variables remained significantly associated with survival: age, extent of surgery, epilepsy, and HIF-1α expression. No significant association was observed between hemoglobin levels (low <120 g/L in females or <140 g/L in males vs. high ≥120 or ≥140 g/L) and survival or HIF-1α/HIF-1β expression. *Conclusions:* In this retrospective study of patients with glioblastoma, four variables—age, extent of surgery, HIF-1α expression, and epilepsy—were significant prognostic factors for survival. Hemoglobin levels were not significantly associated with survival or HIF-1α expression. Although hypoxia is a well-recognized component of the glioblastoma microenvironment, more research is needed to understand the pathogenesis of onset tumor hypoxia and treatment implication.

**Keywords:** glioblastoma; prognostic markers; HIF; hemoglobin

## 1. Introduction

Glioblastoma, the most common type of malignant brain tumor [1], is characterized by intratumoral heterogeneity, an invasive growth pattern, and poor response to treatment [2]. In the last two decades, the most significant advance in the treatment of glioblastoma was the addition of postoperative chemoradiation (Stupp's protocol) [3]. Survival outcomes are still poor, with a median survival of approximately 15 months [4] and 5-year survival rates of around 5% [2].

Confirmed prognostic factors for survival in glioblastoma included performance status, age, extent of resection, postoperative chemoradiotherapy, MGMT methylation, and IDH (isocitrate dehydrogenase) 1/2 mutation [5]. In addition, several other potential prognostic factors have been identified, including hemoglobin level [6,7], epilepsy [8,9], and subventricular zone involvement [10]. However, the true prognostic value of these parameters is uncertain due to the inconsistent results reported to date.

Tumor hypoxia is a common feature in glioblastoma, mainly due to abnormal neovascularization. Hypoxic stress triggers the accumulation of hypoxia-inducible factor 1 alfa (HIF-1$\alpha$) and its partner HIF beta (HIF-1$\beta$), which initiate the transcription of hundreds of genes involved in the regulation of angiogenesis, glycolysis, autophagy, motility and invasion, and resistance to chemotherapy and radiation. HIF-1$\alpha$ expression is associated with poor prognosis in gliomas [11]. The low hemoglobin level before radiotherapy was proved to be the negative prognostic factor of survival of glioblastoma [7], but this finding is not consistent across all studies. A decrease in hemoglobin levels can contribute to the upregulation of HIF expression in glioblastoma, further promoting tumor growth and aggressiveness. Understanding the interplay between hemoglobin levels and HIF expression in glioblastoma may provide valuable insights into potential therapeutic strategies to target hypoxia and improve patient outcomes. Although there are some publications dealing with hemoglobin level and HIF expression in several tumor types [12,13], publications on glioblastoma are lacking.

In this context, the aim of the present study was to evaluate clinical and molecular factors—particularly HIF-1$\alpha$ and hemoglobin levels—to determine whether these variables are associated with survival in patients with glioblastoma. A secondary aim was to determine whether hemoglobin levels were associated with HIF-1$\alpha$ expression.

## 2. Patients and Methods

### 2.1. Patients

Patients diagnosed and treated for glioblastoma between 2012 and 2021 at our hospital (Hospital Hradec Kralove, Czech Republic) were evaluated for possible study inclusion; 324 patients were treated during the study period. Of these, 136 had complete clinical data, including tissue samples.

The patients' demographic and clinical variables were obtained from clinical records, as follows: age, sex, data of diagnosis, type of procedure (gross total resection [GTR], subtotal resection [STR], or biopsy alone), clinical course, and imaging, laboratory, and histopathological data. Other data obtained from the medical records included: comorbid medical conditions, Karnofsky performance status (KPS), epilepsy (yes/no), tumor size, and hemoglobin levels.

Table 1 shows the main characteristics of the cohort. The cohort included 136 patients (94 males, 42 females). The median age was 59 years (range: 26–83). The type of surgery was distributed as follows: GTR ($n$ = 54; 39.7%), STR ($n$ = 64; 47.1%), and biopsy alone ($n$ = 8; 13.2%). In nearly all cases ($n$ = 135; 99.3%), the tumor was located in the supratentorial region. Of these, 125 (92%) were located in the superficial area (cortical or subcortical area) and 10 (7.4%) in deep anatomical structures (basal ganglia and/or corpus callosum). At diagnosis, 18 patients (13.2%) presented multifocal spread.

Ninety-three patients (68.4%) received standard temozolomide (75/mg$^2$) administered concomitantly with conventional radiation therapy (maximum dose, 60 Gy). Palliative radiotherapy was performed in 31 patients (22.8%). The patients ($n$ = 12, 8.8%) with an ECOG (Eastern Cooperative Oncology Group) performance status >3 received best supportive care.

Patients were stratified according to the peak preoperative hemoglobin levels. Hemoglobin levels were classified into low (<120 g/L in females, <140 g/L in males) and high ($\geq$120 or $\geq$140 g/L, respectively).

**Table 1.** Basic statistical characteristics of the involved patients.

| Variable | Number of Patients (%) |
|---|---|
| **Total** | 136 (100%) |
| **Sex** | |
| Male | 94 (69.1%) |
| Female | 42 (30.9%) |
| **Age (mean 59.3 year)** | |
| <70 year | 110 (80.9%) |
| ≥70 year | 26 (19.1%) |
| **KPS, mean** | |
| ≥70 | 87 (64%) |
| <70 | 49 (36%) |
| **Surgery** | |
| GTR | 54 (39.7%) |
| STR | 64 (47.1%) |
| Biopsy alone | 18 (13.2%) |
| **Peak hemoglobin level** * | |
| Low | 62 (45.6%) |
| High | 74 (54.4%) |
| Epilepsy | |
| Yes | 56 (41.1%) |
| No | 80 (58.8%) |
| **Localization** ** | |
| Superficial | 125 (92%) |
| Deep | 11(8%) |
| **Postoperative treatment** | |
| Chemoradiotherapy | 93 (68%) |
| Radiotherapy | 31 (23%) |
| Best supportive care | 12 (9%) |

**Abbreviations:** KPS: Karnofsky performance score; GTR: gross total resection (radical surgery); STR: subtotal resection. * The cut-off for low vs. high hemoglobin levels in females was <120 g/L vs. ≥120 g/L. In males, the cut-off was <140 g/L vs. ≥140 g/L. ** Superficial localization = cortical or subcortical area; deep localization = deep anatomical structures (basal ganglia or corpus callosum).

### 2.2. *Histochemical Analyses*

All histological samples were formalin-fixed and paraffin-embedded following standard procedures at the hospital laboratory. Review of all available archival material, including immunohistochemical slides, was performed by a single experienced neuropathologist prior to study inclusion. Tissue blocks were cut into 2-mm-thick sections for immunohistochemical analysist.

We reviewed the literature to identify histochemical parameters potentially associated with hypoxia signaling [14–16], which included the following: IDH, CD44 (family of transmembrane glycoproteins), HIF-1α, HIF–1β, platelet-derived growth factor receptors (PDFGR), p16, p53, vimentin (VIM), SRY-Box Transcription Factor 11 (SOX11), D240 (podoplanin), MER proto-oncogene tyrosine kinase (MERTK), mesenchyme Homeobox 2 (MEOX2), oligodendrocyte transcription factor 2 (OLIG2), *signal transducer and activator of transcription 3 (STAT3)*, epidermal growth factor receptor (EGFR), CD34 (transmembrane phosphoglycoprotein), and *glial fibrillary acidic protein (GFAP)*.

Staining was carried out with the Ventana Benchmark Ultra immunostainer (Ventana/Roche, Tucson, AZ, USA) or the DAKO Omnis stainer (Agilent, Santa Clara, CA, USA). The Ventana OptiView DAB IHC kit (or DAKO EnVision Flex kit) was used for visualization: both methods use an avidin–biotin complex method with horseradish peroxidase as an enzyme and DAB (3,3′-diaminobenzidine) as chromogen. All slides were counterstained with hematoxylin, and positive controls were used on the slides, as appropriate.

The immunohistochemistry results were assessed by a single observer (HIF-1α, HIF-1β, STAT3) or two observers (remaining markers) blinded to the clinical and pathological data. First, the observer determined the percentage of positive tumor cells and staining intensity, which was rated on a scale from 1 to 3 (mild, moderate, or strong). Tumors in which <1% of cells were positive were considered negative (Figure 1). A modified H-score was then calculated by multiplying the most common staining intensity by the percentage of positive cells (range: 0 to 300). Consensus meeting with both observers was held for discrepant cases to attain a consensus score.

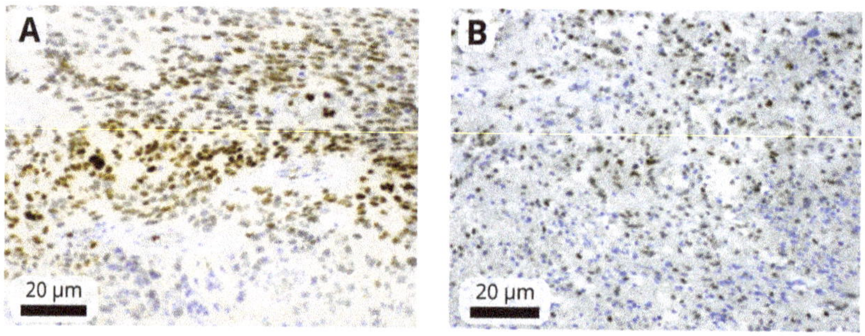


**Figure 1.** Representative images of moderate-to-strong nuclear immunoreactivity of HIF-1α (**A**) and weak-to-moderate immunoreactivity of HIF-1β (**B**) in glioblastoma cells. Images were obtained at magnification of 200×. The black scale bars indicate 20 μm.

### 2.3. *Statistical Analysis*

The following variables were evaluated to check for an association with survival outcomes: age at first surgery, sex, KPS at diagnosis, surgery (radical vs. nonradical), hemoglobin levels at surgery, tumor localization (deep vs. superficial), tumor (T) size (<10 vs. ≥10 mm), epilepsy (yes vs. no), postoperative radiochemotherapy (yes/no), and the mutation status of the following: IDH, CD44, HIF-1α, HIF–1β, p53, p16, PDFGR, VIM, SOX11, D240, MERTK, MEOX2, OLIG2, EGFR, STAT3, CD34, and GFAP.

Survival was calculated as the time from surgery to date of death. For censored cases, survival was defined as the time from surgery until the last medical visit.

Survival data were analyzed using the Kaplan–Meier method with 95% confidence intervals (CI). The statistical significance of the Kaplan–Meier models was evaluated by log-rank Mantel–Cox (MC) and Tarone–Ware (TW) tests of the overall model $\chi^2$. All variables were tested in a univariate Cox regression model and in multivariate Cox models. The overall statistical significance of the Cox models was estimated from the total $\chi^2$. Identifiability of the models was monitored by controlling the maximal values in the correlation matrix of the regression coefficients. The top-scoring statistical models were tested by random bootstrapping (1000 repetitions) to estimate model robustness or bias-corrected accelerated CIs. Only the statistically significant variables (type I error, $p < 0.05$) were considered when interpreting the results. Bivariate non-parametric correlation analyses were carried out (Spearman $\rho$ and Kendal $\tau$ criteria) to assess the mutual independence of the variables: epilepsy and tumor size, radicality of resection, HIF-1α, STAT3, IDH, and age. All statistical analyses were performed with the IBM-SPSS Statistical software program (v. 29.0).

## 3. Results

Median survival in the full cohort ($n$ = 136) was 17.6 months (range: 0–119.4). The overall survival (OS) rate at one and three years was 47.8% ($n$ = 65) and 12.5% ($n$ = 17), respectively. Currently (as of 8 March 2024), four patients remain alive (60.8, 31.8, 38.0, and 38.8 months from surgery, respectively).

Survival was significantly correlated ($p < 0.05$; Kaplan–Meier) with the following variables: age (<70 vs. ≥70 years), KPS (≥70 vs. <70), extent of surgery (radical vs.

nonradical), tumor depth (superficial vs. invasion of central structures), tumor size (<10 vs. ≥10 mm), epilepsy (yes vs. no), postoperative chemoradiotherapy (yes vs. no), and IDH mutation (yes vs. no) (Table 2, Figure 2).

**Table 2.** Results of the Kaplan–Meier analyses.

| Factor | *Subclass* | $N_C$ | $N_E$ | $EOS_{med}$ [Months] | $^{0.95}CI$ [Months] | $_{MC}\chi^2$ (*p*) | $_{TW}\chi^2$ (*p*) |
|---|---|---|---|---|---|---|---|
| Age | <70 year | 3 | 104 | 13.630 | 10.810–16.450 | 11.169 (<0.001) | 12.057 (<0.001) |
| | ≥70 year | 1 | 28 | 6.130 | 2.438–9.822 | | |
| Sex | Male | 4 | 90 | 11.530 | 7.986–15.074 | 0.044 (0.833) | 0.039 (0.843) |
| | Female | 0 | 42 | 11.930 | 8.578–15.282 | | |
| KPS | ≥70 | 4 | 101 | 14.730 | 12.638–16.822 | 28.788 (<0.001) | 37.463 (<0.001) |
| | <70 | 0 | 30 | 4.370 | 2.357–6.383 | | |
| Surgery * | Radical | 4 | 50 | 17.570 | 14.689–20.451 | 21.940 (<0.001) | 26.693 (<0.001) |
| | Non-radical | 0 | 80 | 11.770 | 9.453–14.087 | | |
| Hb ** | Low | 3 | 58 | 10.770 | 7.851–13.689 | 1.437 (0.231) | 1.711 (0.191) |
| | High | 1 | 73 | 13.370 | 9.855–16.885 | | |
| Location | Peritumoral | 4 | 120 | 11.930 | 9.202–14.658 | 5.850 (0.016) | 5.514 (0.019) |
| | Deep | 0 | 10 | 5.470 | 0.775–10.165 | | |
| T-size | <10 mm | 4 | 56 | 16.930 | 14.197–19.663 | 17.757 (<0.001) | 22.673 (<0.001) |
| | ≥10 mm | 0 | 69 | 7.230 | 5.638–8.822 | | |
| Epilepsy | Yes | 2 | 54 | 20.300 | 15.411–25.189 | 20.968 (<0.001) | 20.071 (<0.001) |
| | No | 2 | 77 | 9.830 | 6.704–12.956 | | |
| Treatment | Yes | 3 | 90 | 14.900 | 13.036–16.764 | 24.995 (<0.001) | 27.615 (<0.001) |
| | No | 1 | 42 | 6.370 | 3.672 | | |
| IDH | Mutated | 2 | 9 | 26.900 | 0.000–63.481 | 12.406 (<0.001) | 9.324 (<0.002) |
| | Wild | 2 | 122 | 11.500 | 8.951–14.049 | | |
| HIF-1α | ≥4 | 0 | 9 | 5.230 | 4.646–5.814 | 4.241 (0.039) | 4.076 (0.043) |
| | <4 | 3 | 119 | 11.770 | 9.118–14.422 | | |

**Abbreviations:** Hb: hemoglobin; KPS: Karnofsky performance status; EOS: estimated overall median survival; $^{0.95}CI$: lower and upper bound of the 95% confidence interval of hazard ratio; $_{MC}\chi^2$: values of the Mantel–Cox chi-square test and the corresponding probabilities; $_{TW}\chi^2$: values of the Tarone–Ware chi-square test and the corresponding probabilities; $N_C$: number of the censored cases; $N_E$: number of the cases that experienced the terminal event. * Surgery was classified as radical or non-radical (biopsy or subtotal resection). ** Peak preoperative hemoglobin levels were classified as low (<120 g/L in females, <140 g/L in males) or high (≥120 or ≥140 g/L, respectively).

On the Cox univariate analyses, survival was significantly ($p < 0.05$) associated with the following variables: age, KPS, extent of surgery, tumor depth, tumor size, epilepsy, postoperative chemoradiotherapy, IDH mutations, CD44, HIF-1α, HIF-1β, vimentin, and PDFGF. Since the HIF-1α marker is a well-known predictor of OS in patients with GBM, a histogram showing the distribution of the HIF-1α levels for short and long OS revealed in this study is provided in Figure 3. None of the other variables involved in this study were found statistically significant.

On the multivariate Cox regression analysis only age, extent of surgery, epilepsy, and HIF-1α expression were significantly associated with survival (Table 3).

Hemoglobin levels (low vs. high) did not correlate with survival of HIF-1α or HIF-1β expression (Kaplan–Meier).

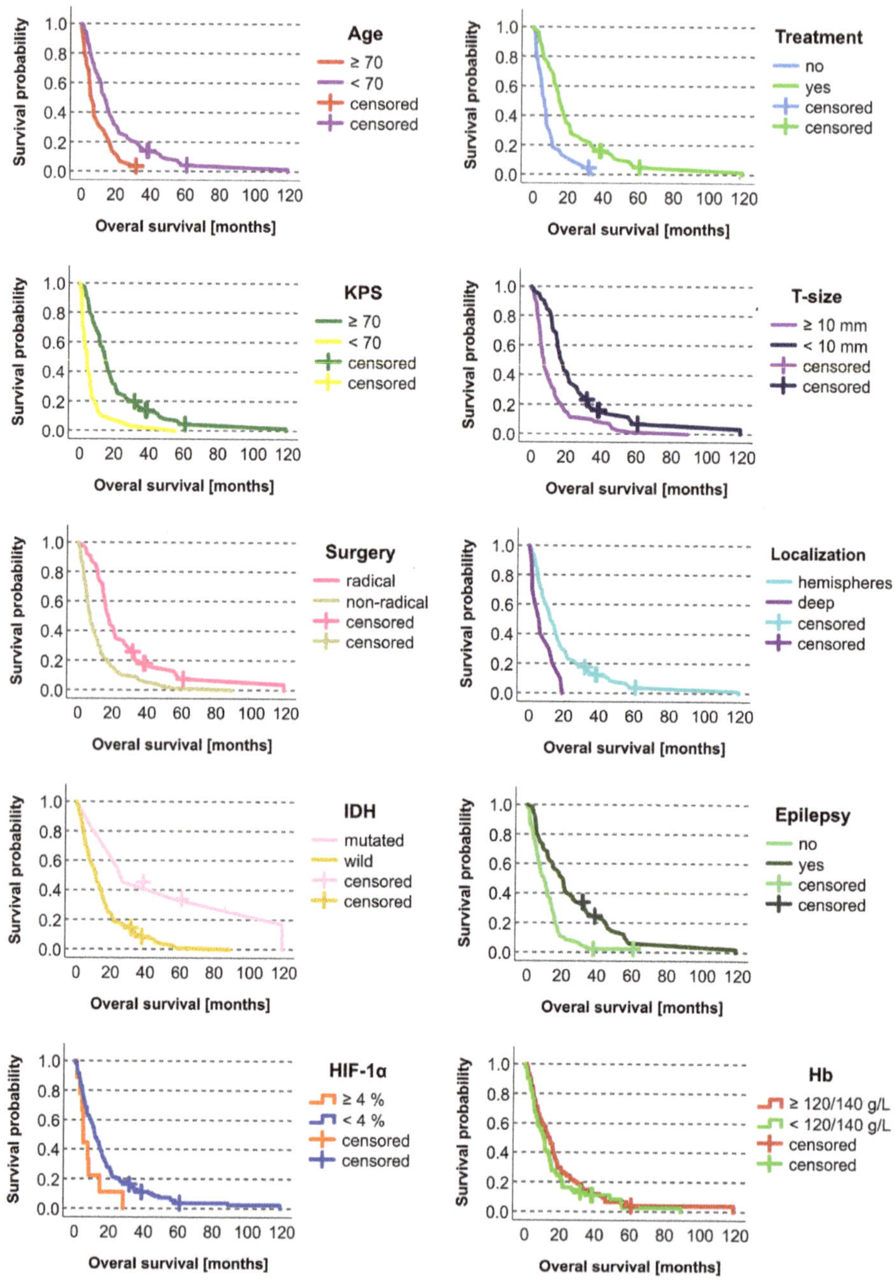


**Figure 2.** Kaplan–Meier curves for statistically significant categorical prognostic factors in the studied cohort of 136 patients with glioblastoma. The retrospective cohort included 132 deceased patients and four who were alive at the time of the study evaluation. The living patients were included as censored cases.

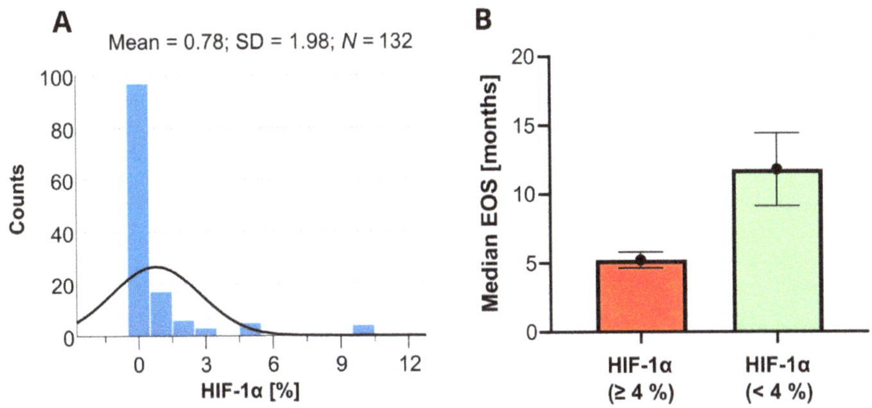


**Figure 3.** Histogram of the available HIF-1α data determined within the present study (**A**) and the corresponding expected median overall survivals (EOS) with the 95% confidence limits as predicted by the Kaplan–Meyer analysis (**B**). The black line in the histogram represents the expected normal distribution.

**Table 3.** Multivariate Cox regression analysis of the most significant categorical and continuous factors in a single model.

| Factor | β | SE | Wald | $p$ | HR | $^{0.95}CI_{LB}$ | $^{0.95}CI_{UB}$ |
|---|---|---|---|---|---|---|---|
| Age(con) | 0.032 | 0.012 | 7.333 | 0.007 | 1.032 | 1.009 | 1.056 |
| Surgery(cat) | −0.874 | 0.363 | 5.811 | 0.016 | 2.396 | 1.177 | 4.877 |
| Epilepsy(cat) | 0.599 | 0.257 | 5.439 | 0.020 | 0.549 | 0.332 | 0.909 |
| HIF-1α(con) | 0.104 | 0.053 | 3.872 | 0.049 | 1.110 | 1.000 | 1.231 |
| HIF1β(con) | −0.008 | 0.006 | 2.095 | 0.148 | 0.992 | 0.980 | 1.003 |
| Localization(cat) | 0.563 | 0.435 | 1.679 | 0.195 | 1.756 | 0.749 | 4.117 |
| KPS(cat) | −0.321 | 0.311 | 1.072 | 0.301 | 1.379 | 0.750 | 2.535 |
| PDGFR(con) | −0.004 | 0.004 | 1.027 | 0.311 | 0.996 | 0.988 | 1.004 |
| IDH1(cat) | −0.406 | 0.505 | 0.646 | 0.422 | 1.500 | 0.558 | 4.033 |
| T-size(cat) | −0.220 | 0.378 | 0.338 | 0.561 | 1.246 | 0.594 | 2.613 |
| Treatment(cat) | −0.163 | 0.280 | 0.338 | 0.561 | 0.850 | 0.491 | 1.471 |
| CD44(con) | 0.002 | 0.005 | 0.134 | 0.715 | 1.002 | 0.991 | 1.013 |
| Vimk(con) | 0.001 | 0.006 | 0.025 | 0.873 | 1.001 | 0.989 | 1.013 |

**Abbreviations:** Con: continuous variable; Cat: categorical variable; β: regression coefficient of a variable in the Cox proportional hazard regression model; SE: standard error of β coefficient; Wald: value of the Wald test; $p$: probability of the Wald test; HR: hazard ratio (i.e., exp(β)) for a one-unit change of a variable; $^{0.95}CI_{LB}$ and $^{0.95}CI_{UB}$: lower and upper bound of the 95% confidence interval of HR; KPS: Karnofsky performance status.

On the bivariate analysis between HIF-1α and p53 and p16, neither the Spearman ρ nor the Kendal τ coefficients were statistically significant.

We tested possible correlation between epilepsy with tumor size, extent of surgery, HIF-1α, and IDH using a non-parametrical Spearman $\rho$ model. A younger age (under 60 years), smaller tumors before treatment (<10 mm), nonradical surgery, and IDH mutations were statistically significant factors. Additional Cox multivariate analyses can be found in Supplementary Materials.

## 4. Discussion

In this retrospective study, we evaluated numerous clinical and molecular factors—including HIF-1$\alpha$ and hemoglobin levels—to determine their potential association with survival in patients with glioblastoma. On the univariate analysis, numerous variables were significantly associated with survival. However, on the multivariate analysis, only four variables—age, extent of surgery, epilepsy, and HIF-1$\alpha$ expression—remained significant. There was no correlation between hemoglobin levels and survival or HIF-1$\alpha$ expression, which suggests that hemoglobin is not a prognostic factor for survival in this patient population.

In glioblastoma, hypoxia is closely associated with tumor aggressiveness, radiation resistance, chemoresistance, and poor prognosis [14]. Several studies have assessed the role of hemoglobin levels in patients with glioblastoma, with contradictory findings. Céfaro et al. [7] found that hemoglobin levels $\leq$12 g/L (versus > 12 g/L) was associated with significantly worse survival outcomes (12 vs. 23 months). Lutterbach et al. [5] also found that low hemoglobin levels were significantly associated with poor treatment outcomes. In a previous study [17], our group found that hemoglobin levels >12 g/L were associated with longer median survival compared to levels $\leq$12 g/L (12.1 vs. 7.9 months, respectively). More recently (2020), Kismar-Elbaz et al. [6] found that low hemoglobin levels (<12 g/dL [females] or <14 g/dL [males]) was a negative prognostic factor for survival, leading the authors to conclude that measures should be taken to normalize preoperative hemoglobin levels and red cell distribution width to improve prognosis. Notwithstanding the findings of those studies, other studies have not found any association between hemoglobin levels and prognosis in these patients [18,19]. In our previous study [17], we hypothesized that low levels of hemoglobin would increase HIF expression. However, in the present study, hemoglobin levels had no impact on HIF expression, nor were they associated with survival outcomes.

The mechanism of action of anemia on treatment outcomes in patients with glioblastoma remains unclear. In fact, only a few studies have assessed the influence of anemia on HIF expression. Winter et al. [12] evaluated HIF-1$\alpha$ expression in surgically-treated head and neck cancer patients and found that high HIF-1$\alpha$/HIF-2$\alpha$ expression was an independent negative prognostic factor for survival. However, no correlation was observed between HIF-1$\alpha$ or HIF-2$\alpha$ expression and hemoglobin levels, leading those authors to conclude that activation of HIF and its signaling pathways is independent of hemoglobin levels, including hypoxia-independent mechanisms. On the other hand, Dallas et al. [13] found that patients with cervical cancer treated by radiotherapy and hemoglobin levels <11 g/dL showed elevated HIF-1$\alpha$ expression compared to patients with hemoglobin levels >12.5 g/L ($p$ = 0.04).

A meta-analysis carried out to assess the role of HIF expression in glioblastoma found that HIF-1$\alpha$ expression was associated with high grade glioma and worse survival [11]. Sfifou et al. [20] recently found that 12-month survival outcomes were better in patients with negative HIF-1$\alpha$ expression and positive IDH1 when compared to patients with HIF-1$\alpha$-positive, IDH1-negative disease. Potharaji et al. [21] found that strong HIF-1$\alpha$ expression was an independent prognostic factor for poor survival. In that study, the subgroup with the worst prognosis had strong expression of both HIF-1$\alpha$ and telomerase reverse transcriptase. In our study, HIF-1$\alpha$ expression was a strong, independent predictor of shorter survival. HIF transcription factors constitute the master regulators of the hypoxia adaptive response. Hypoxia occurs in GBM due to increased cell proliferation and tumor growth resulting in tumor neovascularization, which is insufficient and leads to hypoxia, acidosis, necrosis and intratumoral edema. Some genetic alteration (activation of epidermal growth factor receptor, the loss of tumor suppressor function—p53, PTEN) can increase HIF expression. Hypoxia is a main promotor of glioblastoma invasion, radio- and chemoresistace, and survival [22]. Further research on HIF factor expression regulations is a key factor for the development of targeted biological therapy directed at HIF [23].

Although our study showed a significant association between epilepsy and better survival, the true prognostic value of epilepsy remains unclear due to the heterogenous results reported in previous studies. For example, Dobran et al. [9] found that epilepsy at presentation was an independent prognostic factor for longer survival, but only in patients younger than 60 years. In the study by Berendsen et al. [8], epilepsy was positively correlated with longer survival. Interestingly, there were no survival differences between patients with or without antiepileptic medication, which suggests that the medication had no impact on survival. In a more recent study by that same research group [24], epilepsy was an independent prognostic factor for better survival in glioblastoma patients. In that study, epileptogenic glioblastoma was correlated with decreased hypoxia/HIF-1α/STAT5b signaling compared to non-epileptogenic glioblastomas. In our study, epilepsy was significantly correlated with tumor size (i.e., smaller tumors were associated with a higher probability of epilepsy), younger age, nonradical surgery, and IDH mutations. Other authors have also found a relationship between epilepsy and IDH mutations [25,26]. By contrast, no such association was observed in the study by Berendesen et al. [8].

Although we evaluated numerous different histochemical parameters potentially linked with hypoxia signaling, glioblastoma invasiveness, and disease progression, the only histochemical variable that was significantly associated with survival on the multivariate analysis was HIF-1α expression.

*Strengths and Limitations*

The main limitation of this study is its retrospective study design. By contrast, an important strength is its large sample size.

## 5. Conclusions

In this retrospective study, four variables—age, extent of surgery, expression of HIF-1α, and epilepsy—were significant prognostic factors for survival in patients with glioblastoma. Importantly, hemoglobin levels were not correlated with survival or HIF-1α expression. Commonly used treatment modalities to correct anemia in cancer patients improve the quality in life and may improve their prognosis and response to treatment particularly when radiation therapy is planned. However, there is still no strong evidence that hemoglobin level in glioblastoma patient have an impact on survival. It may be caused by a higher influence of other prognostic markers, be they clinical (age, extent of surgery) or molecular (HIF-1α, IDH), or the tumor microenvironment. Although hypoxia is a well-recognized component of the glioblastoma microenvironment, more research is needed to better understand the pathogenesis of its onset and the treatment implications.

**Supplementary Materials:** The following supporting information can be downloaded at https://www.mdpi.com/article/10.3390/biomedicines12061210/s1. Table S1: Results of the Cox multivariate analyses focused on the genetic and other factors with continuous values.

**Author Contributions:** Conceptualization, Z.P., J.P., and E.R.; Methology, H.V., V.S., P.K. (Petra Kasparova), and B.V.; Surgery, M.B. and P.K. (Petr Krupa); Radiotherapy, D.P. and V.N.; Statistics, R.D. All authors have read and agreed to the published version of the manuscript.

**Funding:** The research was funded by research grant: AZV 20-03-00360.

**Institutional Review Board Statement:** The study was conducted according to the guidelines of the Declaration of Helsinki and approved by Institutional Review Board of University Hradec Kralove, Charles University, 201906S2596P.

**Informed Consent Statement:** Informed consent of the all surviving subjects involved in the study was obtained.

**Data Availability Statement:** All data mentioned within the manuscript are available upon request to corresponding author.

**Acknowledgments:** We thank Bradley Londres for professional English language editing. We thank Jiri Soukup for their evaluation of biopsy samples and great help during the entire research process.

**Conflicts of Interest:** The authors declare no conflicts of interest.

## References

1. Patel, A.P.; Fisher, J.L.; Nichols, E.; Abd-Allah, F.; Abdela, J.; Abdelalim, A.; Abraha, H.N.; Agius, D.; Alahdab, F.; Alam, T.; et al. Global, regional, and national burden of brain and other CNS cancer, 1990–2016: A systematic analysis for the Global Burden of Disease Study 2016. *Lancet Neurol.* **2019**, *18*, 376–393. [CrossRef] [PubMed]
2. Oronsky, B.; Reid, T.R.; Oronsky, A.; Sandhu, N.; Knox, S.J. A Review of Newly Diagnosed Glioblastoma. *Front. Oncol.* **2021**, *10*, 574012. [CrossRef] [PubMed]
3. Stupp, R.; Mason, W.P.; Van Den Bent, M.J.; Weller, M.; Fisher, B.; Taphoorn, M.J.; Belanger, K.; Brandes, A.A.; Marosi, C.; Bogdahn, U.; et al. Radiotherapy plus concomitant and adjuvant temozolomide for glioblastoma. *N. Engl. J. Med.* **2005**, *352*, 987–996. [CrossRef] [PubMed]
4. Stupp, R.; Hegi, M.E.; Mason, W.P.; Van Den Bent, M.J.; Taphoorn, M.J.; Janzer, R.C.; Ludwin, S.K.; Allgeier, A.; Fisher, B.; Belanger, K.; et al. Effects of radiotherapy with concomitant and adjuvant temozolomide versus radiotherapy alone on survival in glioblastoma in a randomised phase III study: 5-year analysis of the EORTC-NCIC trial. *Lancet Oncol.* **2009**, *10*, 459–466. [CrossRef] [PubMed]
5. Lutterbach, J.; Sauerbrei, W.; Guttenberger, R. Multivariate analysis of prognostic factors in patients with glioblastoma. *Strahlenther. Onkol.* **2003**, *179*, 8–15. [CrossRef] [PubMed]
6. Kaisman-Elbaz, T.; Elbaz, Y.; Merkin, V.; Dym, L.; Noy, A.; Atar-Vardi, M.; Bari, R.; Turiel, S.; Alt, A.; Zamed, T.; et al. Hemoglobin Levels and Red Blood Cells Distribution Width Highlights Glioblastoma Patients Subgroup With Improved Median Overall Survival. *Front. Oncol.* **2020**, *10*, 432. [CrossRef]
7. Céfaro, G.A.; Genovesi, D.; Vinciguerra, A.; Trignani, M.; Taraborrelli, M.; Augurio, A.; Buonaguidi, R.; Galzio, R.J.; Di Nicola, M. Prognostic impact of hemoglobin level and other factors in patients with high-grade gliomas treated with postoperative radiochemotherapy and sequential chemotherapy based on temozolomide: A 10-year experience at a single institution. *Strahlenther. Onkol.* **2011**, *187*, 778–783. [CrossRef] [PubMed]
8. Berendsen, S.; Varkila, M.; Kroonen, J.; Seute, T.; Snijders, T.J.; Kauw, F.; Spliet, W.G.; Willems, M.; Poulet, C.; Broekman, M.L.; et al. Prognostic relevance of epilepsy at presentation in glioblastoma patients. *Neuro-oncology* **2016**, *18*, 700–706. [CrossRef]
9. Dobran, M.; Nasi, D.; Chiriatti, S.; Gladi, M.; di Somma, L.; Iacoangeli, M.; Scerrati, M. Prognostic Factors in Glioblastoma: Is There a Role for Epilepsy? *Neurol. Med.-Chir.* **2018**, *58*, 110–115. [CrossRef]
10. Mistry, A.M.; Hale, A.T.; Chambless, L.B.; Weaver, K.D.; Thompson, R.C.; Ihrie, R.A. Influence of glioblastoma contact with the lateral ventricle on survival: A meta-analysis. *J. Neurooncol.* **2017**, *131*, 125–133. [CrossRef]
11. Liu, Q.; Cao, P. Clinical and prognostic significance of HIF-1$\alpha$ in glioma patients: A meta-analysis. *Int. J. Clin. Exp. Med.* **2015**, *8*, 22073–22083.
12. Winter, S.C.; Shah, K.A.; Han, C.; Campo, L.; Turley, H.; Leek, R.; Corbridge, R.J.; Cox, G.J.; Harris, A.L. The relation between hypoxia-inducible factor (HIF)-1$\alpha$ and HIF-2$\alpha$ expression with anemia and outcome in surgically treated head and neck cancer. *Cancer* **2006**, *107*, 757–766. [CrossRef]
13. Taubert, H.; Kappler, M.; Holzapfel, D.; Zorn, E.; Holzhausen, H.J.; Haensgen, G. Prognostic value of HIF-1$\alpha$ expression in patients with definive radiotherapy for cervical cancer. *Strahlenther. Und Onkol.* **2008**, *184*, 169–174.
14. Soukup, J.; Gerykova, L.; Rachelkar, A.; Hornychova, H.; Bartos, M.C.; Krupa, P.; Vitovcova, B.; Pleskacova, Z.; Kasparova, P.; Dvorakova, K.; et al. Diagnostic Utility of Immunohistochemical Detection of MEOX2, SOX11, INSM1 and EGFR in Gliomas. *Diagnostics* **2023**, *13*, 2546. [CrossRef]
15. Vitovcova, B.; Skarkova, V.; Rudolf, K.; Rudolf, E. Biology of Glioblastoma Multiforme—Exploration of Mitotic Catastrophe as a Potential Treatment Modality. *Int. J. Mol. Sci.* **2020**, *21*, 5324. [CrossRef] [PubMed]
16. Li, Y.; Zhao, L.; Li, X.F. Hypoxia and the Tumor Microenvironment. *Technol. Cancer Res. Treat.* **2021**, *20*, 153303382110363. [CrossRef] [PubMed]
17. Karel Odrazka, M.D.; Petera, J.; Kohlova, T.; Dolezel, M.; Vaculikova, M.; Zouhar, M.; Malek, V.; Hobza, V.; Latr, I.; Nemecek, S.; et al. Prognostic impact of hemoglobin level prior to radiotherapy on survival in patients with glioblastoma. *Strahlenther. Onkol.* **2003**, *179*, 615–619. [CrossRef]
18. Lally, B.E.; Colasanto, J.M.; Fischer, J.J.; Knisely, J.P.S. Is there an optimal hemoglobin level for patients with glioblastoma multiforme? *Cancer J. Sudbury Mass.* **2004**, *10*, 391–396. [CrossRef] [PubMed]
19. Fiorentino, A.; Fusco, V. Elderly patients affected by glioblastoma treated with radiotherapy: The role of serum hemoglobin level. *Int. J. Neurosci.* **2013**, *123*, 133–137. [CrossRef]
20. Sfifou, F.; Hakkou, E.M.; Bouaiti, E.A.; Slaoui, M.; Errihani, H.; Al Bouzidi, A.; Abouqal, R.; El Ouahabi, A.; Cherradi, N. Correlation of immunohistochemical expression of HIF-1alpha and IDH1 with clinicopathological and therapeutic data of moroccan glioblastoma and survival analysis. *Ann. Med. Surg.* **2021**, *69*, 102731. [CrossRef]

21. Potharaju, M.; Mathavan, A.; Mangaleswaran, B.; Patil, S.; John, R.; Ghosh, S.; Kalavakonda, C.; Ghosh, M.; Verma, R.S. Clinicopathological Analysis of HIF-1alpha and TERT on Survival Outcome in Glioblastoma Patients: A Prospective, Single Institution Study. *J. Cancer* **2019**, *10*, 2397–2406. [CrossRef] [PubMed]
22. Monteiro, A.R.; Hill, R.; Pilkington, G.J.; Madureira, P.A. The role of hypoxia in glioblastoma invasion. *Cells* **2017**, *6*, 45. [CrossRef] [PubMed]
23. Wang, G.; Wang, J.J.; Fu, X.L.; Guang, R.; To, S.S.T. Advances in the targeting of HIF-1α and future therapeutic strategies for glioblastoma multiforme. *Oncol. Rep.* **2017**, *37*, 657–670. [CrossRef] [PubMed]
24. Berendsen, S.; Spliet, W.G.; Geurts, M.; Van Hecke, W.; Seute, T.; Snijders, T.J.; Bours, V.; Bell, E.H.; Chakravarti, A.; Robe, P.A. Epilepsy Associates with Decreased HIF-1α/STAT5b Signaling in Glioblastoma. *Cancers* **2019**, *11*, 41. [CrossRef] [PubMed]
25. Li, Y.; Shan, X.; Wu, Z.; Wang, Y.; Ling, M.; Fan, X. IDH1 mutation is associated with a higher preoperative seizure incidence in low-grade glioma: A systematic review and meta-analysis. *Seizure* **2018**, *55*, 76–82. [CrossRef]
26. Neal, A.; Kwan, P.; O'Brien, T.J.; Buckland, M.E.; Gonzales, M.; Morokoff, A. IDH1 and IDH2 mutations in postoperative diffuse glioma-associated epilepsy. *Epilepsy Behav.* **2018**, *78*, 30–36. [CrossRef]

*biomedicines*

MDPI

*Article*

# Identification of CD44 as a Reliable Biomarker for Glioblastoma Invasion: Based on Magnetic Resonance Imaging and Spectroscopic Analysis of 5-Aminolevulinic Acid Fluorescence

**Akihiro Inoue [1,*], Takanori Ohnishi [2], Masahiro Nishikawa [1], Hideaki Watanabe [1], Kosuke Kusakabe [1], Mashio Taniwaki [3], Hajime Yano [4], Yoshihiro Ohtsuka [1], Shirabe Matsumoto [1], Satoshi Suehiro [1], Daisuke Yamashita [1], Seiji Shigekawa [1], Hisaaki Takahashi [5], Riko Kitazawa [3], Junya Tanaka [4] and Takeharu Kunieda [1]**

1 Department of Neurosurgery, Ehime University School of Medicine, 454 Shitsukawa, Toon, Ehime 791-0295, Japan; ma.nishikawa1985@gmail.com (M.N.); whideaki@m.ehime-u.ac.jp (H.W.); kk_0145@yahoo.co.jp (K.K.); y.ohtsuka0818@gmail.com (Y.O.); shirabem13@gmail.com (S.M.); satoshi.suehiro@gmail.com (S.S.); yamadai551208@gmail.com (D.Y.); shigekaw@m.ehime-u.ac.jp (S.S.); kuny@m.ehime-u.ac.jp (T.K.)

2 Department of Neurosurgery, Washoukai Sadamoto Hospital, 1-6-1 Takehara, Matsuyama, Ehime 790-0052, Japan; tohnishi@m.ehime-u.ac.jp

3 Division of Diagnostic Pathology, Ehime University Hospital, 454 Shitsukawa, Toon, Ehime 791-0295, Japan; hechota@gmail.com (M.T.); riko.kitazawa@gmail.com (R.K.)

4 Department of Molecular and Cellular Physiology, Ehime University School of Medicine, 454 Shitsukawa, Toon, Ehime 791-0295, Japan; hajime-y@m.ehime-u.ac.jp (H.Y.); jtanaka@m.ehime-u.ac.jp (J.T.)

5 Division of Pathophysiology, Faculty of Pharmaceutical Sciences, Hokuriku University, Kanazawa, Ishikawa 920-1181, Japan; h-takahashi@hokuriku-u.ac.jp

* Correspondence: iakihiro@m.ehime-u.ac.jp; Tel.: +81-89-960-5338

**Abstract:** Recurrent glioblastoma multiforme (GBM) is largely attributed to peritumoral infiltration of tumor cells. As higher CD44 expression in the tumor periphery correlates with higher risk of GBM invasion, the present study analyzed the relationship between CD44 expression and magnetic resonance imaging (MRI)-based invasiveness of GBM on a large scale. We also quantitatively evaluated GBM invasion using 5-aminolevulinic acid (5-ALA) spectroscopy to investigate the relationship between CD44 expression and tumor invasiveness as evaluated by intraoperative 5-ALA intensity. Based on MRI, GBM was classified as high-invasive type in 28 patients and low-invasive type in 22 patients. High-invasive type expressed CD44 at a significantly higher level than low-invasive type and was associated with worse survival. To quantitatively analyze GBM invasiveness, the relationship between tumor density in the peritumoral area and the spectroscopic intensity of 5-ALA was investigated. Spectroscopy showed that the 5-ALA intensity of infiltrating tumor cells correlated with tumor density as represented by the Ki-67 staining index. No significant correlation between CD44 and degree of 5-ALA-based invasiveness of GBM was found, but invasiveness of GBM as evaluated by 5-ALA matched the classification from MRI in all except one case, indicating that CD44 expression at the GBM periphery could provide a reliable biomarker for invasiveness in GBM.

**Keywords:** glioblastoma; invasion; CD44; 5-aminolevulinic acid; fluorescence spectroscopy; glioma stem cell; biomarker

## 1. Introduction

Glioblastoma multiforme (GBM) is the most malignant brain tumor, characterized by a highly invasive nature and inter- and intratumor heterogeneity. These features make GBM a tumor with poor prognosis due to diffuse infiltration of tumor cells into normal peritumoral brain tissues. In particular, tumor cells existing in the peritumoral area, where gadolinium (Gd) enhancement is not detected on magnetic resonance imaging (MRI), create a big obstacle to radical excision of the tumor, resulting in infiltrating tumor cells being left in place. The current therapy in the primary GBM consists of maximal tumor resection with minimal brain injury, the extended local radiotherapy with 60 Gy, and the concomitant chemotherapy with temozolomide. In addition to non-radicality in tumor resection, various factors make GBM extremely poor tumors. These include the fact that GBM cells show a resistance to chemotherapy due to elevated extracellular exclusion of chemotherapeutic agents and the non-disrupted blood–brain barrier and a resistance to radiotherapy due to upregulation of repair enzymes for DNA damages besides a hypoxic environment surrounding tumor cells. Glioma stem-like cells (GSCs), which constitute a small population in GBM but have abilities to make self-proliferation, multilineage differentiation to neurons and astrocytes, and rapid tumorigenesis in a mouse xenograft model, could participate in all these forms of resistance to the therapy. If GSCs are included in the infiltrating tumor cells, they will become a key cause of early tumor recurrence for GBM [1–6]. Specifically, we found that GSCs expressing high cluster of differentiation (CD) 44, which were established by the primary culture of tumor cells obtained from the tumor periphery of highly CD44-expressing GBM patients, can present a high invasive feature [7,8].

CD44 is a transmembrane glycoprotein expressed in various cells, such as embryonic stem cells, differentiated cells and cancer cells, including GBM [9,10]. CD44 is a multifunctional receptor molecule with important roles not only in mediating cell-to-cell and cell-to-extracellular matrix interactions, but also in transducing intracellular signaling pathways that promote cell migration, invasion, proliferation, and tumorigenesis in malignant tumors (Supplementary Figure S1) [11]. Our studies demonstrated that GBM with higher expression of CD44 in the tumor periphery than in the core correlated well with the highly invasive-type GBM as defined on magnetic resonance imaging (MRI), associated with poorer survival of patients. Conversely, lower expression of CD44 in the tumor periphery corresponded with low invasive-type GBM and longer survival [7]. In those studies, the ratio of CD44 expression in the tumor periphery to expression in the tumor core (P/C ratio) could be evaluated as a marker of GBM invasiveness. CD44 expression in the tumor periphery alone was not associated with the invasiveness of GBM. Why only the P/C ratio of CD44 expression correlated with GBM invasiveness remains unclear. To confirm the correlation between CD44 expression and invasiveness of GBM, the present study investigated in an increasing number of patients with GBM and analyzed whether CD44 offered the most reliable indicator for predicting GBM invasiveness.

Elucidating whether CD44 expression correlates with the degree of actual invasiveness of GBM apart from MRI is another critical issue. To achieve supra-total resection of malignant glioma, fluorescence-guided surgery (FGS) has recently been adopted for glioma surgery using image-guided neuro-navigation systems [6,12]. This technique has been developed to enhance the extent of GBM resection, and supra-total resection of GBM has been demonstrated to contribute to good patient prognosis [13,14]. Among the various photosensitizers available, 5-aminolevulinic acid (5-ALA), an amino acid and natural precursor in heme biosynthesis, has been widely used as a fluorescence-inducing molecule in intraoperative FGS for malignant glioma [15]. Many studies have reported that the intensity of fluorescence from 5-ALA-derived protoporphyrin IX (PpIX) in brain tumors correlates with tumor cell grade and malignancy, with both sensitivity and specificity exceeding 80% [6,16,17]. Fluorescence intensity of 5-ALA-derived PpIX is usually evaluated by visually assessed strength of red color. Tumors showing fluorescence with a strong red color correspond to malignant tumors with high grade (GBM). Tumors presenting almost no red color correspond to low-grade glioma, and tumors showing a pale red or pinkish color correspond to grade-III malignant glioma. Photodynamic diagnosis using 5-ALA appears useful to visualize tumor cells not recognizable under white light in the operative field, but such photodynamic diagnosis sometimes fails to detect tumor cells that do not show red-color fluorescence and cannot quantitatively evaluate the fluorescence intensity of tumor cells. To overcome these problems, the present study introduced spectroscopy for 5-ALA-derived PpIX fluorescence using a violet laser diode (VLD) to enhance the visualization of tumor cells and allow for quantitative evaluation of 5-ALA fluorescence intensity from infiltrating glioma cells. We evaluated the degree of invasiveness of GBM by analyzing the relationship between spectroscopic 5-ALA fluorescence signal intensity and tumor proliferation rate (Ki-67 staining index [SI]), which is considered to represent the density of infiltrating tumor cells in the peritumoral area of GBM [5]. We then investigated whether the P/C ratio of CD44 expression correlated with the degree of invasiveness of GBM as evaluated by spectroscopic 5-ALA fluorescence intensity. These studies might identify potential biomarkers for evaluating the degree of GBM invasiveness. Identification of a novel marker for GBM invasion will provide a useful tool for not only elucidating the molecular mechanisms underlying glioma invasion, but also developing new strategies for the effective treatment of GBM.

## 2. Materials and Methods

This study was approved by the Ethics Committee for Clinical Research of Ehime University Hospital (approval no. 2211011). All procedures were performed in accordance with the ethical standards of the 1964 Declaration of Helsinki and its later amendments.

### 2.1. Patients and Study Design

Fifty patients with histologically verified GBM treated under the same protocol in the Department of Neurosurgery at Ehime University Hospital between April 2014 and January 2021 were enrolled in the present study. Informed consent was obtained from each participant after receiving explanations of the potential risks of positron emission tomography (PET) and MRI, surgical procedure, radio-chemotherapy, and oral administration of 5-ALA in patients undergoing 5-ALA-guided surgery. Neurological findings were assessed preoperatively, immediately after surgery, and 3 months postoperatively. The same neurosurgeon and neuroradiologist analyzed the clinical records and radiological examinations of these patients on admission, within 72 h and 3 months after surgery. Other requirements for enrollment included: Karnofsky performance status (KPS) score $\geq$60; tumor resection more than biopsy (gross total resection (GTR), 100% resection of tumor volume; subtotal resection (STR), $\geq$95% but <100% resection of tumor volume; or partial resection (PR), <95% resection of tumor volume); and bevacizumab, a monoclonal antibody targeting vascular endothelial growth factor (VEGF), therapy at tumor recurrence. The treatment protocol comprised tumor resection followed by radiotherapy (60 Gy) and chemotherapy with temozolomide (TMZ), in accordance with the Stupp regimen [2]. Ki-67 SI was evaluated by immunohistochemistry and used to estimate the number of proliferative tumor cells in tumor tissue samples. In all 50 patients, mRNA expression of CD44 was examined in the periphery and core of the tumor, to clarify what patterns of CD44 expression correlated most significantly with GBM invasiveness. Then, among the 50 patients, the most recent 21 patients who had been operated on under the guidance of spectroscopic 5-ALA fluorescence were entered into the present study to quantitatively analyze the degree of GBM invasiveness. Finally, the correlation between CD44 expression in the tumor periphery and tumor invasiveness as evaluated by spectroscopic analysis of 5-ALA fluorescence was investigated.

### 2.2. Imaging Studies and Analysis

MRI examinations were achieved using a 3-T scanner (Achieva; Philips, Best, the Netherlands) with a standard head coil. Axial, coronal, and sagittal T1-weighted images were obtained in 2 mm slice thickness before and after intravenous administration of gadolinium-diethylenetriamine pentaacetic acid (0.1 mmol/kg). Based on the features observed on MRI, patients were divided into a high-invasive (HI) type and a low-invasive (LI) type. Four features were used to define the invasive type, including morphology of the tumor margin, enhancement of the tumor margin wall, peritumoral edema, and central necrosis (Table 1; Figure 1). In cases presenting equivocal features of invasiveness on MRI, methionine (Met)-PET was performed. Details of the criteria for these phenotypes in GBM on MRI and PET have been described previously [5].

**Table 1.** Criteria for judging invasiveness in GBM based on MRI findings.

| Features on MRI | High-Invasive Type | Low-Invasive Type |
|---|---|---|
| Morphology of tumor margin | irregular, thin wall | demarcated, thick wall |
| Enhancement of tumor margin wall | heterogeneous, weak-moderate | homogeneous, intense |
| Peritumoral brain edema | diffuse, extensive | focal, localized |
| Morphology of central necrosis | irregular, unevenly located | round, located at or near center |

GBM, glioblastoma multiforme. MRI, magnetic resonance imaging.

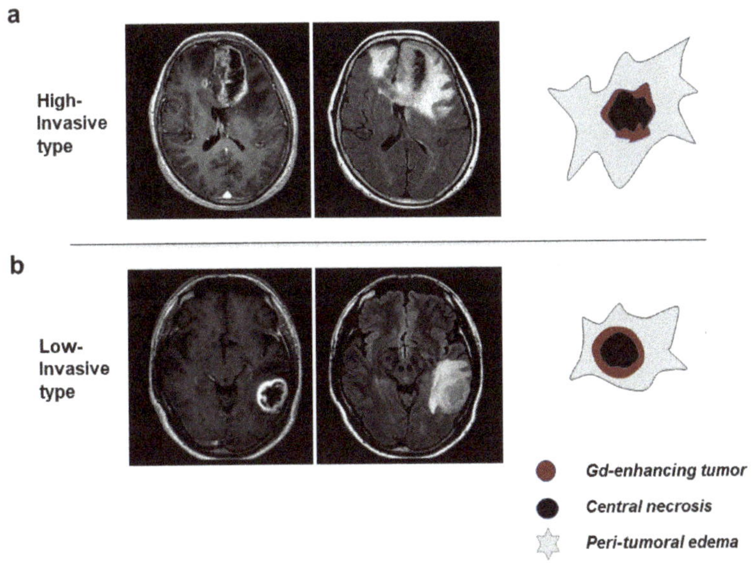


**Figure 1.** Phenotypes of invasiveness in glioblastoma multiforme (GBM) based on the evaluation of imaging features on magnetic resonance imaging (MRI). GBM was classified into high-invasive (HI) and low-invasive (LI) phenotypes using four features on MRI (Table 1). (**a**) MRI (**left**: gadolinium-enhanced T1-weighted imaging; right: FLAIR) and illustration (**far right**) of HI-type GBM. (**b**) MRI and illustration of LI-type GBM.

### *2.3. Image-Guided Navigation Surgery and Selective Tissue Sampling*

Tumor resection was performed by echo-linked navigation-guided microsurgery using MRI and Met-PET fusion images and fence-post catheter techniques. Tumor samples were separately obtained from the core and periphery of the tumor using surgical techniques, as previously described [7,8]. In addition, tumor tissues were obtained from several different areas in the GBM, including areas of SUVmax on Met-PET, and areas that showed positive 11C-Met uptake but no Gd-enhancement in the tumor periphery.

### *2.4. Immunohistochemistry*

Tumor samples were fixed in buffered formalin and embedded in paraffin according to our regular histological examination. Hematoxylin and eosin (HE)-stained sections and immunohistochemical studies were obtained from all patients. In the immunohistochemical studies, deparaffinized 4 μm tissue sections from paraffin blocks were hydrated in a graded alcohol series and underwent heat-activated antigen retrieval. After blocking endogenous peroxidase activity, tissue was incubated with anti-Ki-67 mouse monoclonal antibody, clone MIB-1 code GA626 (1:400) (DAKO, Santa Clara, CA, USA). Sections were then washed and incubated with biotinylated secondary antibody for 30 min at room temperature. Reaction complexes were visualized with diaminobenzidine and counterstained with hematoxylin. Appropriate negative and positive control experiments were run for all antibodies tested.

### *2.5. RNA Isolation and Quantitative Real-Time Reverse Transcription Polymerase Chain Reaction (qRT-PCR)*

Total RNA was extracted from both the core and periphery of each tumor sample using ISOGEN (Nippon Gene, #319-90211, Tokyo, Japan) according to the attached instructions. Complementary DNA (cDNA) was synthesized using ReverTra Ace qPCR RT Master Mix with a gDNA remover kit (Toyobo, #FSQ-301, Osaka, Japan). The qPCR analysis was performed using Fast Start Universal SYBR Green Master Mix (Roche Diagnostic Japan, #04887352001, Tokyo, Japan) with an MJ Mini instrument (BioRad, Hercules, CA, USA). All gene-specific mRNA expression values were normalized by the expression level of the GAPDH housekeeping (reference) gene encoding glyceraldehyde-3-phosphate dehydrogenase. Quantification of gene expression was performed using $\Delta$Ct values, defined as the difference between the target and reference gene Ct values. All primer sequences were as described previously [7].

### *2.6. Quantitative Evaluation of 5-ALA Fluorescence Intensity by Spectroscopic Analysis*

In 21 of the 50 patients, 5-ALA fluorescence intensity was quantitatively evaluated by spectroscopic analysis. Patients received oral administration of 5-ALA (Alabel Oral; Nobelpharma Co., #877290, Tokyo, Japan) at 20 mg/kg body weight 3 h before elective craniotomy as part of the routine procedures for fluorescence-guided resection of malignant glioma. The signal intensity of 5-ALA fluorescence from the resected tumor and peritumoral area after tumor resection was measured using an optic fiber probe. We performed spectroscopic analysis of 5-ALA fluorescence intensity using a violet laser diode (VLD) (Alcedo LS-VLD: VLD-EX; SBI Pharmaceuticals Co., Tokyo, Japan) and evaluated fluorescence intensity as the signal intensity of 5-ALA fluorescence (arbitrary units; a.u.).

### *2.7. Statistical Methods*

Values are described in mean $\pm$ standard deviation. Data were compared using a two-tailed Student's *t*-test (unpaired). Comparisons of more than two groups were conducted using two-tailed one-way analysis of variance with the Tukey post hoc test. Kaplan–Meier plots were formed to estimate unadjusted time-to-event variables. Spearman's correlation analysis was performed to examine correlations for non-parametric data. Values of $p < 0.05$ were considered significant. All analyses were performed using Office Excel 2016 software (Microsoft, Redmond, WA, USA) and Easy R (EZR) version 1.54 software (Saitama Medical Center, Jichi Medical University, Saitama, Japan) [18].

## 3. Results

### *3.1. Patient Characteristics*

The 50 patients enrolled in this study comprised 37 men and 13 women. The follow-up period was for nine years between April 2014 and March 2023. All patients were newly diagnosed cases receiving primary therapy. The mean age was 65.3 years (range, 19–86 years). No significant differences were found in terms of age or sex. The median KPS score was 80 (range, 60–100). The mean body mass index (BMI) was 21.0 kg/m$^2$ (range, 14.5–25.9 kg/m$^2$). All patients underwent craniotomy for tumor resection, followed by radiotherapy (60 Gy) and chemotherapy with temozolomide, in accordance with the Stupp protocol [1,2]. The extent of resection was evaluated by volumetric analysis on MRI before and after surgery, as previously described [19]. GTR was achieved in 31 patients (62.0%), STR in 5 patients (10.0%), and PR in 14 patients (28.0%). Histopathological evaluation verified that all tumors were GBM and that 45 tumors lacked mutation in the gene encoding isocitrate dehydrogenase-1 (IDH-1). Five tumors showed mutation in the gene encoding IDH-1. The presence of hotspot mutations in IDH1 (R132) was analyzed by

Sanger sequencing. Consequently, 45 tumors corresponded to "GBM, IDH-wild type" and 5 tumors were classified as "astrocytoma IDH-mutant" according to the World Health Organization classification of central nervous system tumors in 2021 [20]. The methylation status of the O6-methylguanine-DNA methyltransferase (MGMT) promoter was analyzed by quantitative methylation-specific PCR after bisulfate modification of genomic DNA, and we used a cutoff of $\geq$1% for MGMT promoter methylation [21]. Methylation was recognized in the MGMT promotor in 22 cases and mean Ki-67 SI in the core of tumors was 33.3% (range, 7.0–60.0%). The characteristics of enrolled patients are summarized in Supplementary Table S1.

### *3.2. Classification of GBM Patients to HI and LI Types on MRI*

Out of fifty patients, 28 were classified as HI type and 22 were classified as LI type based on MRI. Met-PET was performed to confirm invasiveness in patients for whom the invasive type could not be determined based on MRI findings alone. The follow-up period was for nine years between April 2014 and March 2023. No significant differences were seen between HI and LI types of GBM in terms of mean age (64.5 vs. 66.4 years, respectively); sex (M/F: 23/5 vs. 14/8, respectively); median KPS score (70 vs. 80, respectively); and Ki-67 SI (33.5 ± 17.2% vs. 31.9 ± 17.4%, respectively). In terms of tumor resection, resection rate $\geq$95% (GTR or STR) was achieved for 64.3% of HI-GBM and 81.8% for LI-GBM, representing no significant difference. Methylation status of the MGMT promotor and mutation of IDH1 did not differ significantly between HI- and LI-GBMs (Table 2). Survival was compared between patients with HI-type and LI-type GBM. Both progression-free survival (PFS) and overall survival (OS) were much longer in patients with LI-type GBM than in those with HI-type GBM (median PFS: HI 6.7 months, LI 10.8 months, $p = 0.017$; median OS: HI 15 months, LI 28 months, $p < 0.001$) (Figure 2).

**Table 2.** Patient characteristics in 50 glioblastomas classified as high- and low-invasive type based on MRI.

| Tumor Phenotypes | Age (Years) | Sex | KPS (%) | mMGMT | IDH-1 Mut | Ki-67 SI (%) | EOR | |
|---|---|---|---|---|---|---|---|---|
| | (Mean ± SD) | Male/Female | (Median) | +/− | +/− | Mean ± SD | GTR + STR ** (%) | PR (%) |
| High invasive (28 patients) | 64.5 ± 13.4 | 23/5 | 70 | 15/12 * | 2/26 | 33.5 ± 17.2 | 19 (68) | 9 (32) |
| Low invasive (22 patients) | 66.4 ± 13.2 | 14/8 | 80 | 7/14 * | 3/19 | 31.9 ± 17.4 | 18 (82) | 4 (18) |

MRI, magnetic resonance imaging. SD, standard deviation. KPS, Karnofsky performance status. mMGMT, methylation of the O6-methylguanine-DNA methyltransferase promoter. IDH-1 mut, isocitrate dehydrogenase 1 mutation. MGMT(m), methylation of O6-methylguanine-DNA methyltransferase. Ki-67 SI, Ki-67 staining index. EOR, extent of resection. GTR, gross total resection. STR, subtotal resection. PR, partial resection. * One patient in both phenotypes did not show a definite result. ** Resection more than 95%.

Invasive type was determined by specific features on MRI. Both progression-free survival (PFS) and overall survival (OS) were significantly longer in patients with an LI-type GBM than in those with HI-type GBM (median PFS: LI vs. HI, 10.8 months (M) vs. 6.7 months, $p = 0.017$; median OS: LI vs. HI, 28.0 months vs. 15.0 months, $p < 0.001$). Inv: Invasive.

### *3.3. Expression of CD44 mRNA in the Core and Periphery of GBM and CD44 Expression as Evaluated by P/C ratio in HI- and LI-Type GBM*

Expression of CD44 mRNA tended to be lower in the tumor core of HI-type GBM than in LI-type GBM, but no significant difference was seen between types (Figure 3a). In contrast, both CD44 expression in the tumor periphery and P/C ratio were significantly higher in HI-type GBM than in LI-type GBM (Figure 3b,c). In addition, the P/C ratio of CD44 expression showed much higher sensitivity and specificity for predicting invasiveness of GBM (100% and 83.3%, respectively) than CD44 expression in the tumor periphery

(85.7% and 79.2%, respectively). Survival was compared between patients with a high-CD44 expression type of GBM and a low-CD44 expression type of GBM. Both PFS and OS were significantly longer in the patients with a low-CD44 expression type of GBM than those with a high-CD44 expression type of GBM (median (m) PFS: Low P/C ratio vs. High P/C ratio, 9.5 months vs. 7.0 months, $p = 0.02$; mOS: Low P/C ratio vs. High P/C ratio, 25.0 months vs. 14.0 months, $p = 0.0000025$) (Figure 4).

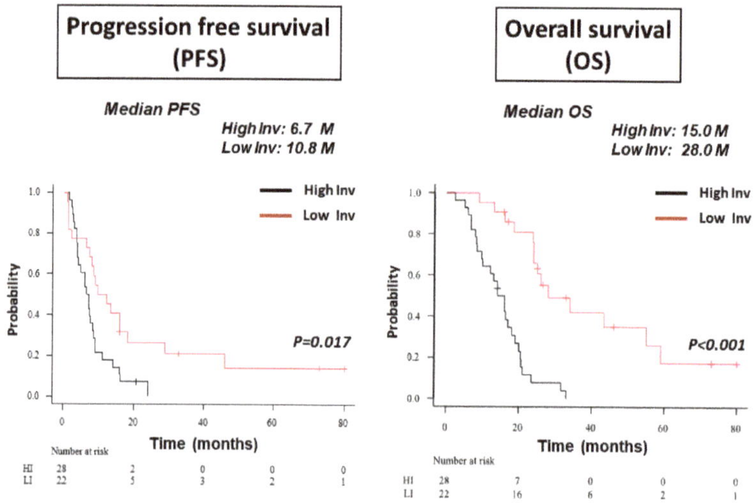


**Figure 2.** Kaplan–Meier survival curves showing progression-free survival (PFS) and overall survival (OS) in 50 patients with GBM, including HI-type GBM (28 patients) and LI-type GBM (22 patients).

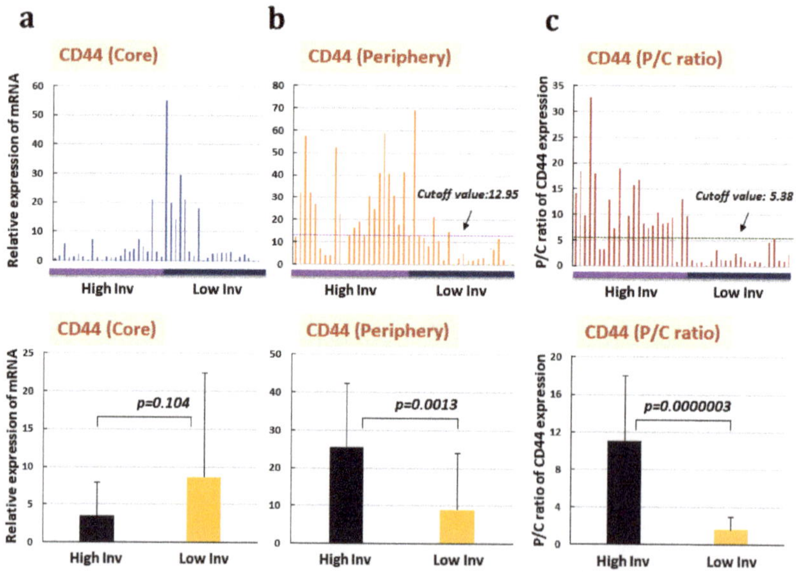


**Figure 3.** mRNA expression of CD44 in the tumor core and tumor periphery in 50 patients with GBM. Upper panel shows expression patterns of CD44 mRNA in GBM classified into HI-type tumor and LI-type tumor on MRI. Lower panel displays differences in types of GBM invasiveness between HI- and LI-type tumors. (**a**) CD44 expression in tumor core. (**b**) CD44 expression in tumor periphery. (**c**) CD44 expression as represented by P/C ratio (ratio of CD44 expression in tumor periphery to

(expression in tumor core). No significant difference in CD44 expression in the tumor core is seen between HI- and LI-type tumors (HI vs. LI: 3.55 ± 4.36 vs. 8.67 ± 13.7, $p$ = 0.104). In contrast, in the tumor periphery, CD44 expression in HI-type tumors is significantly higher than that in LI-type tumors (HI vs. LI: 25.56 ± 16.83 vs. 9.14 ± 14.93, $p$ = 0.0013). P/C ratios of CD44 expression disclose a much greater difference in CD44 expression between HI- and LI-type tumors than CD44 expression in the tumor periphery alone (HI vs. LI: 11.06 ± 7.0 vs. 1.62 ± 1.35, $p$ = 0.0000003). These results coincide with sensitivity and specificity for predicting HI-type tumors (P/C ratio: 100% sensitivity, 83.8% specificity; expression in tumor periphery alone: 85.7% sensitivity, 79.2% specificity). (Cutoff values are displayed in CD44 expression patterns in the upper panel.)

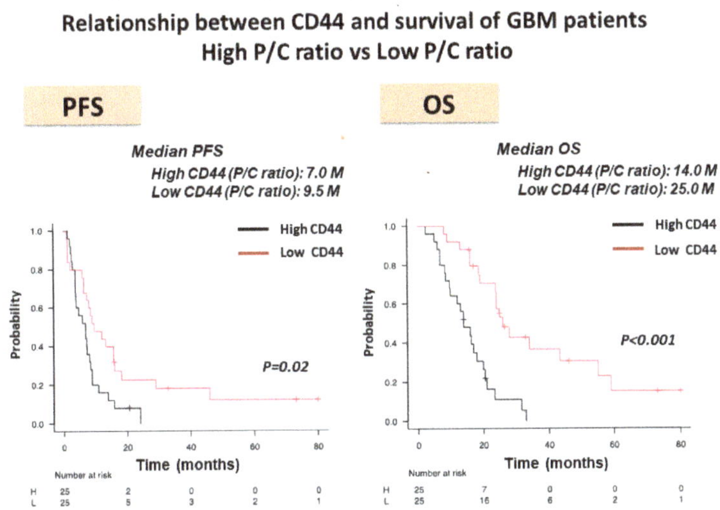


**Figure 4.** Kaplan–Meier survival curves showing progression-free survival (PFS) and overall survival (OS) in 50 patients with GBM, including GBM with high CD44 expression (High P/C ratio) (25 patients) and GBM with low CD44 expression (Low P/C ratio) (25 patients). CD44 expression types were determined by the cutoff values of 5.38. Both PFS and OS were significantly longer in the patients with a low-CD44 expression type of GBM than those with a high-CD44 expression type of GBM (median (m) PFS: Low P/C ratio vs. High P/C ratio, 9.5 months (M) vs. 7 M, $p$ = 0.02; mOS: Low P/C ratio vs. High P/C ratio, 25.0 M vs. 14.0 M, $p$ = 0.0000025).

### 3.4. Quantitative Analysis of Spectroscopic 5-ALA Fluorescence Intensity in the Invasion Area

In 21 patients, to evaluate the strength of GBM invasion, we applied spectroscopy of 5-ALA fluorescence using a VLD with a wavelength of 400 nm. As the area illuminated by the VLD was small, like a spotlight, fluorescence signals were obtained from several sites in target tissues. Since the spectroscopic signal intensities of 5-ALA fluorescence varied within the same target areas, the intensity of 5-ALA fluorescence was expressed not by individual values but by the range of fluorescence intensities in each target, which were divided into four groups according to the strength of signal intensities. These included groups of strong high fluorescence intensity (HFI-S; ≥5000 a.u.), moderately high intensity (HFI-M; ≥3000 a.u. but <5000 a.u.), weak high intensity (HFI-W; ≥1000 a.u., but <3000 a.u.), and low fluorescence intensity (LFI; <1000 a.u.). These intensity levels of each group corresponded to tumor density represented as the number of proliferative tumor cells (Ki-67 SI) (Figure 5a). The HFI-S group showed a large number of tumor cells with a mean Ki-67 SI of 32.8% (range, 20.0–73.3%), the HFI-M group revealed a moderate number of tumor cells with a mean Ki-67 SI of 18.4% (range, 10–40.0%), and the HFI-W group showed a few infiltrating tumor cells with a mean Ki-67 SI of 1.82% (range, 1.0–3.5%) (Figure 5b; Table 3). With 5-ALA signal intensity <1000 a.u. (LFI), almost no tumor cells were detected in peritumoral tissue, showing a Ki-67 SI of 0.27% (range, 0–1.0%).

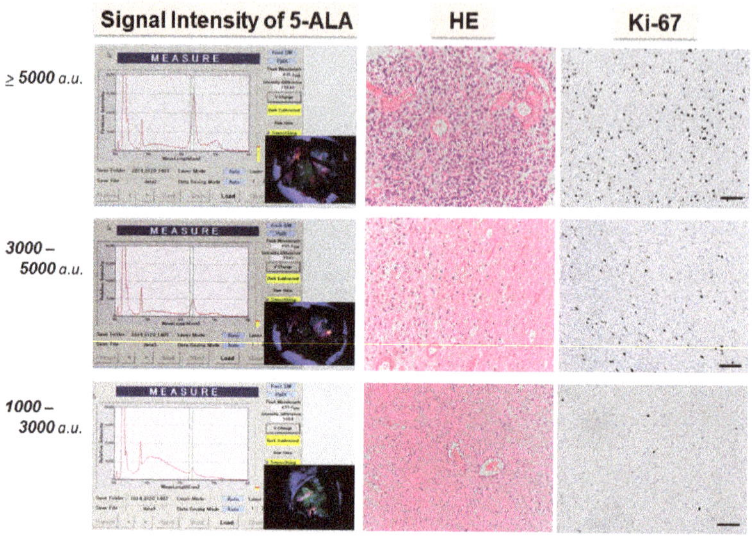


(**a**)

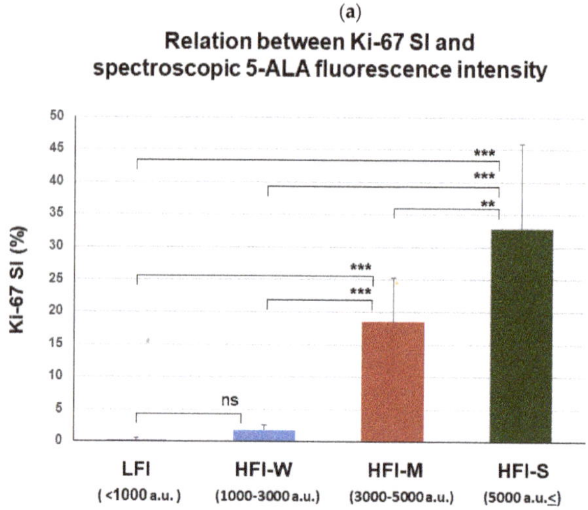


(**b**)

**Figure 5.** Spectroscopic intensity of 5-ALA fluorescence and its pathology in the target area. (**a**) Left panel: Spectroscopy of 5-ALA fluorescence showing the fluorescence spectrum and fluorescence intensity of the peak at a wavelength of around 636 nm and intraoperative views showing tumors with various shades of red color according to the strength of ALA-fluorescence signal intensity. Right panel: Histological images of hematoxylin and eosin staining and immunohistochemical staining with Ki-67 corresponding to strength of signal intensity shown in the left panels. **Upper**: Tissues at the tumor border show signal intensity ≥5000 a.u. containing numerous tumor cells and a mean Ki-67 LI of 32.8%. **Middle**: Tissues containing medium numbers of tumor cells with a mean Ki-67 SI of 18.4% under a signal intensity of 3000–5000 a.u. **Lower**: Tissues containing few tumor cells with a mean Ki-67 SI of 1.82% under a signal intensity of 1000–3000 a.u. In tissues showing signal intensity <1000 a.u., tumor cells are almost absent. Magnification, ×100. Scale bar, 100 μm. (**b**) A bar graph presenting the relationship between signal intensities of 5-ALA fluorescence and tumor density as represented by Ki-67 SI. The graph demonstrates that intensities of 5-ALA depend on the number of proliferating tumor cells as represented by Ki-67 SI. ** $p < 0.01$, *** $p < 0.001$, ns: not significant.

**Table 3.** P/C ratio for CD44, Ki-67 SI at peritumoral areas with three different levels of 5-ALA intensity, and 5-ALA intensity at the final tumor resection stage and invasion types of GBM as evaluated by MRI and 5-ALA spectroscopy in 21 patients with GBM.

| Patient no. | P/C Ratio | Ki-67 SI (%) | | | Ki-67 SI Ratio | | Fluorescence Intensity (a.u.) at Final Resection (Peritumoral Area) | Type of Invasiveness | |
|---|---|---|---|---|---|---|---|---|---|
| | (CD44) | A0 (≥5000) | A1 (3000–5000) | A2 (1000–3000) | A1/A0 | A2/A0 | | on MRI | on 5-ALA |
| | | | | | | | | | High (s) |
| A1 | 1.3 | 73.3 | 40 | <1 | 0.55 | 0.014 | <1000 | L | L |
| A2 | 14.2 | 18.4 | 18.8 | 3.5 | 1.00 | 0.180 | 3000–5000 | H | H (m) |
| A3 | 18.3 | 44.9 | 18.6 | 2.8 | 0.41 | 0.062 | 1000–3000 | H | H (w) |
| A4 | 9.8 | 27.1 | 15.3 | 2 | 0.56 | 0.070 | 3000–5000 | H | H (m) |
| A5 | 0.7 | 26 | 20.5 | <1 | 0.79 | 0.038 | <1000 | L | L |
| A6 | 46.5 | 26.1 | 13.3 | 2.5 | 0.51 | 0.096 | 1000–3000 | H | H (w) |
| A7 | 16.3 | 23.2 | 17.2 | 2.3 | 0.74 | 0.099 | 1000–3000 | H | H(w) |
| A8 | 2 | 30 | 20 | <1 | 0.67 | 0.030 | <1000 | L | L |
| A9 | 16.8 | 22.9 | 16.3 | 3 | 0.71 | 0.130 | 3000–5000 | H | H (m) |
| A10 | 0.8 | 20 | 10 | 1 | 0.50 | 0.050 | 3000–5000 | L | * H (m) |
| A11 | 1 | 20 | 18 | 0.5 | 0.90 | 0.025 | <1000 | L | L |
| A12 | 8.1 | 30 | 13 | 2.5 | 0.43 | 0.080 | 3000–5000 | H | H (m) |
| A13 | 7.5 | 40 | 22 | 2 | 0.55 | 0.050 | 1000–3000 | H | H (w) |
| A14 | 8.1 | 44.9 | 20.2 | 2 | 0.45 | 0.045 | 1000–3000 | H | H (w) |
| A15 | 0.9 | 45 | 30 | 0.5 | 0.67 | 0.010 | ≥5000 | L | L |
| A16 | 10.3 | 40 | 15 | 2 | 0.38 | 0.050 | 1000–3000 | H | H (w) |
| A17 | 9.2 | 26 | 12 | 2 | 0.46 | 0.076 | ≥5000 | H | H (s) |
| A18 | 8.4 | 35 | 15 | 2 | 0.43 | 0.080 | ≥5000 | H | H (s) |
| A19 | 2.4 | 20 | 14 | 0.8 | 0.70 | 0.040 | <1000 | L | L |
| A20 | 0.03 | 48 | 25 | 1 | 0.69 | 0.034 | <1000 | L | L |
| A21 | 15.5 | 33 | 14 | 2.5 | 0.42 | 0.075 | ≥5000 | H | H (s) |

P/C ratio, periphery/core ratio. CD44, cluster of differentiation 44. Ki-67 SI, Ki-67 staining index. 5-ALA, 5-aminolevulinic acid. GBM, glioblastoma multiforme. MRI, magnetic resonance imaging. no., number. a.u., arbitrary units. H, high invasive. L, low invasive. H(w), weakly high invasive. H(m), moderately high invasive. H(s), strongly high invasive. * reclassified from L to H on 5-ALA.

### *3.5. Evaluation of Invasiveness of GBM by Spectroscopic Analysis of 5-ALA Fluorescence Intensity*

To examine the degree of invasiveness in GBM, after contrast-enhancing tumors were resected, the amount of residual tumor cells infiltrating the peritumoral area was assessed by spectroscopic 5-ALA fluorescence intensity in the tissue of the wall of the resection cavity (Supplementary Figure S2). The peritumoral area showing a 5-ALA fluorescence intensity ≥5000 a.u. was defined as area "A0", the area showing an intensity of ≥3000 to <5000 a.u. was "A1", and the area showing an intensity of ≥1000 to <3000 a.u. was "A2". The peritumoral area presenting 5-ALA intensity <1000 a.u. was regarded as having almost no tumor cells (Figure 6a). Six tumors showed 5-ALA intensity <1000 a.u. in peritumoral brain tissue at the final resection stage. These tumors were therefore regarded as LI-type GBM. The remaining 15 tumors presenting 5-ALA intensity ≥1000 a.u. at the final resection stage were classified as HI-type GBM (Table 3). The ratios of Ki-67 SI in the "A1" and "A2" areas to those in the A0 area were calculated to allow for comparisons with Ki-67 SI, with Ki-67 SI in the A0 area defined as 1. Then, using the A1/A0 and A2/A0 ratios, attenuation curves for tumor density were created for LI- and HI-type GBM (Figure 6b). Both A1/A0 ratio and A2/A0 ratio of Ki-67 SI showed significant differences between LI- and HI-type tumors (HI vs. LI at A1: $0.548 \pm 0.168$ vs. $0.717 \pm 0.118$, $p = 0.038$; HI vs. LI at A2: $0.077 \pm 0.04$ vs. $0.03 \pm 0.01$, $p = 0.012$). A cutoff of 0.55 for A1/A0 ratio offered 100% sensitivity and 93.1% specificity, while a cutoff of 0.04 for A2/A0 ratio offered 100% sensitivity and 60% specificity (Figure 6c). These results showed that LI-type GBM had cutoff Ki-67 SI ratios of A1/A0 ratio ≥0.55 and A2/A0 ratio <0.04. When the 21 GBMs were re-classified using these cutoff values, one patient (Patient 10) who had been classified with LI-type GBM based on MRI was reclassified as having HI-type tumor on 5-ALA fluorescence spectroscopy. As a result, 7 patients were classified as having LI-type GBM and 14 patients were classified as having HI-type GBM. The 14 HI-type GBMs were further divided into three subgroups by fluorescence intensity at the final resection stage. Six tumors showing fluorescence intensity ≥1000 a.u. to <3000 a.u. were defined as weakly high-invasive (HIw) tumors, five tumors with intensity ≥3000 a.u. to <5000 a.u. were defined as moderately high-invasive (HIm) tumors, and three tumors with intensity ≥5000 a.u. were defined as strongly high-invasive (HIs) tumors. Patients with low-invasive GBM based on 5-ALA showed significantly longer PFS and OS than patients with high-invasive GBM, including all subgroups. Median PFS was: HIs, 4.0 months; HIm, 9.0 months; HIw, 5.9 months; and LI, 15.7 months ($p$ = 0.0014). Median OS was: HIs, 8.3 months; HIm, 17.0 months; HIw, 8.4 months; and LI, 26.5 months ($p$ = 0.0068) (Figure 6d).

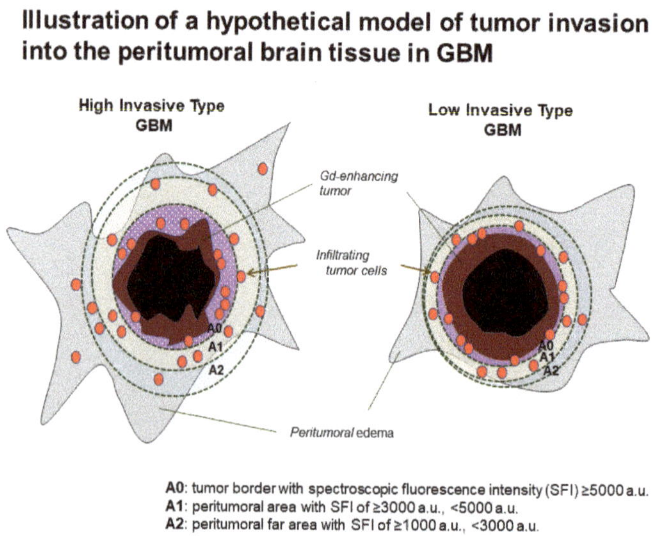


(a)

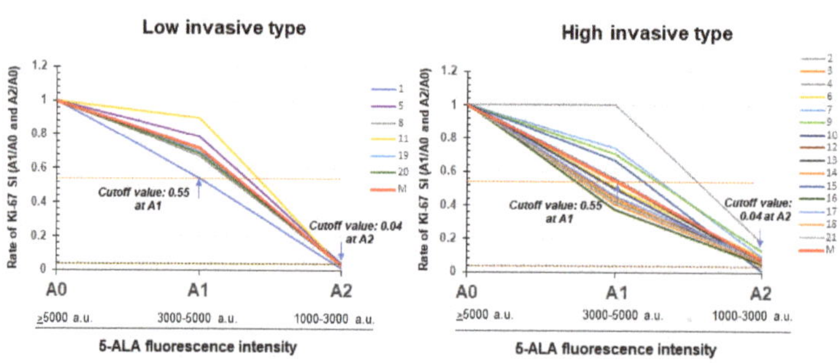


(b)

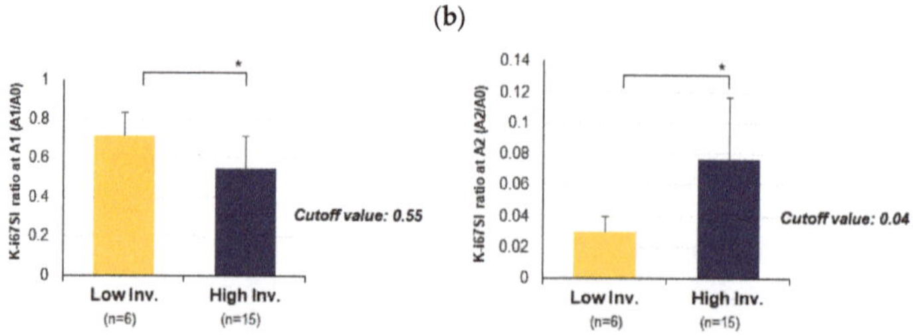


(c)

**Figure 6.** *Cont.*

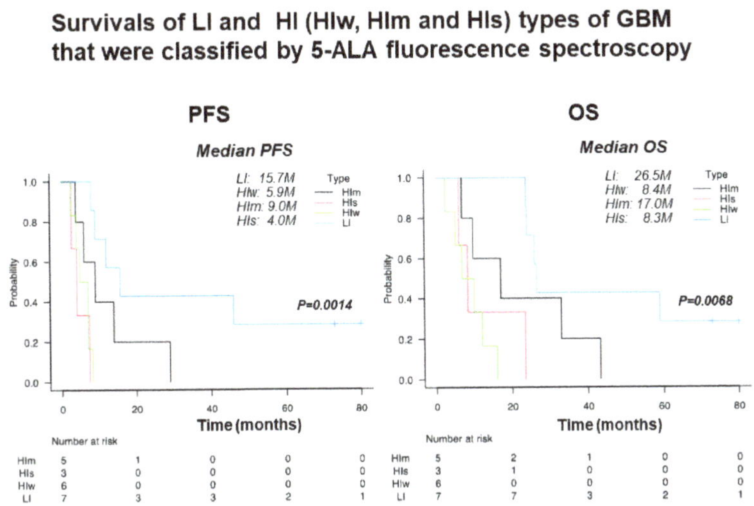


(d)

**Figure 6.** Quantitative evaluation of invasive types in GBM on 5-ALA fluorescence spectroscopy. (**a**) Illustration of a hypothetical model of tumor invasion into the peritumoral brain tissue in HI-type GBM and LI-type GBM. At the tumor border zone, massive numbers of tumor cells crowd the area around the tumor mass. This area is defined as A0. Around area A0, areas are expected to present less signal intensity of 5-ALA than A0. These areas are divided into two areas, with signal intensities of 3000–5000 a.u. (A1) and 1000–3000 a.u. (A2). Tissue samples were obtained from each area and histological examinations including immunostaining of Ki-67 were performed. (**b**) Attenuation curves of Ki-67 SI ratios at A1 and A2 compared to A0 in the group with fluorescence intensity <1000 a.u. at final tumor resection (six patients, corresponding to LI-type GBM) (**left graph**) and in the group with intensity ≥1000 a.u. at final tumor resection (15 patients, corresponding to HI-type GBM) (**right graph**). In LI-type GBM, all six patients show A1/A0 ratios greater than or equal to the cutoff value of 0.55 and A2/A0 ratios less than the cutoff value of 0.04. In HI-type GBM, 14 patients present A2/A0 ratios ≥0.04, except one patient (Patient 15). This patient displays an A1/A0 ratio of 0.67, and so was classified with LI type tumour, the same as the invasive type evaluated on MRI. Although Patient 10 had been classified with LI-type tumor on MRI, both Ki-67 SI ratios of A1/A0 and A2/A0 showed values corresponding to HI-type tumor. (**c**) Bar graphs showing differences in Ki-67 SI ratio at A1 (A1/A0) (**left**) and at A2 (A2/A0) (**right**) between HI-type GBM and LI-type GBM. Values at A1 are significantly higher in LI-type GBM than in HI-type GBM (LI vs. HI: 0.717 ± 0.118 vs. 0.548 ± 0.168, $p = 0.038$). The cutoff of 0.55 offers 100% sensitivity and 60% specificity. Values at A2 are significantly higher in HI-type GBM than in LI-type GBM (HI vs. LI: 0.077 ± 0.04 vs. 0.03 ± 0.01, $p = 0.0115$), with a cutoff of 0.04 offering 100% sensitivity and 93.1% specificity. (**d**) Kaplan–Meier survival curves show PFS and OS in 21 patients with GBM types classified as HI (14 patients, weakly high-invasive (HIw): 6, moderately high-invasive (HIm): 5, strongly high-invasive (HIs): (3) and LI (7 patients), for which invasive types were determined by spectroscopic analysis of 5-ALA fluorescence). Both PFS and OS were significantly longer in patients with an LI-type GBM than in those with HI-type GBM (median PFS: LI vs. HIw, HIm, and HIs: 15.7 months vs. 5.9 months, 9.0 months, and 4.0 months, $p = 0.0014$; median OS: LI vs. HIw, HIm, and HIs: 26.5 months vs. 8.4 months, 17.0 months, and 8.3 months, respectively, $p = 0.0068$). * $p < 0.05$.

### *3.6. Relationship between P/C Ratio of CD44 Expression and Spectroscopic Signal Intensity of 5-ALA Fluorescence in the Tumor Periphery of GBM*

To examine whether CD44 expression in the tumor periphery offers a useful marker to predict invasiveness of GBM, the relationship between P/C ratio of CD44 expression and tumor invasiveness as evaluated by spectroscopic 5-ALA fluorescence intensity was investigated. Tumor invasiveness as determined by the strength of 5-ALA fluorescence intensity in the peritumoral area at final resection correlated well with the A2/A0 ratio of Ki-67 SI at the area A2, except for the relationship between HIw and HIm (Figure 7a). On the other hand, analysis of the relationship between CD44 expression and invasiveness based on 5-ALA fluorescence revealed that tumors in the HI group with high fluorescence intensity ≥1000 a.u. showed significantly higher P/C ratio of CD44 expression than patients in the LI group with fluorescence intensity <1000 a.u. (Figure 7b). However, among high-invasive groups, P/C ratio of CD44 did not correlate with degree of invasiveness among subgroups (Figure 7b). On the other hand, in all 21 GBM patients, P/C ratio of CD44 expression correlated with the A2/A0 ratio of Ki-67 SI at A2 in the peritumoral area (Figure 7c).

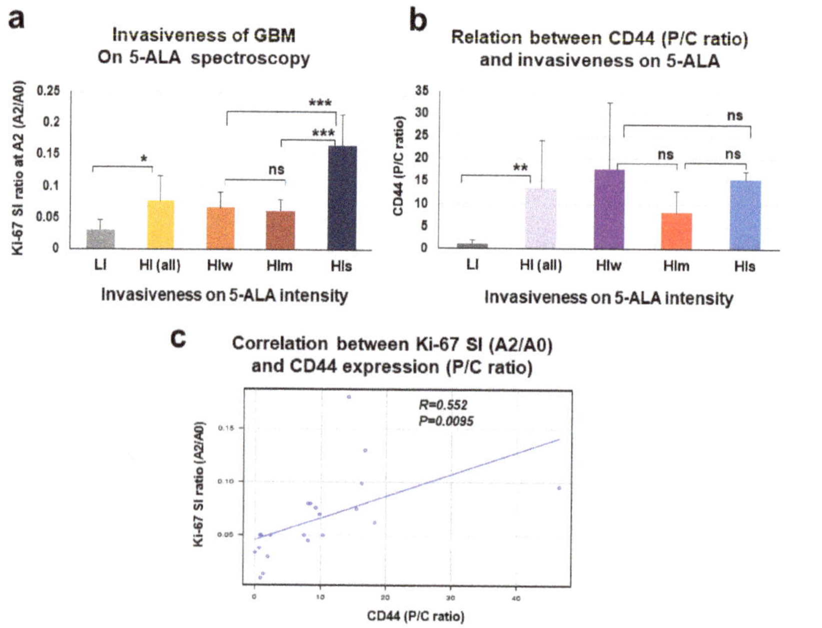


**Figure 7.** Relation between the degree of tumor invasiveness evaluated by 5-ALA fluorescence intensity and P/C ratio of CD44 expression. (**a**) Relationship between tumor invasiveness measured by 5-ALA fluorescence intensity and tumor density represented by Ki-67 SI ratio at A2 area (A2/A0). LI-type tumors showing fluorescence intensity <1000 a.u. present significantly lower Ki-67 SI ratio than all HI-type tumors showing fluorescence intensity ≥1000 a.u. (LI vs. HI: 0.031± 0.017 vs. 0.077± 0.04, $p = 0.0115$). Among HI-type tumors, HIs (>5000 a.u.) shows significantly higher Ki-67 SI ratio than HIw (1000–3000 a.u.) and HIm (3000–5000 a.u.), but HIw and HIm do not show any difference in Ki-67 SI (A2/A0) (HIs vs. HIw: 0.165 ± 0.05 vs. 0.067 ± 0.024, $p = 0.00026$, HIs vs. HIm: 0.165 ± 0.05 vs. 0.062 ± 0.017, $p = 0.00014$, HIw vs. HIm: 0.067 ± 0.024 vs. 0.062 ± 0.017, $p = 0.975$). (**b**) Relationship between CD44 expression (P/C ratio) and tumor invasiveness measured by 5-ALA fluorescence intensity. HI-type tumors with fluorescence intensity ≥1000 a.u. express significantly higher CD44 than LI-type tumors showing fluorescence intensity <1000 a.u. (HI vs. LI: 13.56 ± 10.59 vs. 1.19 ± 0.8, $p = 0.0067$). However, no significant difference is seen among HI-type tumor groups. (**c**) Spearman's regression analysis shows a positive correlation between P/C ratio of CD44 expression and Ki-67 SI ratio at A2 (A2/A0). * $p < 0.05$, ** $p < 0.01$, *** $p < 0.001$, ns: not significant.

## 4. Discussion

High invasiveness is a characteristic feature of GBM and a major obstacle to surgical eradication, leading to early recurrence of GBM. The development of novel therapeutic methods to regulate the invasion of GBM would greatly contribute to improved outcomes for patients with GBM. To control the high invasiveness of GBM, understanding not only the molecular mechanisms underlying GBM invasion, but also the degree of invasiveness in each patient with GBM is crucial. To date, potential markers for numerically predicting the invasiveness of GBM have not been identified. Our previous studies reported that higher expression of CD44 in the tumor periphery compared to the tumor core in GBM correlates with tumor invasiveness as defined from MRI [7]. CD44 is a multi-functional transmembrane glycoprotein that not only mediates cell adhesion and cell migration, but also transduces various signaling pathways of intra- and intercellular events [11,22], thus activating a variety of cellular processes, such as migration, invasion, proliferation, and metastasis in cancer cells, including in GBM [23]. CD44 is also known as a stem cell marker of many cancer cells [24,25]. The migratory and invasive activities of cancer stem cells are highly enhanced by the high expression of CD44 [26], promoting local tumor recurrence and metastasis.

In the present study, we demonstrated that GBMs expressing CD44 much higher in the tumor periphery than in the tumor core (as represented by a higher P/C ratio) were most strongly associated with HI-type GBM. CD44 expression in the tumor periphery also correlated with HI-type GBM, but both specificity and sensitivity in differentiating HI-type tumors from LI-type tumors were much greater with evaluation by P/C ratio of CD44 expression than by CD44 expression in the tumor periphery alone. The co-existence of differentiated non-stem glioma cells expressing CD44 may be the cause of such differences. Taking the ratio of CD44 expression in the periphery to that in the core (P/C ratio) could more clearly represent the degree of CD44 expression in GSCs present within the invasive front of the tumor periphery, as described previously [7]. The present study may have revealed another reason for the differences in CD44 expression reflected by P/C ratio and levels in the periphery alone. In HI-type GBM, CD44 expression tended to be much lower in the tumor core than in the tumor periphery (Figure 3a). Consequently, evaluation of CD44 expression in the tumor periphery using the P/C ratio may reflect the specificity of CD44 expression in the periphery much better than that of CD44 expression in the tumor periphery alone. Our previous study found that GSCs from tumor tissues in the periphery of HI-type GBMs, presenting a high P/C ratio of CD44, expressed higher CD44 in stem-like cells and much higher invasion than GSC from GBM with a low P/C ratio of CD44 expression [7,8]. These results may indicate that glioma cells, particularly GSCs highly expressing CD44, actually exist in the tumor periphery. This may be a reason why CD44 expression in the tumor periphery is higher in HI-type GBM than in LI-type GBM, even if one's evaluating by either CD44 expression in the periphery alone or CD44 expression by P/C ratio. The P/C ratios of CD44 expression also correlated negatively with OS in all GBM patients, including both HI and LI types. This indicates that the P/C ratio of CD44 expression may provide an indicator for differentiating between HI- and LI-type GBM based on a cutoff P/C ratio for CD44 expression. However, analyses of CD44 expression to date have not been able to evaluate how extensively tumors invade the peritumoral area. To respond to this question, we investigated whether CD44 expression as represented by P/C ratio actually corresponded with the degree of invasiveness in GBM patients using intraoperative 5-ALA fluorescence spectroscopy.

Accumulating studies have demonstrated that a fair number of infiltrating tumor cells exist in the peritumoral area beyond the contrast-enhanced main tumor mass [7,27]. These infiltrating tumor cells, particularly GSCs, are thought to represent sources for early recurrence of GBM. However, neither the density nor the extent of tumor cells infiltrating into the peritumoral area can be depicted by standard morphological imaging modalities, such as CT and MRI. In contrast, Met-PET has been shown to be useful for delineating the extent of tumor invasion from solid GBM masses [28,29]. We reported that the extent of GBM infiltration could be estimated by Met-PET analysis [5]. In that study, we demonstrated that infiltrating tumor cells expand to the peritumoral area while decreasing the number of proliferative tumor cells in accordance with decreases in the level of tumor-to-contralateral normal brain tissue ratio (TNR) in Met-PET. Compared to the present study, Met-PET displayed a mean Ki-67 SI of 39.3% at SUVmax (TNR > 2.0), 23.6% at the area of TNR 1.4 and 1.85% at the area of TNR 1.2, respectively. Accordingly, areas A0, A1, and A2 as defined in the present study may correspond to areas of TNR > 2.0, 1.4 and 1.2 on Met-PET, respectively. As almost no tumor cells were identified in the area of TNR 1.2 on Met-PET, areas showing a 5-ALA fluorescence intensity signal <1000 a.u. would indicate areas without tumor invasion. These results suggest that spectroscopic analysis of 5-ALA fluorescence intensity may take the place of Met-PET in evaluating the invasiveness of GBM.

FGS, particularly using 5-ALA, has been a common procedure for achieving safe maximal resection of GBM cells [17,30]. FGS enables visualization of tumor cells and effective resection of tumor cells that emit a distinct red fluorescence from PpIX. However, infiltrating tumor cells cannot be always detected by photodynamic visualization using conventional excitation methods. In contrast, 5-ALA fluorescence spectroscopy allows the neurosurgeon to detect infiltrating tumor cells in the invasion area. When infiltrating tumor cells were irradiated with laser light in the target area, tumor cells that had not been detected by conventional light were prompted to emit red fluorescence and simultaneously the fluorescent spectrum was depicted on the monitoring display. The fluorescence intensity of PpIX appeared on the display as a peak intensity at a wavelength of 636 nm. The 5-ALA fluorescence spectroscopy not only enhanced the visualization of tumor cells, but also enabled quantitative analysis of the fluorescence intensities of infiltrating tumor cells in the tumor periphery of GBM.

In the present study, we analyzed the spectroscopic intensity of 5-ALA fluorescence to evaluate the extent of tumor invasiveness in the peritumoral area of GBM. As spectroscopic fluorescence signals were derived from small spots of the target area, signals were obtained from several spots. As a result, 5-ALA fluorescence intensity was presented as a range of intensities because intensity signals from several spots in each target area showed a large degree of variability. To evaluate degrees of tumor invasiveness based on 5-ALA fluorescence intensity, we created attenuation curves for tumor density with intensity represented by Ki-67 SI ratios at areas A1 and A2 compared to area A0 (A1/A0 and A2/A0, respectively). The curves showed two patterns according to HI and LI types. Decreases in the number of tumor cells from area A0 to area A1 were less in LI-type tumors than in HI-type tumors. In contrast, the reduction in the number of tumor cells from area A1 to area A2 was less in HI-type tumors than in LI-type tumors. Consequently, LI tumors tended to show a graph with a sharp decline, whereas most HI-type tumors presented an almost linear graph. In the present study, all HI-type tumors presented a Ki-67 SI ratio at area A2 (A2/A0) ≥0.04. When cutoff values of A1/A0 ratio and A2/A0 ratio of Ki-67 SI were applied to differentiate HI- and LI-type tumors, only one patient was reclassified from LI-type tumor in the classification based on MRI to HI-type tumor based on 5-ALA classification (Patient 10). The remaining 20 patients showed the same invasion types as the classification of tumor invasiveness based on MRI. This indicates that criteria determining invasive types from MRI may be commonly used for showing invasiveness in GBM. As a result, in the classification based on 5-ALA fluorescence intensity, seven patients showed LI type and 14 patents showed HI type. Patients with HI-type tumor were further divided into three groups by the strength of fluorescence intensity at area A2. These included subgroups showing intensity >1000 to ≤3000 a.u. (HIw), >3000 to ≤5000 a.u. (HIm), and >5000 a.u. (HIs). Survival curves of patients with tumor invasiveness determined on 5-ALA fluorescence presented the same patterns as patients classified by MRI—that is, the group with the LI-type tumor on 5-ALA showed much longer PFS and OS than those with the HI-type tumor in any of the three subgroups of HIw, HIm, and HIs. These results indicate that the spectroscopic analysis of 5-ALA fluorescence may be applicable for determining the invasiveness of GBM. Consequently, we investigated the relationship between degree of tumor invasiveness based on 5-ALA intensity and levels of CD44 expression (P/C ratio). Although the correlation between the two types of invasiveness (low or high) and CD44 expression (P/C ratio) as evaluated by 5-ALA spectroscopy was quite similar to that from MRI, no relationship was seen among HI-type tumors. In contrast, Ki-67 SI ratios at area A2 (A2/A0 ratio) revealed a positive correlation with P/C ratios of CD44 expression. These results may indicate that the P/C ratio of CD44 expression could better predict the invasive types of GBM, but the degree to which levels of CD44 expression correspond to the degree of GBM invasiveness was not determined. Although the fluorescence intensity of 5-ALA on spectroscopy could not be evaluated by individual numerical values of intensity, the method clarified that almost no tumor cells exist in the peritumoral area where spectroscopy shows a fluorescence intensity <1000 a.u. These data imply that this area may represent a histological border of invasive GBM, and resection of the tumor up to this area while avoiding impairment of normal brain function allows more effective supra-total resection of GBM. From the present study, CD44 expression (represented by P/C ratio) and spectroscopy of 5-ALA fluorescence were recognized as potential biomarkers for predicting GBM invasiveness. However, these predictive methods have several disadvantages. One is that these methods cannot evaluate tumor invasiveness until the tumor is being resected. Preoperative evaluation of tumor invasiveness must therefore still rely on imaging modalities such as MRI and Met-PET. Another disadvantage is that current evaluation of tumor invasiveness cannot always consider the existence of GSCs infiltrating the peritumoral area. Concerning this problem in terms of CD44 expression, we are analyzing CD44 isoforms specifically expressed in GSC, as many cancer stem cells express specific CD44 isoforms and the isoforms thus represent biomarkers for each cancer [31,32]. As for 5-ALA spectroscopy, we have identified a novel enhancer of 5-ALA fluorescence that increases 5-ALA fluorescence in not only differentiated glioma cells, but also GBM stem-like cells (submitted). In the future, utilizing these methods of evaluation with CD44 and 5-ALA fluorescence spectroscopy may identify more useful markers for quantitatively and conventionally predicting tumor invasiveness in GBM. In addition, essential roles of CD44 in the cellular process of tumor invasion may be clarified and reveal the molecular mechanisms underlying GBM invasion.

This study was conducted by analyzing data from a relatively small cohort of patients. This may reflect the difficulty of enrolling a sufficient number of GBM cases from a single center and achieving adequate laboratory evaluations, including Met-PET, for diagnosis. Extensive analysis with a larger sample size is required to obtain further definitive conclusions.

The present study demonstrates that high CD44 expression in the tumor periphery correlated with HI-type GBM, in which tumor invasiveness was evaluated by the specific features on MRI, whereas CD44 expression in the tumor core tended to be lower in HI-type tumor. Consequently, the ratio of CD44 expression in the tumor periphery to that in the tumor core (P/C ratio) was found to be more effective for differentiating between LI- and HI-type tumors. To clarify the relationship between P/C ratio of CD44 expression and actual invasiveness of GBM, invasion was quantitatively evaluated by spectroscopic analysis of 5-ALA fluorescence. Tumor invasiveness was evaluated from the spectroscopic intensity signal of 5-ALA fluorescence, which was demonstrated to correlate with tumor density as represented by Ki-67 SI. As tumors showing an intensity signal <1000 a.u. at the final resection stage displayed no tumor cells in the area, these were considered LI-type tumors. In contrast, tumors presenting a fluorescent intensity signal ≥1000 a.u. were included as HI-type tumors. The HI-type tumors with tumor invasiveness determined by 5-ALA intensity showed significantly greater P/C ratios of CD44 expression than LI-type tumors. In addition, P/C ratios of CD44 expression in all 21 GBM patients correlated with the strength of 5-ALA fluorescence at final resection, representing the strength of tumor invasiveness. These results indicate that the level of CD44 expression (as represented by P/C ratio) was associated with the degree of actual invasiveness of GBM and could provide a reliable biomarker for predicting GBM invasiveness. These findings are of great importance for surgical planning and may help with decision-making processes in the treatment of GBM.

**Supplementary Materials:** The following supporting information can be downloaded at: https://www.mdpi.com/article/10.3390/biomedicines11092369/s1, Figure S1: Illustration showing surgical procedures for tumor resection in the peritumoral area under guidance of spectroscopic 5-ALA fluorescence; Figure S2. Illustration showing surgical procedures for tumor resection in the peritumoral area under guidance of spectroscopic 5-ALA fluorescence; Table S1: Characteristics of 50 patients with GBM, including 21 patients who were also enrolled in the study of 5-ALA spectroscopy.

**Author Contributions:** Conceptualization, A.I. and T.O.; methodology, A.I.; software, A.I.; validation, A.I., T.O. and M.N.; formal analysis, A.I.; investigation, A.I., M.N., H.W., K.K., M.T., H.Y., Y.O., S.M., S.S. (Satoshi Suehiro), D.Y., S.S. (Seiji Shigekawa) and H.T.; resources, A.I.; data curation, A.I.; writing—original draft preparation, A.I.; writing—review and editing, T.O.; visualization, A.I.; supervision, H.T., R.K., J.T. and T.K.; project administration, A.I.; funding acquisition, A.I. All authors have read and agreed to the published version of the manuscript.

**Funding:** This work was supported by the Japan Society for the Promotion of Science (JSPS) KAKENHI Grant Number JP 22K09208.

**Institutional Review Board Statement:** This study was approved by the Ethics Committee for Clinical Research of Ehime University Hospital (approval no. 2211011) prior to initiation, and was performed in accordance with the ethical standards as laid down in the 1964 Declaration of Helsinki and its later amendments.

**Informed Consent Statement:** Informed consent was obtained from all individual participants included in the study.

**Data Availability Statement:** All data used for analysis are presented in the tables in this article.

**Acknowledgments:** The authors would like to express their gratitude to Taichi Furumochi, Yasuhiro Shiraishi, Yoshihiro Takimoto and Mizusa Harada of the Department of Neurology, Ehime University Hospital, Japan; and Eiji Katayama, Takuya Kondo and Satsuki Myoga of the Department of Pathology, Ehime University Hospital, Japan, for their help in obtaining pathological and radiological findings. We also thank Rei Shirogane at SBI Pharmaceuticals Co. for the invaluable advice regarding the technical background of quantitative evaluation of 5-ALA fluorescence intensity by spectroscopic analysis during the revision process. This work was supported by the Japan Society for the Promotion of Science (JSPS) KAKENHI Grant Number JP 22K09208.

**Conflicts of Interest:** The authors declare that there are no conflict of interest regarding the publication of this paper. None of the authors has any financial interest in this work.

## References


1. Stupp, R.; Mason, W.P.; van den Bent, M.J.; Weller, M.; Fisher, B.; Taphoorn, M.J.; Belanger, K.; Brandes, A.A.; Marosi, C.; Bogdahn, U.; et al. Radiotherapy plus concomitant and adjuvant temozolomide for glioblastoma. *N. Engl. J. Med.* **2005**, *352*, 987–996. [CrossRef] [PubMed]
2. Stupp, R.; Hegi, M.E.; Mason, W.P.; van den Bent, M.J.; Taphoorn, M.J.; Janzer, R.C.; Ludwin, S.K.; Allgeier, A.; Fisher, B.; Belanger, K.; et al. Effects of radiotherapy with concomitant and adjuvant temozolomide versus radiotherapy alone on survival in glioblastoma in a randomised phase III study: 5-year analysis of the EORTC-NCIC trial. *Lancet Oncol.* **2009**, *10*, 459–466. [CrossRef]
3. Munthe, S.; Petterson, S.A.; Dahlrot, R.H.; Poulsen, F.R.; Hansen, S.; Kristensen, B.W. Glioma cells in the tumor periphery have a stem cell phenotype. *PLoS ONE* **2016**, *11*, e0155106.
4. Angelucci, C.; D'Alessio, A.; Lama, G.; Binda, E.; Mangiola, A.; Vescovi, A.L.; Proietti, G.; Masuelli, L.; Bei, R.; Fazi, B.; et al. Cancer stem cells from peritumoral tissue of glioblastoma multiforme: The possible missing link between tumor development and progression. *Oncotarget* **2018**, *9*, 28116–28130. [CrossRef] [PubMed]
5. Inoue, A.; Ohnishi, T.; Kohno, S.; Ohue, S.; Nishikawa, M.; Suehiro, S.; Matsumoto, S.; Ozaki, S.; Fukushima, M.; Kurata, M.; et al. Met-PET uptake index for total tumor resection: Identification of 11C-methionine uptake index as a goal for total tumor resection including infiltrating tumor cells in glioblastoma. *Neurosurg. Rev.* **2021**, *44*, 587–597. [CrossRef]
6. Shimizu, K.; Tamura, K.; Hara, S.; Inaji, M.; Tanaka, Y.; Kobayashi, D.; Sugawara, T.; Wakimoto, H.; Nariai, T.; Ishii, K.; et al. Correlation of Intraoperative 5-ALA-Induced Fluorescence Intensity and Preoperative 11C-Methionine PET Uptake in Glioma Surgery. *Cancers* **2022**, *14*, 1449. [CrossRef]

7. Nishikawa, M.; Inoue, A.; Ohnishi, T.; Kohno, S.; Ohue, S.; Matsumoto, S.; Suehiro, S.; Yamashita, D.; Ozaki, S.; Watanabe, H.; et al. Significance of glioma stem-like cells in the tumor periphery that express high levels of CD44 in tumor invasion, early progression, and poor prognosis in glioblastoma. *Stem Cells Int.* **2018**, *23*, 5387041. [CrossRef]
8. Nishikawa, M.; Inoue, A.; Ohnishi, T.; Yano, H.; Ozaki, S.; Kanemura, Y.; Suehiro, S.; Ohtsuka, Y.; Kohno, S.; Ohue, S.; et al. hypoxia-induced phenotypic transition from highly invasive to less invasive tumors in glioma stem-like cells: Significance of CD44 and osteopontin as therapeutic targets in glioblastoma. *Trans. Oncol.* **2021**, *14*, 101137. [CrossRef]
9. Jijiwa, M.; Demir, H.; Gupta, S.; Leung, C.; Joshi, K.; Orozco, N.; Huang, T.; Yildiz, V.O.; Shibahara, I.; de Jesus, J.A.; et al. CD44v6 regulates growth of brain tumor stem cells partially through the AKT-mediated pathway. *PLoS ONE* **2011**, *6*, e242217. [CrossRef]
10. Chen, C.; Zhao, S.; Kamad, A.; Freeman, J.W. The biology and role of CD44 in cancer progression: Therapeutic implications. *J. Hematol. Oncol.* **2018**, *11*, 64.
11. Hassn, M.M.; Syafruddin, S.E.; Mohtar, M.A.; Syahir, A. CD44: A multifunctional mediator of cancer progression. *Biomolecules* **2021**, *11*, 1850. [CrossRef] [PubMed]
12. Ewelt, C.; Nemes, A.; Senner, V.; Wolfer, J.; Brokinkel, B.; Stummer, W.; Holling, M. Fluorescence in neurosurgery: Its diagnostic and therapeutic use. Review of the literature. *J. Photochem. Photobiol. B* **2015**, *148*, 302–309. [CrossRef] [PubMed]
13. Schneider, M.; Ilic, I.; Potthoff, A.L.; Hamed, M.; Schafer, N.; Velten, M.; Guresir, E.; Herrlinger, U.; Borger, V.; Vatter, H.; et al. Safety metric profiling in surgery for temporal glioblastoma: Lobectomy as a supra-total resection regime preserves perioperative standard quality rates. *J. Neuro-Oncol.* **2020**, *149*, 455–461. [CrossRef] [PubMed]
14. Baik, S.H.; Kim, S.Y.; Na, Y.C.; Cho, J.M. Supratotal resection of glioblastoma: Better survival outcome than gross total resection. *J. Pers. Med.* **2023**, *13*, 383. [CrossRef]
15. Stummer, W.; Pichlmeier, U.; Meinel, T.; Wiestler, O.D.; Zanella, F.; Reulen, H.J.; ALA-Glioma Study Group. Fluorescence-guided surgery with 5-aminolevulinic acid for resection of malignant glioma: A randomised controlled multicentre phase III trial. *Lancet Oncol.* **2006**, *7*, 392–401. [CrossRef]
16. Ferraro, N.; Barbarite, E.; Albert, T.R.; Berchmans, E.; Shah, A.H.; Bregy, A.; Ivan, M.E.; Brown, T.; Komotar, R.J. The role of 5-aminolevulinic acid in brain tumor surgery: A systematic review. *Neurosurg. Rev.* **2016**, *39*, 545–555. [CrossRef]
17. Eatz, T.A.; Eichberg, D.G.; Lu, V.M.; Di, L.; Komotar, R.J.; Ivan, M.E. Intraoperative 5-ala fluorescence-guided resection of high-grade glioma leads to greater extent of resection with better outcomes: A systematic review. *J. Neurooncol.* **2022**, *156*, 233–256. [CrossRef]
18. Kanda, Y. Investigation of the freely available easy-to-use software 'EZR' for medical statistics. *Bone Marrow Transplant.* **2013**, *48*, 452–458. [CrossRef]
19. Ohue, S.; Kohno, S.; Inoue, A.; Yamashita, D.; Matsumoto, S.; Suehiro, S.; Kumon, Y.; Kikuchi, K.; Ohnishi, T. Surgical results of tumor resection using tractography-integrated navigation-guided fence-post catheter techniques and motor-evoked potentials for preservation of motor function in patients with glioblastomas near the pyramidal tracts. *Neurosurg. Rev.* **2015**, *38*, 293–306. [CrossRef]
20. Louis, D.N.; Perry, A.; Wesseling, P.; Brat, D.J.; Cree, I.A.; Figarella-Branger, D.; Hawkins, C.; Ng, H.K.; Pfister, S.M.; Reifenberger, G.; et al. The 2021 WHO Classification of Tumors of the Central Nervous System: A summary. *Neuro Oncol.* **2021**, *23*, 1231–1251. [CrossRef]
21. Okita, Y.; Nonaka, M.; Shofuda, T.; Kanematsu, D.; Yoshioka, E.; Kodama, Y.; Mano, M.; Nakajima, S.; Kanemura, Y. (11)C-methinine uptake correlates with MGMT promoter methylation in nonenhancing gliomas. *Clin. Neurol. Neurosurg.* **2014**, *125*, 212–216. [CrossRef] [PubMed]
22. Xu, H.; Niu, M.; Yuan, X.; Wu, K.; Liu, A. CD44 as a tumor biomarker and therapeutic target. *Exp. Hematol. Oncol.* **2020**, *9*, 36. [CrossRef] [PubMed]
23. Mooney, K.L.; Choy, W.; Sidhu, S.; Pelargos, P.; Bui, T.T.; Voth, B.; Barnette, N.; Yang, I. The role of CD44 in glioblastoma multiforme. *J. Clin. Neurosci.* **2016**, *34*, 1–5. [CrossRef]
24. Al-Hajj, M.; Wicha, M.S.; Benito-Hemandez, A.; Morrison, S.J.; Clarke, M.F. Prospective identification of tumorigenic breast cancer cells. *Proc. Natl. Acad. Sci. USA* **2003**, *100*, 3983–3988. [CrossRef]
25. Takaishi, S.; Okumura, T.; Tu, S.; Wang, S.S.; Shibata, W.; Vigneshwaran, R.; Gordon, S.A.; Shimada, Y.; Wang, T.C. Identification of gastric cancer stem cells using the cell surface marker CD44. *Stem Cells* **2009**, *27*, 1006–1020. [CrossRef]
26. Chen, J.E.; Wu, Q.; Guryanova, O.A.; Huang, Z.; Hung, Q.; Ricch, J.N. Elevated invasive potential of glioblastoma stem cells. *Biochem. Biophys. Res. Commun.* **2011**, *406*, 643–648. [CrossRef]
27. Miwa, K.; Shinoda, J.; Yano, H.; Okumura, A.; Iwama, T.; Nakashima, T.; Sakai, N. Discrepancy between lesion distributions on methionine PET and MR images in patients with glioblastoma multiforme: Insight from a PET and MR fusion image study. *J. Neurol. Neurosurg. Psychiatry* **2004**, *75*, 1457–1462. [CrossRef]
28. Kracht, L.W.; Miletic, H.; Busch, S.; Jacobs, A.H.; Voges, J.; Hoevels, M.; Klein, J.C.; Herholz, K.; Heiss, W.D. Delineation of brain tumor extent with [11C]L-methionine positron emission tomography: Local comparison with stereotactic histopathology. *Clin. Cancer Res.* **2004**, *10*, 7163–7170. [CrossRef] [PubMed]
29. Lucas, J.T., Jr.; Serrano, N.; Kim, H.; Li, X.; Snyder, S.E.; Hwang, S.; Li, Y.; Hua, C.H.; Broniscer, A.; Merchant, T.E.; et al. 11C-Methionine positron emission tomography delineates non-contrast enhancing tumor regions at high risk for recurrence in pediatric high-grade glioma. *J. Neurooncol.* **2017**, *132*, 163–170. [CrossRef]

30. Mirza, A.B.; Christodoulides, I.; Lavradort, J.P.; Giamouriadis, A.; Vastani, A.; Boardman, T.; Ahmed, R.; Noman, I.; Murphy, C.; Devi, S.; et al. 5-aminolevulinic acid-guided resection improves the overall survival of patients with glioblastoma-a comparative cohort study of 343 patients. *Neuro-Oncol. Adv.* **2021**, *3*, 1–11.
31. Taniguchi, D.; Saeki, H.; Nakashima, Y.; Kudou, K.; Nakanishi, R.; Kubo, N.; Ando, K.; Oki, E.; Oda, Y.; Maehara, Y. CD44v9 is associated with epithelial-mesenchymal transition and poor outcomes in esophageal squamous cell carcinoma. *Cancer Med.* **2018**, *7*, 6258–6268. [CrossRef] [PubMed]
32. Todaro, M.; Gaggianesi, M.; Catalano, V.; Benfante, A.; Iovino, F.; Biffoni, M.; Apuzzo, T.; Sperduti, I.; Volpe, S.; Cocorullo, G.; et al. CD44v6 is a marker of constitutive and reprogrammed cancer stem cells driving colon cancer metastasis. *Cell Stem Cell* **2014**, *14*, 342–356. [CrossRef] [PubMed]

biomedicines

MDPI

Article

# Comparison of MRI Sequences to Predict *IDH* Mutation Status in Gliomas Using Radiomics-Based Machine Learning

**Dilek N. G. Kasap, Nabila Gala Nacul Mora, David A. Blömer, Burak Han Akkurt, Walter Leonhard Heindel, Manoj Mannil *,† and Manfred Musigmann †**

University Clinic for Radiology, University of Münster, University Hospital Münster, Albert-Schweitzer-Campus 1, DE-48149 Münster, Germany; dilek.kasap@uni-muenster.de (D.N.G.K.); nabilagala.naculmora@ukmuenster.de (N.G.N.M.); musigma@uni-muenster.de (M.M.)

* Correspondence: mannil@uni-muenster.de

† These authors contributed equally to this work.

**Abstract:** Objectives: Regarding the 2021 World Health Organization (WHO) classification of central nervous system (CNS) tumors, the isocitrate dehydrogenase (*IDH*) mutation status is one of the most important factors for CNS tumor classification. The aim of our study is to analyze which of the commonly used magnetic resonance imaging (MRI) sequences is best suited to obtain this information non-invasively using radiomics-based machine learning models. We developed machine learning models based on different MRI sequences and determined which of the MRI sequences analyzed yields the highest discriminatory power in predicting the *IDH* mutation status. Material and Methods: In our retrospective IRB-approved study, we used the MRI images of 106 patients with histologically confirmed gliomas. The MRI images were acquired using the T1 sequence with and without administration of a contrast agent, the T2 sequence, and the Fluid-Attenuated Inversion Recovery (FLAIR) sequence. To objectively compare performance in predicting the *IDH* mutation status as a function of the MRI sequence used, we included only patients in our study cohort for whom MRI images of all four sequences were available. Seventy-one of the patients had an *IDH* mutation, and the remaining 35 patients did not have an *IDH* mutation (*IDH* wild-type). For each of the four MRI sequences used, 107 radiomic features were extracted from the corresponding MRI images by hand-delineated regions of interest. Data partitioning into training data and independent test data was repeated 100 times to avoid random effects associated with the data partitioning. Feature preselection and subsequent model development were performed using Random Forest, Lasso regression, LDA, and Naïve Bayes. The performance of all models was determined with independent test data. Results: Among the different approaches we examined, the T1-weighted contrast-enhanced sequence was found to be the most suitable for predicting *IDH* mutations status using radiomics-based machine learning models. Using contrast-enhanced T1-weighted MRI images, our seven-feature model developed with Lasso regression achieved a mean area under the curve (AUC) of 0.846, a mean accuracy of 0.792, a mean sensitivity of 0.847, and a mean specificity of 0.681. The administration of contrast agents resulted in a significant increase in the achieved discriminatory power. Conclusions: Our analyses show that for the prediction of the *IDH* mutation status using radiomics-based machine learning models, among the MRI images acquired with the commonly used MRI sequences, the contrast-enhanced T1-weighted images are the most suitable.

**Keywords:** glioma; *IDH* mutation status; machine learning; artificial intelligence; neuroimaging; MRI; radiomics

## 1. Introduction

Gliomas, a diverse group of central nervous system (CNS) tumors, present an intricate challenge in the realm of neuro-oncology due to their diverse clinical behaviors and complex molecular underpinnings [1]. Accurate characterization and classification of gliomas are crucial for guiding treatment decisions and predicting patient outcomes [2]. With the release of the 2021 World Health Organization (WHO) classification guidelines [3], the landscape of glioma classification underwent a profound transformation [4–6]. Central to this classification are molecular markers, notably the isocitrate dehydrogenase (*IDH*) and alpha thalassemia/mental retardation syndrome X-linked (*ATRX*) mutations, which have emerged as pivotal elements in characterizing gliomas and informing clinical decision making [1,7,8].

*IDH*-mutant gliomas tend to initially present as lower histologic grades with better outlooks, boasting a median survival of over 12 years [9]. However, they frequently transform to higher grades as the disease progresses. On the other hand, *IDH* wild-type gliomas typically manifest as glioblastomas (GBMs), which are the most prevalent and highly aggressive grade IV brain tumors according to the WHO classification [3]. Despite aggressive treatments involving multiple approaches, the average survival for individuals with *IDH* wild-type gliomas is only 12 to 15 months [10,11].

Additionally, the presence or absence of *ATRX* mutations and of the *1p/19q* co-deletion status further refines our ability to dissect glioma heterogeneity [12]. The *ATRX* gene holds significant relevance in various types of tumors, particularly adult-type diffuse glioma [13]. This gene is pivotal in the process of chromatin remodeling and the preservation of genomic stability [14].

As the diagnostic landscape of gliomas evolves, non-invasive methods for assessing molecular markers are becoming increasingly important. Magnetic resonance imaging (MRI) is the preferred method in the diagnosis and characterization of gliomas, offering valuable insights into their structure and function [15,16]. Specific MRI sequences such as T1-weighted native (T1w native), T1-weighted with contrast enhancement (T1w CE), T2-weighted (T2w), and Fluid-Attenuated Inversion Recovery (FLAIR) are commonly employed for this purpose [17,18]. It is worth noting that magnetic resonance spectroscopy has emerged as a promising tool for predicting the *IDH* mutation status in gliomas, contributing to the advancing diagnostic precision in glioma classification. Suh et al. and Bauer et al. investigated the role of 2-Hydroxyglutarate as a predictor of the *IDH* mutation in gliomas with very promising results [19,20]. In recent years, radiomics, a cutting-edge field at the intersection of medical imaging and data science, has emerged as a promising approach to extract valuable information from medical images, including magnetic resonance imaging, and transform it into clinically meaningful insights [21,22]. Radiomics, at its core, harnesses the information embedded within medical images, by extracting a multitude of quantitative features [21,22]. These radiomic features encompass a wide array of data, including textural, morphological, and functional characteristics, which have the potential to offer unique insights into glioma subtypes and their underlying molecular alterations [23,24]. The combination of radiomics based on MRI and machine learning offers numerous potential applications in neuroradiological diagnostics. For example, Alves et al. investigated the differentiability of inflammatory lesions and brain tumors using such methods [25]. Radiomics, when coupled with machine learning techniques, has also demonstrated its potential in predicting various molecular markers and clinical outcomes in diverse malignancies [26–28]. In this context, machine learning refers to the use or development of algorithms that are able to independently recognize patterns in data and subsequently draw their own conclusions based on the previously learned relationships. In glioma research, radiomics-based machine learning models have shown promise in non-invasively predicting molecular alterations using MRI data [29,30]. By integrating radiomics analysis with molecular information, there is an opportunity to refine and enhance the precision of glioma classification.

In alignment with our study, current research in the field of neuroimaging has investigated the efficacy of radiomics models in non-invasively predicting mutation status in gliomas [26,31–35]. Choi et al. employed a fully automated segmentation model based on Convolutional Neural Networks (CNNs) for multi-parametric MRI images, demonstrating its effectiveness in predicting the *IDH* mutation status [31]. W.Y. Huang et al., using radiomics analyses, found the radiomics model based on T1w CE to perform better than models based on other sequences in predicting tumor grade and *IDH1* status of gliomas [34]. Additionally, L. Han et al. conducted an MRI texture analysis, emphasizing the suitability of the T1w CE sequence for predicting *IDH1* mutations [35]. Conversely, some studies suggest the T2w sequence to be the most suitable for predicting the *IDH* mutation status [26,33].

We are interested in predicting the *IDH* mutation status, i.e., distinguishing between *IDH*-mutated and *IDH* wild-type gliomas using radiomics-based machine learning. However, the specific aim of our current study is not to predict *IDH* mutation status with the highest possible accuracy that can somehow be achieved using machine learning models. Rather, we are interested to determine which of the commonly used MRI sequences is best suited for this diagnostic task. For this purpose, we compared the results obtained with different machine learning algorithms using native and contrast-enhanced T1-weighted MRI images, T2-weighted MRI images, and MRI images acquired with the FLAIR sequence.

## 2. Materials and Methods

Our study was performed in compliance with the Declaration of Helsinki and approved by the local ethics committee (Ärztekammer Westfalen-Lippe, ÄKWL 2021-596-f-S). Due to its retrospective nature, written informed consent was waived. We aimed to analyze which of the commonly used MRI sequences is best suited to determine *IDH* mutation status non-invasively using radiomics-based machine learning models. As already explained, *IDH* mutation status is one of the most important markers for the classification of diffuse gliomas in adults according to the 2021 WHO classification of CNS tumors. We searched our database for patients diagnosed with glioma between October 2007 and December 2021. In our study cohort, only patients with (1) native, T1-weighted; (2) contrast-enhanced (CE), T1-weighted; (3) T2-weighted; and (4) FLAIR sequence MRI images were included. We only included patients for whom MRI images of all four sequences mentioned were available to compare the performance achievable with the different sequences in terms of predictability of *IDH* mutation status as objectively as possible. The MRI sequences utilized in our study were obtained from various hospitals and radiology practices, resulting in a combination of images captured using both 1.5 Tesla and 3 Tesla MRI machines. Despite the variability in magnetic field strength, we ensured uniformity across all included patients by employing consistent imaging protocols. This standardization of imaging protocols allowed for a comprehensive analysis of radiomic features across different MRI sequences, irrespective of the magnetic field strength utilized during the imaging process.

Initially, 107 patients were included in our study. We excluded one of these 107 patients due to unknown *IDH* mutation status. No further patients were excluded. The final cohort of 106 patients contained 71 patients with an *IDH* mutation and 35 patients without an *IDH* mutation (*IDH* wild-type). The demographic characteristics of the data used to compare the MRI sequences in terms of predictability of *IDH* mutation status are summarized in Table 1.

**Table 1.** Demographic characteristics of patients included in the study cohort, separately for training data and independent test data as well as for the total data.

| | Training Data | Independent Test Data | Total Data |
|---|---|---|---|
| Number of patients | 85 | 21 | 106 |
| Gender (in %) | | | |
| Female | 45.09 | 46.05 | 45.28 |
| Male | 54.91 | 53.95 | 54.72 |
| Mean age (in years) | 43.52 | 42.85 | 43.39 |
| *IDH* status (in %) | | | |
| Mutated | 67.06 | 66.67 | 66.98 |
| Non-mutated (wild-type) | 32.94 | 33.33 | 33.02 |

### 2.1. Radiomics

Segmentation of the entirety of each tumor was semi-automatically performed by two radiology residents in their 3rd year of residency and two medical students with experience in MRI segmentation. Additionally, all images were visually assessed by a board-certified neuroradiologist with 11 years of experience. We utilized the 3D slicer open-source software platform (version 4.10, October 2018, www.slicer.org, accessed on 1 December 2021) and the Segmentation Wizard plugin. Expert consensus from two radiologists with five and ten years of experience in neuroimaging was used in anatomically difficult segmentations.

For each of the four MRI sequences used, we determined a total of 107 radiomic features using the PyRadiomisc package. The package is available as an implementable plugin for the 3D slicer platform. The 107 radiomic features were extracted by hand-delineated regions of interest (ROI) from the MRI images of each patient. They were assigned to the seven different feature classes: "first-order statistics" (18 features), "shape-based features" (14 features), "gray-level co-occurrence matrix features" (24 features), "gray-level run-length matrix features" (16 features), "gray-level size-zone matrix features" (16 features), "neighboring gray-tone difference matrix features" (5 features), and "gray-level dependence matrix features" (14 features). The numerical values in the brackets indicate the respective number of features belonging to the corresponding feature class. The longlist used for model development initially contained all 107 radiomic features, the age of the patients at the time of diagnosis, and their gender. Redundant features were subsequently removed from the longlist using a 95% correlation filter. All features were standardized.

### 2.2. Statistical Analysis

We performed our statistical analysis using R software (version 4.1.2, November 2021). The 106 patients included in our final study cohort were assigned randomly to a training group and an independent test group, using a stratified 4:1 ratio with a balanced distribution of *IDH*-mutated and *IDH* non-mutated patients between these two groups (see Table 1). We assigned the data/MRI images generated with the four sequences T1w native, CE, T2w, and FLAIR in each case identically to the respective training groups or independent test groups to be able to compare the results obtained with these four different sequences as objectively as possible. A total of 80% of the total data (i.e., the training data) were used for the feature preselection and subsequent model development. We performed feature preselection and model development using four different machine learning algorithms: Random Forest, Lasso (least absolute shrinkage and selection operator) regression, linear discriminant analysis (LDA), and Naïve Bayes. We used 10-fold cross-validations to optimize the hyperparameters included in the models. The models themselves were optimized by maximizing the area under the curve (AUC) of the receiver operator characteristic (ROC). Finally, the achieved model performance was tested using the independent test data (i.e., using the remaining 20% of the total data). The discriminatory power of our models was determined in terms of AUC, accuracy, sensitivity, specificity, positive predictive value (PPV), and negative predictive value (NPV). With respect to our analyses, sensitivity describes the proportion of correctly predicted cases with an *IDH* mutation. Accordingly, specificity describes the fraction of correctly predicted cases without an *IDH* mutation (*IDH* wild-type). The positive predictive value indicates the proportion of correctly predicted cases with an *IDH* mutation relative to all predicted cases with an *IDH* mutation. Finally, the negative predictive value describes the rate of correct predictions of cases without an *IDH* mutation (*IDH* wild-type) relative to all predictions of cases without *IDH* mutation.

Because the performance achievable with the models depends on the number of features included in the models and to rule out possible overfitting, we started by constructing each model multiple times with an increasing number of features. We always started with univariate models. Only the most important features were included in the models. These were previously determined based on the training data using the "varImp" function in R (varImp = variable importance). The varImp function determines the significance of the individual features by determining the model performance with and without a specific feature. The significance of the individual features is subsequently determined based on a comparison of the respective performance results. We identified the optimal and final number of features to include in the models by determining which number of features achieved the highest model performance in relation to the independent test data. Our approach minimized the risk of possible overfitting.

Both the exact model compositions and the discriminatory power obtained depended slightly on the specific division of the data into training and independent test data used. Our aim was to eliminate such slight random effects associated with data partitioning as much as possible and thus to demonstrate the robustness of our approach and its results. We therefore performed the data partitioning and subsequent model development a total of 100 times, each time using a new partitioning of the data into training and independent test data. All performance values were calculated based on the associated independent test datasets and as averages of these 100 cycles. A detailed description of the methodology we used can be found in Musigmann et al. [36]. For better understanding, we have additionally described the entire methodological approach in a flowchart (see Figure 1).

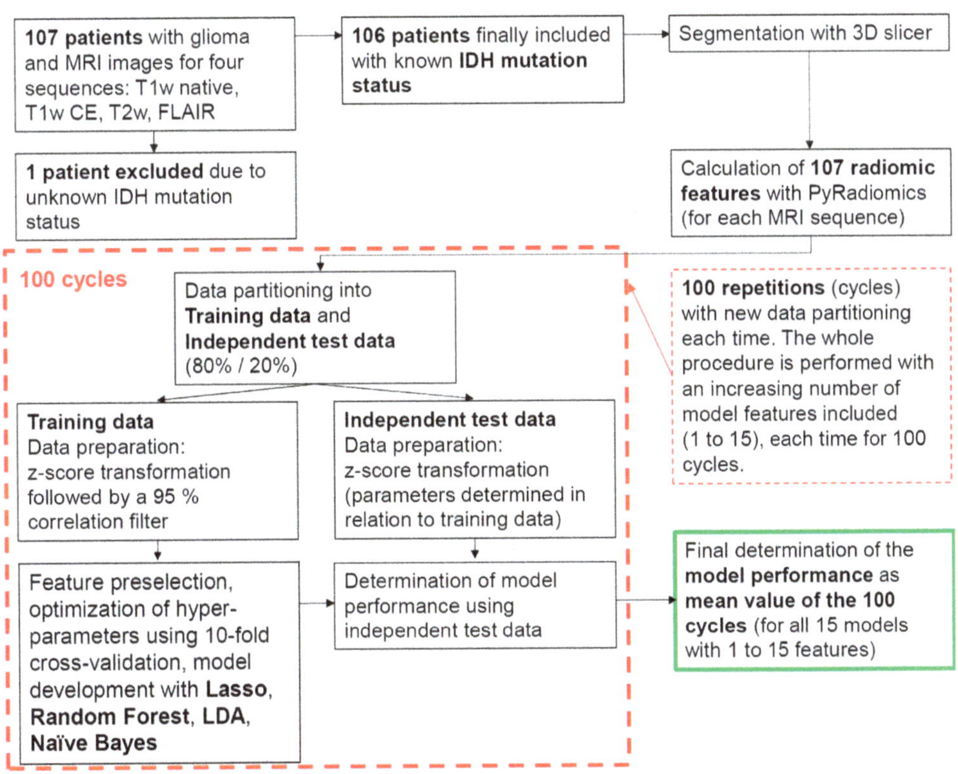


**Figure 1.** Flowchart describing the methodological approach. Four machine learning algorithms to predict *IDH* mutation status were tested: Lasso regression, Random Forest, LDA, and Naive Bayes. All models were developed with an increasing number (1 to 15) of features and for each of the four MRI sequences analyzed (T1 native, T1 CE, T2, and FLAIR). Each of these models was developed 100 times with new data partitioning each time and tested with independent test data afterward. The final model performances were determined as the average values of the 100 runs in each case.

It should be noted that the described procedure with 100 repetitions can, in principle, lead to up to 100 slightly different models. Therefore, based on the model variant that yielded the highest discriminatory power overall, we additionally created two final models that contained, in a fixed manner, only the features that were most frequently selected in the associated 100 runs previously conducted. The first of these two models contained 10 features. This is the number of features belonging to the model that exhibited the highest discriminatory power of all our tested model variants. In the second model, only the features that were selected in at least 50% of the previously performed 100 runs were considered. This applied to the seven most important features. The procedure used to develop these two models with 7 and 10 included features, respectively, was the same as the approach described previously. This means that again 100 different data partitions were used, and all discriminatory power values were calculated as mean values of the 100 repetitions and with respect to the respective independent test data.

## 3. Results

### 3.1. Results of Comparing MRI Sequences to Predict IDH Mutation Status

As explained earlier, we tested four machine learning algorithms to predict *IDH* mutation status: Random Forest, Lasso regression, LDA, and Naive Bayes. Each of these four algorithms was performed with an increasing number of features included in the models. At this point, however, it should be pointed out once again that the specific aim of this study was not to predict *IDH* mutation status with the highest possible accuracy that can somehow be achieved using radiomics-based machine learning models. The aim of this study was to determine which of the commonly used MRI sequences is best suited for this purpose. We used MRI images acquired with the T1 sequence (with and without contrast administration), the T2 sequence, and the FLAIR sequence. Based on the four machine learning algorithms we tested (each algorithm was tested with MRI images acquired with each of the four different MRI sequences), we obtained the best results using Lasso regression. In comparison, we received slightly worse results with the Random Forest algorithm. The worst results were obtained using LDA and the Naive Bayes algorithm. Figure 2 shows the discriminatory power obtained with our models using Lasso regression in terms of AUC (top left), accuracy (top right), sensitivity and specificity (bottom left), and, finally, positive and negative predictive values (bottom right) as a function of the number of features included in the models. The figures show the average values determined with the respective independent test data. All results were calculated based on 100 different data partitions of the data into training and independent test data. Comparing the results obtained with the four different MRI sequences, it is evident that *IDH* mutation status can be most accurately predicted using T1-weighted contrast-enhanced MRI images. The T2-weighted and the native T1-weighted MRI images are found to be the worst for this task. Finally, the results obtained with the MRI images generated using the FLAIR sequence are in the middle range.

In the case of the three sequences T1w CE, T2w, and Flair, the inclusion of additional features in the model leads to an increase in discriminatory power approximately up to the ninth feature. When using the native T1-weighted MRI images, the discriminatory power increases only up to the sixth feature. If more than nine (six) features are used, the discriminatory power no longer increases significantly or even decreases again. This indicates the beginning of an overfitting. However, regardless of the number of final features included in the models, T1-weighted contrast-enhanced MRI images almost always yielded the highest discriminatory power in predicting the *IDH* mutation status.

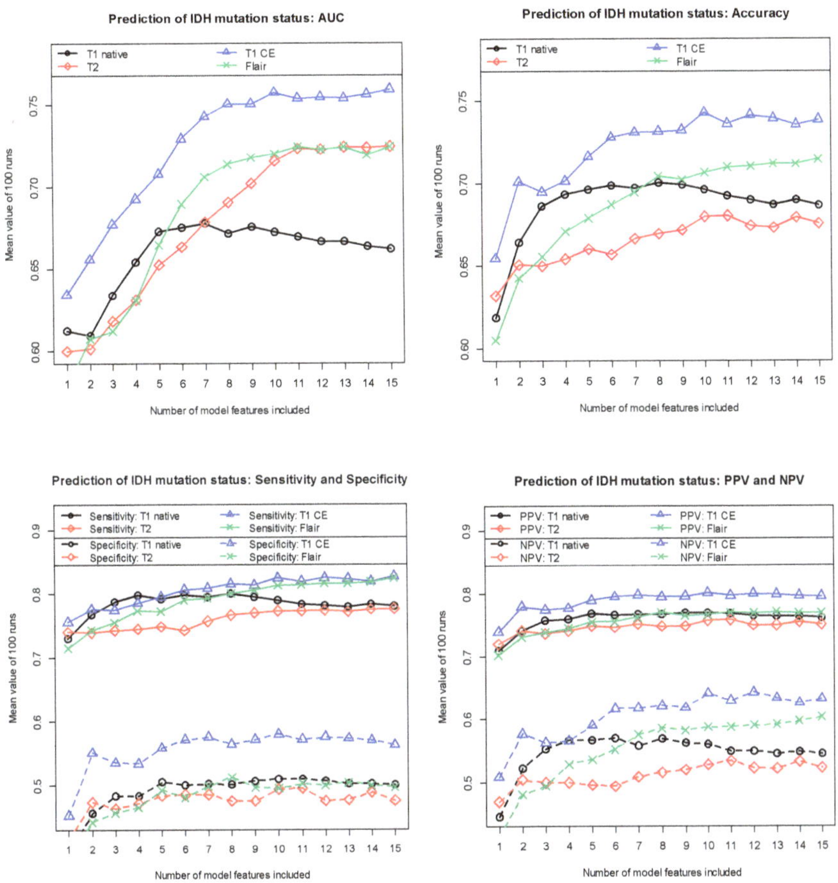


**Figure 2.** Mean discriminatory power (using 100 runs) in the prediction of *IDH* mutation status in gliomas using Lasso regression based on MRI images of four different sequences. AUC = area under the curve; PPV and NPV = positive and negative predictive values. All results were calculated using independent test data.

At this point, one might ask whether a different machine learning approach other than the Lasso regression we used would not have resulted in a different MRI sequence other than the T1-CE sequence we found. Three remarks are worth mentioning in this context:

(1) As described, we tested not only Lasso regression but also Random Forest, LDA, and Naive Bayes for predicting the *IDH* mutation status. With all four machine learning algorithms, the contrast-enhanced T1-weighted MRI images were invariably and unequivocally found to be the most suitable for predicting the *IDH* mutation status. The images generated using the FLAIR sequence produced the second-best results on average. Accordingly, using the native T1-weighted or the T2-weighted MRI images yielded the worst results on average.

(2) In addition, we also calculated and compared the univariate discriminatory power for all features included in the longlist for each of the four MRI sequences tested. These values do not depend on the machine learning algorithm subsequently used. Here, several features derived from the contrast-enhanced T1-weighted MRI images also proved to be the most discriminative factors. Some of the features derived based on the FLAIR sequence also showed promising univariate discriminatory power. In contrast,

the features derived from the native T1-weighted MRI images showed the lowest discriminatory power on average, which is again consistent with Figure 2. Of course, these results do not mandatorily apply to a multivariate approach as well, but they can be interpreted as an additional indication that contrast-enhanced T1-weighted images are likely to provide the best results with respect to our diagnostic question.

(3) Finally, we also developed models using Lasso regression, where we simultaneously provided the features of all four MRI sequences for feature selection. Similar to the results of our univariate analyses, features determined with contrast-enhanced T1-weighted MRI images and with the FLAIR sequence again proved to be particularly important. However, the models developed using the features of all four sequences simultaneously did not show a higher discriminatory power on average than the models based solely on the contrast-enhanced T1-weighted MRI images.

We would like to briefly point out here that our analyses not only determined the most appropriate MRI sequence for predicting the *IDH* mutation status but also identified the added value of contrast administration with respect to this diagnostic question. The added value obtained using a contrast agent can be calculated directly using Figure 2 as the difference of the curves determined for the T1-weighted native and contrast-enhanced MRI images. It is obvious that the administration of the contrast agent significantly increased the discriminatory power.

### *3.2. Results of Comparing MRI Sequences to Predict IDH Mutation Status Using Fixed Features*

As explained earlier, due to the 100 different partitions into training and independent test data, followed by new feature selection and model development in each cycle, the final 100 models may in principle be composed differently. The question is how different the final models really are in terms of their specific feature composition. Therefore, for each feature that was included in at least one of the 100 models developed with Lasso regression and the contrast-enhanced T1-weighted MRI images, we determined the number of 100 runs in which the feature was selected. Table 2 lists the frequencies of the most frequently selected features. The numbers indicate the respective frequency of selection of the associated feature during the 100 cycles. The results were calculated for the 10-feature models.

**Table 2.** Most important features and their frequencies of selection during 100 runs. Calculations were performed for the 10-feature models using Lasso regression based on the contrast-enhanced T1-weigthed MRI images.

| Level of Importance | Feature Name | Number of Runs Included |
|---|---|---|
| 1 | age_at_diagnosis | 99 |
| 2 | original_shape_Flatness | 97 |
| 3 | original_gldm_LargeDependenceHighGrayLevelEmphasis | 86 |
| 4 | original_shape_LeastAxisLength | 78 |
| 5 | original_firstorder_Kurtosis | 61 |
| 6 | original_glszm_LargeAreaHighGrayLevelEmphasis | 55 |
| 7 | original_glcm_DifferenceAverage | 52 |
| 8 | original_ngtdm_Contrast | 43 |
| 9 | original_shape_Maximum2DDiameterColumn | 35 |
| 10 | original_glszm_ZonePercentage | 30 |
| 11 | original_glrlm_LongRunHighGrayLevelEmphasis | 29 |
| 12–14 | original_glcm_Imc1 | 25 |
| 12–14 | original_glszm_LowGrayLevelZoneEmphasis | 25 |
| 12–14 | original_ngtdm_Strength | 25 |
| 15 | original_glcm_ClusterShade | 23 |

It turns out that the 100 models are very similar in their composition of features. The two features "age_at_diagnosis" and "original_shape_Flatness" are almost always included. Two further features were selected in more than three out of four runs (in 86% and 78% of all runs, respectively). Three additional features were included in at least more than every second model (in 61%, 55%, and 52% of all runs). All other features listed in Table 2 were already selected in less than 50% of all runs and quickly lost importance. This fact is consistent with the observation that for the T1-CE sequence, discriminatory power increases only slightly from about the seventh feature onward and not at all from the tenth feature onward (see Figure 2).

We were interested in the extent to which the performance achieved with our models depended on the exact composition of the individual models. Therefore, we developed two additional models with fixed features. The first model contains only those features which were selected in at least 50% of all runs, i.e., the first seven features listed in Table 2. The second model includes the 10 most important features in a fixed manner (i.e., "age_at_diagnosis" to "original_glszm_ZonePercentage", see Table 2). Based on the independent test data and the contrast-enhanced T1-weighted MRI images, we received the highest discriminative power with our 10-feature models. These two models were again developed 100 times with different data partitions, and the performance was subsequently determined using the independent test data. This time, however, the features included were fixed within the 100 runs.

The classification results for predicting *IDH* mutation of these 7- and 10-feature models are summarized in Table 3. According to the method used to calculate the results shown in Figure 2, the columns labeled "Different features" contain the results for the models with newly determined relevant features in each of the 100 runs. The columns labeled "Fixed features", on the other hand, contain the results for the models calculated with the fixed features according to Table 2. All results were again calculated with independent test data and as mean values of 100 runs. The numbers in brackets in the table indicate the 95% confidence intervals of the results. Comparing the 7-feature and the 10-feature models, it is evident that the additional three features yield only a very moderate increase in discriminatory power (see also Figure 2). The models show high performance in predicting the *IDH* mutation status. The 7-feature models yield the following average performance values (different features/fixed features): mean AUC = 74.3%/84.6%, mean accuracy = 73.1%/79.2%, mean sensitivity = 80.9%/84.7%, mean specificity = 57.6%/68.1%, mean PPV = 79.8%/84.8%, and mean NPV = 61.8%/71.0%. Our described methodology with 100 (or even more) runs is well suited to compare different approaches and to investigate the stability of the respective model approaches. The models with the fixed features show a slightly higher discriminatory power compared to the models with different features. The comparison of the widths of the individual confidence intervals indicates that the approach with the fixed features also leads to more stable results. Overall, however, both approaches (with different and fixed features) yield high discriminatory power in predicting the *IDH* mutation status.

**Table 3.** Mean discriminatory power (using 100 runs) in the prediction of *IDH* mutation status in gliomas using Lasso regression based on contrast-enhanced T1-weighted MRI images. The models include 7 and 10 features. Values in brackets indicate 95% confidence intervals. All results were calculated using independent test data based on models with newly determined features for each run, on the one hand, and fixed features according to Table 2, on the other.

| | Models with 7 Features | | Models with 10 Features | |
|---|---|---|---|---|
| | **Different Features** | **Fixed Features** | **Different Features** | **Fixed Features** |
| AUC | 0.743 [0.490:0.933] | 0.846 [0.634:0.984] | 0.757 [0.511:0.939] | 0.842 [0.630:0.984] |
| Accuracy | 0.731 [0.549:0.905] | 0.792 [0.596:0.905] | 0.743 [0.549:0.905] | 0.795 [0.619:0.905] |
| Sensitivity | 0.809 [0.643:1.000] | 0.847 [0.643:1.000] | 0.824 [0.643:1.000] | 0.844 [0.643:1.000] |
| Specificity | 0.576 [0.143:0.857] | 0.681 [0.286:1.000] | 0.580 [0.218:0.925] | 0.696 [0.286:1.000] |
| PPV | 0.798 [0.657:0.929] | 0.848 [0.680:1.000] | 0.801 [0.667:0.962] | 0.855 [0.714:1.000] |
| NPV | 0.618 [0.311:1.000] | 0.710 [0.355:1.000] | 0.641 [0.333:1.000] | 0.716 [0.429:1.000] |

## 4. Discussion

*IDH* mutation status has gained paramount importance in glioma classification since the 2021 WHO guidelines were released. It is not just a marker but a key determinant of clinical behavior and patient outcome. The accuracy of identifying *IDH* mutations non-invasively through MRI sequences holds immense clinical value.

While biopsy remains the gold standard for molecular diagnosis, it is often hindered by associated risks, including mortality, seizure, impaired consciousness, brain swelling, and hemorrhaging [37]. Moreover, in certain clinical scenarios, conducting a comprehensive MRI examination with all recommended sequences may not be feasible. Therefore, our study aimed to provide a non-invasive alternative for *IDH* mutation prediction, particularly catering to situations where a biopsy is impractical or MRI sequences are limited. The results we obtained shed light on the most effective MRI sequence for this crucial diagnostic task and underscore the critical role of contrast-enhanced T1-weighted MRI images.

Based on the results of our study, we specifically propose the use of contrast-enhanced T1-weighted MRI images for the non-invasive prediction of the *IDH* mutation status using machine learning algorithms.

Our patient cohort consisted of 106 individuals, comprising 71 with *IDH* mutations and 35 with *IDH* wild-type gliomas. We employed four common MRI sequences to extract radiomic features encompassing various data classes. To ensure the robustness of our findings, we randomly divided the patients into training and independent test groups through 100 iterations, maintaining a balance between *IDH*-mutated and wild-type cases. We leveraged four machine learning algorithms including Lasso regression, Random Forest, LDA, and Naïve Bayes for feature preselection and model development. Hyperparameters were optimized via 10-fold cross-validation. Model performance was rigorously assessed using independent test data. To mitigate overfitting, we determined the optimal number of features for each model and identified feature importance through the varImp function. In our study, we placed great emphasis on avoiding possible overfitting. Therefore, we developed each model 100 times using different training datasets and determined the average performance values during these 100 runs based on independent test data. We also reduced the number of features included in the final models as much as possible and analyzed how often these features were selected during the 100 runs (i.e., we ended up using only the best-ranked features). The importance of avoiding overfitting in machine learning is also discussed in more detail by Dietterich [38], for example.

In our comprehensive analysis, which involved various machine learning algorithms, Lasso regression consistently emerged as the superior model for predicting the MRI sequence, highlighting its unique diagnostic value for predicting the *IDH* mutation status. Among the tested MRI sequences in our study, contrast-enhanced T1-weighted images consistently demonstrated the highest suitability for predicting the *IDH* mutation status. This sequence consistently exhibited superior performance metrics, including an average AUC of 0.846, an average accuracy of 0.792, an average sensitivity of 0.847, and an average specificity of 0.681, as determined by the seven-feature model developed with Lasso regression. Our analysis also explored the impact of feature selection and model complexity. It was observed that increasing the number of features in the models up to 10 features led to improved performance, after which overfitting became a concern. Additionally, it was found that the administration of contrast agents significantly contributed to the enhancement of discriminatory power, highlighting the significance of this approach. Interestingly, human readers were unable to assess *IDH* mutation status by mere visually assessment in a subset of 20 patients with contrast-enhancing gliomas (sensitivity 100% (95% CI 69.15% to 100.00%) and specificity 0% (95% CI 0.00% to 30.85%)).

Our study also revealed that the FLAIR sequence is moderately suitable for this purpose, while the T1w and T2w native sequences were identified as the least suitable options. Comparing our findings to those in the existing literature (as highlighted in the introduction), it is noteworthy that there has been limited exploration on the identification of the most effective MRI sequence for the non-invasive prediction of the *IDH* mutation status in gliomas. Our study contributes significantly to this gap by demonstrating the superiority of contrast-enhanced T1w MRI images for this purpose.

The varied findings in the current literature underscore the complexity of the relationship between MRI sequences and the prediction of the *IDH* mutation status. The discrepancies across studies could be attributed to several factors. Firstly, the unique microenvironment of gliomas may influence the performance of different MRI sequences. Gliomas exhibit considerable heterogeneity in terms of vascularity and cellularity and the presence of necrotic regions, presenting challenges in accurately capturing these diverse characteristics through imaging modalities. It is possible that specific characteristics of our patient cohort favored the discriminative power of contrast-enhanced T1w images.

Secondly, the choice of imaging modality may be influenced by the radiomic features extracted and the machine learning algorithms employed. Recent advancements in radiomics have introduced various approaches to feature extraction, and the selection of optimal features may vary across studies. Additionally, the choice of machine learning algorithms can significantly impact the performance of predictive models. In our study, the consistent superiority of contrast-enhanced T1w images may be attributed to a combination of feature selection, algorithm choice, and specific radiomic characteristics captured by T1w images.

Despite this diversity, our study contributes to the ongoing discourse on the optimal choice of MRI sequences in glioma research and adds valuable insights into the evolving field of radiomics. As our study aligns with the therapeutic guidelines set by the German Association of Neurology and Neurosurgery, emphasizing the importance of the WHO's classification in determining the standard of care for gliomas, it further supports the integration of non-invasive techniques into routine clinical practice.

Regarding model stability, our study demonstrated that the selected features were robust across multiple runs, suggesting that our findings are reliable and not overly influenced by random effects associated with data partitioning.

As already mentioned in the introduction, the primary aim of our current study was neither to achieve the highest possible discriminatory power in the non-invasive prediction of the *IDH* mutation status using a conventional machine learning algorithm or a deep learning approach, nor to find the most suitable algorithm for this task. Rather, the aim of our study was to find out which of the commonly used MRI sequences is best suited for this important diagnostic task. New, promising medical image segmentation methods could lead to even more precise results for diagnostic questions such as the one considered in this study. Interesting ideas in this regard are semi-supervised medical image segmentation (see Zhao et al. [39]) and transformer-based approaches [40,41].

The molecular and histological characteristics of diffuse gliomas of the adult type are decisive for their classification, treatment strategies, and prognosis. The current standard treatment of gliomas is based on adhering to the guidelines established by the German Association of Neurology and Neurosurgery according to their WHO classification. This therapeutic approach typically involves a multimodal strategy encompassing surgical intervention or tumor reduction; chemotherapy with either procarbazine/lomustine/vincristine (PCV) or temozolomide (TMZ); and radiotherapy [42].

Of particular significance is the role of the *IDH* mutation status in gliomas. Beyond aiding in the initial classification of gliomas, the *IDH* mutation status holds predictive value for treatment responses. Notably, *IDH* mutations have been linked to heightened sensitivity to certain chemotherapeutic agents, such as Temozolomide [43]. This heightened sensitivity can translate into improved response rates and longer progression-free intervals in patients with *IDH*-mutated gliomas. This finding further underscores the critical importance of being able to predict *IDH* mutation status in gliomas.

It is worth noting that in our previous study by Nacul Mora et al. [44], we were able to successfully predict the *ATRX* status in gliomas using radiomics-based machine learning. Notably, T2-weighted MRI sequences emerged as being the most effective in predicting *ATRX* status in gliomas, indicating their pivotal role in molecular classification. This finding aligns with our observations regarding the potential of specific MRI sequences in providing valuable insights into the molecular landscape of gliomas.

Despite the valuable insights gained from our study, several limitations should be acknowledged. Firstly, due to the retrospective nature of our research, some patient data were incomplete, leading to the exclusion of 1 of the 107 patients. Additionally, while our study represents a substantial contribution to the field, further validation in larger cohorts and exploration of avenues for clinical translation are warranted to fully harness the clinical implications of these findings. An interesting way to further validate our results could also be the use of external databases such as the Rembrandt database (REpository for Molecular BRAin Neoplasia DaTa) of The Cancer Imaging Archive (TCIA). Finally, it should also be noted that the parameters used for image acquisition, such as repetition time and echo time, can influence the analysis results to a certain extent [45,46].

The incorporation of advanced MRI techniques, such as diffusion-weighted imaging and perfusion imaging, may further refine the accuracy of molecular predictions.

While our study focused on the efficacy of radiomics-based machine learning models in predicting the *IDH* mutation status using MRI images, we acknowledge the importance of considering the performance comparison with human experts. Unfortunately, our dataset did not include data on predictions made by human experts during the MRI assessment. Therefore, a direct comparison between the performance of our model and that of human experts was not feasible within the scope of this study. However, exploring this comparison in future research would provide valuable insights into the clinical utility of our model and its potential role alongside human expertise in tumor classification.

## 5. Conclusions

Our study sought to determine which of the commonly used MRI sequences, when coupled with radiomics-based machine learning models, holds the greatest potential for non-invasively predicting the *IDH* mutation status. Our retrospective analysis identified contrast-enhanced T1-weighted MRI images as the optimal MRI sequence for this purpose, yielding impressive performance metrics, with an accuracy of up to 79 percent. The administration of contrast agents significantly contributed to this enhanced predictive power. These findings have the potential to influence clinical practice by guiding neuro-oncologists toward a more precise and efficient approach to glioma characterization, ultimately benefitting patients through improved treatment strategies and outcomes.

**Author Contributions:** Conceptualization: M.M. (Manoj Mannil); methodology: M.M. (Manfred Musigmann); software: M.M. (Manfred Musigmann); validation: B.H.A. and M.M. (Manoj Mannil); formal analysis: D.N.G.K.; investigation: D.N.G.K., N.G.N.M. and D.A.B.; data curation: N.G.N.M. and B.H.A.; writing—original draft preparation: D.N.G.K. and M.M. (Manfred Musigmann); writing—review and editing: N.G.N.M.; visualization: M.M. (Manfred Musigmann); supervision: W.L.H. and M.M. (Manoj Mannil); project administration: M.M. (Manoj Mannil). All authors have read and agreed to the published version of the manuscript.Conceptualization: M.M. (Manoj Mannil); methodology: M.M. (Manfred Musigmann); software: M.M. (Manfred Musigmann); validation: B.H.A. and M.M. (Manoj Mannil); formal analysis: D.N.G.K.; investigation: D.N.G.K., N.G.N.M. and D.A.B.; data curation: N.G.N.M. and B.H.A.; writing—original draft preparation: D.N.G.K. and M.M. (Manfred Musigmann); writing—review and editing: N.G.N.M.; visualization: M.M. (Manfred Musigmann); supervision: W.L.H. and M.M. (Manoj Mannil); project administration: M.M. (Manoj Mannil). All authors have read and agreed to the published version of the manuscript.

**Funding:** This research received no external funding.

**Institutional Review Board Statement:** The study was conducted in accordance with the Declaration of Helsinki and approved by the local ethics committee (Ärztekammer Westfalen-Lippe, ÄKWL 2021-596-f-S, approval date: 21 September 2021).

**Informed Consent Statement:** Due to its retrospective nature, written informed consent was waived.

**Data Availability Statement:** Data available on request. The data presented in this study are available on reasonable request from the corresponding author.

**Conflicts of Interest:** The authors declare no conflicts of interest.

## References


1. Nicholson, J.G.; Fine, H.A. Diffuse Glioma Heterogeneity and Its Therapeutic Implications. *Cancer Discov.* **2021**, *11*, 575–590. [CrossRef]
2. Weller, M.; van den Bent, M.; Preusser, M.; Le Rhun, E.; Tonn, J.C.; Minniti, G.; Bendszus, M.; Balana, C.; Chinot, O.; Dirven, L.; et al. EANO guidelines on the diagnosis and treatment of diffuse gliomas of adulthood. *Nat. Rev. Clin. Oncol.* **2021**, *18*, 170–186. [CrossRef]
3. WHO Classification of Tumours Editorial Board. *Central Nervous System Tumours*, 5th ed.; WHO Classification of Tumours Series; International Agency for Research on Cancer: Lyon, France, 2021; Volume 6. Available online: https://publications.iarc.fr/601 (accessed on 2 October 2023).
4. Whitfield, B.T.; Huse, J.T. Classification of adult-type diffuse gliomas: Impact of the World Health Organization 2021 update. *Brain Pathol.* **2022**, *32*, e13062. [CrossRef]
5. Cancer Genome Atlas Research Network; Brat, D.J.; Verhaak, R.G.; Aldape, K.D.; Yung, W.K.; Salama, S.R.; Cooper, L.A.; Rheinbay, E.; Miller, C.R.; Vitucci, M.; et al. Comprehensive, Integrative Genomic Analysis of Diffuse Lower-Grade Gliomas. *N. Engl. J. Med.* **2015**, *372*, 2481–2498. [CrossRef] [PubMed]
6. Louis, D.N.; Perry, A.; Wesseling, P.; Brat, D.J.; Cree, I.A.; Figarella-Branger, D.; Hawkins, C.; Ng, H.K.; Pfister, S.M.; Reifenberger, G.; et al. The 2021 WHO Classification of Tumors of the Central Nervous System: A summary. *Neuro-Oncology* **2021**, *23*, 1231–1251. [CrossRef]
7. Eckel-Passow, J.E.; Lachance, D.H.; Molinaro, A.M.; Walsh, K.M.; Decker, P.A.; Sicotte, H.; Pekmezci, M.; Rice, T.; Kosel, M.L.; Smirnov, I.V.; et al. Glioma Groups Based on *1p/19q*, *IDH*, and TERT Promoter Mutations in Tumors. *N. Engl. J. Med.* **2015**, *372*, 2499–2508. [CrossRef] [PubMed]
8. Solomou, G.; Finch, A.; Asghar, A.; Bardella, C. Mutant *IDH* in Gliomas: Role in Cancer and Treatment Options. *Cancers* **2023**, *15*, 2883. [CrossRef]
9. Franceschi, E.; Tosoni, A.; Bartolini, S.; Minichillo, S.; Mura, A.; Asioli, S.; Bartolini, D.; Gardiman, M.; Gessi, M.; Ghimenton, C.; et al. Histopathological grading affects survival in patients with *IDH*-mutant grade II and grade III diffuse gliomas. *Eur. J. Cancer* **2020**, *137*, 10–17. [CrossRef]
10. Stupp, R.; Mason, W.P.; Van Den Bent, M.J.; Weller, M.; Fisher, B.; Taphoorn, M.J.; Belanger, K.; Brandes, A.A.; Marosi, C.; Bogdahn, U.; et al. Radiotherapy plus concomitant and adjuvant temozolomide for glioblastoma. *N. Engl. J. Med.* **2005**, *352*, 987–996. [CrossRef] [PubMed]
11. Brown, N.F.; Ottaviani, D.; Tazare, J.; Gregson, J.; Kitchen, N.; Brandner, S.; Fersht, N.; Mulholland, P. Survival Outcomes and Prognostic Factors in Glioblastoma. *Cancers* **2022**, *14*, 3161. [CrossRef] [PubMed]
12. Nandakumar, P.; Mansouri, A.; Das, S. The Role of *ATRX* in Glioma Biology. *Front. Oncol.* **2017**, *7*, 236. [CrossRef]
13. Antonelli, M.; Poliani, P.L. Adult type diffuse gliomas in the new 2021 WHO Classification. *Pathologica* **2022**, *114*, 397–409. [CrossRef]
14. Pang, Y.; Chen, X.; Ji, T.; Cheng, M.; Wang, R.; Zhang, C.; Liu, M.; Zhang, J.; Zhong, C. The Chromatin Remodeler *ATRX*: Role and Mechanism in Biology and Cancer. *Cancers* **2023**, *15*, 2228. [CrossRef] [PubMed]
15. Bonm, A.V.; Ritterbusch, R.; Throckmorton, P.; Graber, J.J. Clinical Imaging for Diagnostic Challenges in the Management of Gliomas: A Review. *J. Neuroimaging Off. J. Am. Soc. Neuroimaging* **2020**, *30*, 139–145. [CrossRef]
16. Zaccagna, F.; Grist, J.T.; Quartuccio, N.; Riemer, F.; Fraioli, F.; Caracò, C.; Halsey, R.; Aldalilah, Y.; Cunningham, C.H.; Massoud, T.F. Imaging and treatment of brain tumors through molecular targeting: Recent clinical advances. *Eur. J. Radiol.* **2021**, *142*, 109842. [CrossRef] [PubMed]
17. Kamepalli, H.K.; Kalaparti, V.; Kesavadas, C. Imaging Recommendations for the Diagnosis, Staging, and Management of Adult Brain Tumors. *Ind. J. Med. Pädiatrie Oncol.* **2023**, *44*, 26–38. [CrossRef]
18. Ellingson, B.M.; Wen, P.Y.; Cloughesy, T.F. modified criteria for radiographic response assessment in glioblastoma clinical trials. *Neurotherapeutics* **2017**, *14*, 307–320. [CrossRef]
19. Suh, C.H.; Kim, H.S.; Jung, S.C.; Choi, C.G.; Kim, S.J. 2-Hydroxyglutarate MR spectroscopy for prediction of isocitrate dehydrogenase mutant glioma: A systemic review and meta-analysis using individual patient data. *Neuro-Oncology* **2018**, *20*, 1573–1583. [CrossRef]
20. Bauer, J.; Raum, H.; Kugel, H.; Müther, M.; Mannil, M.; Heindel, W. 2-Hydroxyglutarate as an MR spectroscopic predictor of an *IDH* mutation in gliomas. *Rofo*, 2024; *in press*.
21. Yi, Z.; Long, L.; Zeng, Y.; Liu, Z. Current Advances and Challenges in Radiomics of Brain Tumors. *Front. Oncol.* **2021**, *11*, 732196. [CrossRef]
22. Mayerhoefer, M.E.; Materka, A.; Langs, G.; Häggström, I.; Szczypiński, P.; Gibbs, P.; Cook, G. Introduction to Radiomics. *J. Nucl. Med. Off. Publ. Soc. Nucl. Med.* **2020**, *61*, 488–495. [CrossRef]
23. Moon, S.H.; Kim, J.; Joung, J.G.; Cha, H.; Park, W.Y.; Ahn, J.S.; Ahn, M.J.; Park, K.; Choi, J.Y.; Lee, K.H.; et al. Correlations between metabolic texture features, genetic heterogeneity, and mutation burden in patients with lung cancer. *Eur. J. Nucl. Med. Mol. Imaging* **2019**, *46*, 446–454. [CrossRef] [PubMed]

24. Choi, E.R.; Lee, H.Y.; Jeong, J.Y.; Choi, Y.L.; Kim, J.; Bae, J.; Lee, K.S.; Shim, Y.M. Quantitative image variables reflect the intratumoral pathologic heterogeneity of lung adenocarcinoma. *Oncotarget* **2016**, *7*, 67302. [CrossRef] [PubMed]
25. Alves, A.F.F.; Miranda, J.R.D.A.; Reis, F.; Souza, S.A.S.D.; Alves, L.L.R.; Feitoza, L.D.M.; Castro, J.T.D.S.D.; Pina, D.R.D. Inflammatory Lesions and Brain Tumors: Is It Possible to Differentiate Them Based on Texture Features in Magnetic Resonance Imaging? *J. Venom. Anim. Toxins Incl. Trop. Dis.* **2020**, *26*, e20200011. [CrossRef]
26. He, A.; Wang, P.; Zhu, A.; Liu, Y.; Chen, J.; Liu, L. Predicting *IDH* Mutation Status in Low-Grade Gliomas Based on Optimal Radiomic Features Combined with Multi-Sequence Magnetic Resonance Imaging. *Diagnostics* **2022**, *12*, 2995. [CrossRef]
27. Liu, Y.F.; Shu, X.; Qiao, X.F.; Ai, G.Y.; Liu, L.; Liao, J.; Qian, S.; He, X.J. Radiomics-Based Machine Learning Models for Predicting P504s/P63 Immunohistochemical Expression: A Noninvasive Diagnostic Tool for Prostate Cancer. *Front. Oncol.* **2022**, *12*, 911426. [CrossRef]
28. Xie, D.; Xu, F.; Zhu, W.; Pu, C.; Huang, S.; Lou, K.; Wu, Y.; Huang, D.; He, C.; Hu, H. Delta radiomics model for the prediction of progression-free survival time in advanced non-small-cell lung cancer patients after immunotherapy. *Front. Oncol.* **2022**, *12*, 990608. [CrossRef] [PubMed]
29. Li, J.; Liu, S.; Qin, Y.; Zhang, Y.; Wang, N.; Liu, H. High-order radiomics features based on T2 FLAIR MRI predict multiple glioma immunohistochemical features: A more precise and personalized gliomas management. *PLoS ONE* **2020**, *15*, e0227703. [CrossRef]
30. Li, Y.; Ammari, S.; Lawrance, L.; Quillent, A.; Assi, T.; Lassau, N.; Chouzenoux, E. Radiomics-Based Method for Predicting the Glioma Subtype as Defined by Tumor Grade, *IDH* Mutation, and *1p/19q* Codeletion. *Cancers* **2022**, *14*, 1778. [CrossRef]
31. Choi, Y.S.; Bae, S.; Chang, J.H.; Kang, S.-G.; Kim, S.H.; Kim, J.; Rim, T.H.; Choi, S.H.; Jain, R.; Lee, S.-K. Fully Automated Hybrid Approach to Predict the *IDH* Mutation Status of Gliomas via Deep Learning and Radiomics. *Neuro-Oncology* **2021**, *23*, 304–313. [CrossRef]
32. Zhang, X.; Tian, Q.; Wang, L.; Liu, Y.; Li, B.; Liang, Z.; Gao, P.; Zheng, K.; Zhao, B.; Lu, H. Radiomics Strategy for Molecular Subtype Stratification of Lower-Grade Glioma: Detecting *IDH* and TP53 Mutations Based on Multimodal MRI. *J. Magn. Reson. Imaging* **2018**, *48*, 916–926. [CrossRef]
33. Lu, J.; Xu, W.; Chen, X.; Wang, T.; Li, H. Noninvasive Prediction of *IDH* Mutation Status in Gliomas Using Preoperative Multiparametric MRI Radiomics Nomogram: A Mutlicenter Study. *Magn. Reson. Imaging* **2023**, *104*, 72–79. [CrossRef] [PubMed]
34. Huang, W.; Wen, L.; Wu, G.; Hu, M.; Zhang, C.; Chen, F.; Zhao, J. Comparison of Radiomics Analyses Based on Different Magnetic Resonance Imaging Sequences in Grading and Molecular Genomic Typing of Glioma. *J. Comput. Assist. Tomogr.* **2021**, *45*, 110. [CrossRef] [PubMed]
35. Han, L.; Wang, S.; Miao, Y.; Shen, H.; Guo, Y.; Xie, L.; Shang, Y.; Dong, J.; Li, X.; Wang, W.; et al. MRI Texture Analysis Based on 3D Tumor Measurement Reflects the *IDH1* Mutations in Gliomas—A Preliminary Study. *Eur. J. Radiol.* **2019**, *112*, 169–179. [CrossRef] [PubMed]
36. Musigmann, M.; Nacul, N.G.; Kasap, D.N.; Heindel, W.; Mannil, M. Use Test of Automated Machine Learning in Cancer Diagnostics. *Diagnostics* **2023**, *13*, 2315. [CrossRef] [PubMed]
37. Riche, M.; Amelot, A.; Peyre, M.; Capelle, L.; Carpentier, A.; Mathon, B. Complications after frame-based stereotactic brain biopsy: A systematic review. *Neurosurg. Rev.* **2021**, *44*, 301–307. [CrossRef] [PubMed]
38. Dietterich, T. Overfitting and Undercomputing in Machine Learning. *ACM Comput. Surv.* **1995**, *27*, 326–327. [CrossRef]
39. Zhao, C.; Xiang, S.; Wang, Y.; Cai, Z.; Shen, J.; Zhou, S.; Zhao, D.; Su, W.; Guo, S.; Li, S. Context-Aware Network Fusing Transformer and V-Net for Semi-Supervised Segmentation of 3D Left Atrium. *Expert Syst. Appl.* **2023**, *214*, 119105. [CrossRef]
40. Cao, H.; Wang, Y.; Chen, J.; Jiang, D.; Zhang, X.; Tian, Q.; Wang, M. Swin-Unet: Unet-like Pure Transformer for Medical Image Segmentation. *arXiv* **2021**, arXiv:2105.05537. [CrossRef]
41. Lin, A.; Chen, B.; Xu, J.; Zhang, Z.; Lu, G. DS-TransUNet:Dual Swin Transformer U-Net for Medical Image Segmentation. *arXiv* **2021**, arXiv:2106.06716. [CrossRef]
42. Wick, W.; Bendszus, M.; Goldbrunner, R.; Grosu, A.; Hattingen, E.; Hau, P.; Herrlinger, U.; Kessler, T.; Platten, M.; Pukrop, T. Gliome, S2k-Leitlinie, 2021. In *Leitlinien für Diagnostik und Therapie in der Neurologie*; Deutsche Gesellschaft für Neurologie, Ed.; Georg Thieme Verlag: Stuttgart, Germany, 2021; Available online: www.dgn.org/leitlinien (accessed on 9 October 2023).
43. Lin, L.; Cai, J.; Tan, Z.; Meng, X.; Li, R.; Li, Y.; Jiang, C. Mutant *IDH1* Enhances Temozolomide Sensitivity via Regulation of the ATM/CHK2 Pathway in Glioma. *Cancer Res. Treat.* **2021**, *53*, 367–377. [CrossRef]
44. Nacul Mora, N.G.; Akkurt, B.H.; Kasap, D.; Blömer, D.; Heindel, W.; Mannil, M.; Musigmann, M. Comparison of MRI Sequences to Predict *ATRX* Status Using Radiomics-Based Machine Learning. *Diagnostics* **2023**, *13*, 2216. [CrossRef] [PubMed]
45. Waugh, S.A.; Purdie, C.A.; Jordan, L.B.; Vinnicombe, S.; Lerski, R.A.; Martin, P.; Thompson, A.M. Magnetic Resonance Imaging Texture Analysis Classification of Primary Breast Cancer. *Eur. Radiol.* **2016**, *26*, 322–330. [CrossRef] [PubMed]
46. Mayerhoefer, M.E.; Szomolanyi, P.; Jirak, D.; Berg, A.; Materka, A.; Dirisamer, A.; Trattnig, S. Effects of Magnetic Resonance Image Interpolation on the Results of Texture-Based Pattern Classification: A Phantom Study. *Investig. Radiol.* **2009**, *44*, 405–411. [CrossRef] [PubMed]

biomedicines

MDPI

Article

# Validating Brain Tumor Reporting and Data System (BT-RADS) as a Diagnostic Tool for Glioma Follow-Up after Surgery

Yassir Edrees Almalki [1], Mohammad Abd Alkhalik Basha [2,*], Maha Ibrahim Metwally [2], Nesma Adel Zeed [2], Mohamad Gamal Nada [2], Sharifa Khalid Alduraibi [3], Ahmed A. Morsy [4], Rawda Balata [5], Ahmed Z. Al Attar [5], Mona M. Amer [6], Mohamed Abd El-Aziz Mohamed Farag [7], Sameh Abdelaziz Aly [8], Ahmed M. Abdelkhalik Basha [9] and Enas Mahmoud Hamed [2]

1 Division of Radiology, Department of Internal Medicine, Medical College, Najran University, Najran 61441, Saudi Arabia; yealmalki@nu.edu.sa
2 Department of Diagnostic Radiology, Faculty of Human Medicine, Zagazig University, Zagazig 44519, Egypt; mimetwally@zu.edu.eg (M.I.M.); nonyzeed1984@gmail.com (N.A.Z.); mohammedgamalnada.mn@gmail.com (M.G.N.); enashamedmya@gmail.com (E.M.H.)
3 Department of Radiology, College of Medicine, Qassim University, Buraidah 52571, Saudi Arabia; salduraibi@qu.edu.sa
4 Department of Neurosurgery, Faculty of Human Medicine, Zagazig University, Zagazig 44519, Egypt; dr.ahmed.ali.morsy@gmail.com
5 Department of Clinical Oncology and Nuclear Medicine, Faculty of Human Medicine, Zagazig University, Zagazig 44519, Egypt; drrawdabalata@gmail.com (R.B.); ahmedalata982@gmail.com (A.Z.A.A.)
6 Department of Neurology, Faculty of Human Medicine, Zagazig University, Zagazig 44519, Egypt; monaamer2345@gmail.com
7 Department of Radio-Diagnosis, Faculty of Human Medicine, Al Azhar University, Cairo 11884, Egypt; mabdelaziz43@yahoo.com
8 Department of Diagnostic Radiology, Faculty of Human Medicine, Benha University, Benha 13511, Egypt; drsamehaly75@gmail.com
9 Faculty of General Medicine, Saint Petersburg State University, Egypt Branch, Cairo 11646, Egypt; ahm7edbasha@gmail.com
* Correspondence: mohammad_basha76@yahoo.com

**Abstract:** Gliomas are a type of brain tumor that requires accurate monitoring for progression following surgery. The Brain Tumor Reporting and Data System (BT-RADS) has emerged as a potential tool for improving diagnostic accuracy and reducing the need for repeated operations. This prospective multicenter study aimed to evaluate the diagnostic accuracy and reliability of BT-RADS in predicting tumor progression (TP) in postoperative glioma patients and evaluate its acceptance in clinical practice. The study enrolled patients with a history of partial or complete resection of high-grade glioma. All patients underwent two consecutive follow-up brain MRI examinations. Five neuroradiologists independently evaluated the MRI examinations using the BT-RADS. The diagnostic accuracy of the BT-RADS for predicting TP was calculated using histopathology after reoperation and clinical and imaging follow-up as reference standards. Reliability based on inter-reader agreement (IRA) was assessed using kappa statistics. Reader acceptance was evaluated using a short survey. The final analysis included 73 patients (male, 67.1%; female, 32.9%; mean age, 43.2 ± 12.9 years; age range, 31–67 years); 47.9% showed TP, and 52.1% showed no TP. According to readers, TP was observed in 25–41.7% of BT-3a, 61.5–88.9% of BT-3b, 75–90.9% of BT-3c, and 91.7–100% of BT-RADS-4. Considering >BT-RADS-3a as a cutoff value for TP, the sensitivity, specificity, and accuracy of the BT-RADS were 68.6–85.7%, 84.2–92.1%, and 78.1–86.3%, respectively, according to the reader. The overall IRA was good ($\kappa = 0.75$) for the final BT-RADS classification and very good for detecting new lesions ($\kappa = 0.89$). The readers completely agreed with the statement "the application of the BT-RADS should be encouraged" (score = 25). The BT-RADS has good diagnostic accuracy and reliability for predicting TP in postoperative glioma patients. However, BT-RADS 3 needs further improvements to increase its diagnostic accuracy.

**Keywords:** glioma; magnetic resonance imaging; BT-RADS; follow-up; tumor progression

## 1. Introduction

Glioma is the most common primary brain tumor, with the highest grade, grade IV glioma, or glioblastoma, representing the most common form in adults and having a particularly poor prognosis [1]. There are multimodal treatments, including radiotherapy, chemotherapy, and surgery; however, the results remain inadequate [2,3]. The poor prognosis associated with glioblastoma emphasizes the need for precise monitoring of tumor progression (TP) and treatment response [4,5].

Magnetic resonance imaging (MRI) is essential for assessing brain tumors during preoperative and postoperative periods [5,6]. Brain tumor MRI findings, however, are notably challenging to interpret since true TP may look similar to treatment effects, leading to ambiguous interpretations and difficult management decisions. Consequently, it is challenging for radiologists to assess the treatment response of brain tumors [5,7,8]. There are currently several guidelines for assessing the therapeutic response of brain tumors, including the World Health Organization, Levin, MacDonald, and Response Assessment in Neuro-Oncology (RANO), which are more often used in clinical trials. Due to reasons such as complexity, requirement for repeated measures, significant interobserver variability, and physician misunderstanding, their use in radiology has been limited to date [5,7,9–12]. To address these shortcomings in current practice, a team of neuroradiologists, neuro-oncologists, neurosurgeons, and radiation oncologists created the Brain Tumor Reporting and Data System (BT-RADS) in 2018 [13].

A BT-RADS is a structured glioma surveillance reporting system that uses MR imaging patterns, clinical evaluation, and therapeutic scheduling to give each report a numeric category from 0 to 4 based on the possibility of TP. Each category is associated with a specific management recommendation. Categorization is based on the changes in four MR imaging patterns: enhancing components, FLAIR components, mass effects, and new lesions compared with the most recent prior brain MRI [12–14]. In this scenario, the BT-RADS has been shown to improve the clarity, consistency, and confidence ratings of referral providers in radiology reports and facilitate patient management decisions [15]. Although qualitative and quantitative improvements for the BT-RADS have been reported [14–16], the complete value of the BT-RADS still needs to be evaluated. Subsequently, we carried out this multi-institutional prospective study to assess the diagnostic accuracy and reliability of the BT-RADS in predicting TP in patients with post-treated glioma. Furthermore, we conducted a simple survey to assess readers' agreement with this classification.

## 2. Methods

### 2.1. Ethical Statement

The study was authorized by the Institutional Review Board (approval number: ZU-9857), and all participants signed informed consent. The study was conducted in accordance with the principles of the Declaration of Helsinki.

### 2.2. Eligibility Criteria

Between October 2021 and August 2023, we identified 90 postoperative glioma patients at three academic institutions. The inclusion criteria were (i) age $\geq$ 18 years and (ii) recent total or subtotal high-grade glioma (HGG) resection. Once enrolled, all patients underwent a series of follow-up MRI scans. The exclusion criteria were (i) patients categorized as BT-RADS-0 (n = 5), (ii) patients lost during the follow-up period (n = 7), (iii) patients with poor-quality contrast-enhanced MR images (n = 2), and (iv) patients with postoperative surgical bed ischemic insult (n = 3).

### 2.3. Postoperative Clinical Assessment

All patients were clinically evaluated by neurosurgeons with >10 years of clinical experience. All neurosurgeons were blinded to the imaging findings but assessed the clinical status of patients according to the Neurological Assessment in Neuro-Oncology (NANO) scale [17]. Nine neurological domains were tested for each patient: upper extremity ataxia, visual field, strength, facial strength, gait, language, sensation, behavior, and degree of consciousness. A score range of 0–2 was given to each domain. Finally, neurosurgeons reported that each patient's clinical state was clinically stable, improved, or worsened.

### 2.4. MRI Examination

Brain MRI examinations were conducted using a 1.5-T MRI system (Optima 450 GEM, GE Healthcare, Chicago, IL, USA, and Achieva class IIa, Philips Medical Systems, Best, Netherlands) with an 8-channel head coil. The acquired MRI sequences included axial fluid attenuation inversion recovery (FLAIR) (TE = 130, TR = 8500, TI = 2509, FOV = 24, matrix = 256 × 192, slice thickness = 5, bandwidth = 31.25), axial fast release fast spin echo (FrFSE) T2WI (TE = 102, TR = 4430, FOV = 24, matrix = 256 × 224, slice thickness = 5, bandwidth = 31.25), coronal FrFSE T2WI (TE = 120, TR = 4597, FOV = 22, matrix = 320 × 224, slice thickness = 2, bandwidth = 22.73), and axial pre-contrast and multiplanar post-contrast FSE T1WI (TE = Min Full, TR = 480, FOV = 24, matrix = 192 × 288, slice thickness = 5) after intravenous injection of 0.1 mmol/kg of gadopentetate dimeglumine (2 mL/s).

### 2.5. MRI Image Analysis

Images were transferred to workstations and analyzed using a dedicated platform-extended workstation (Advantage Workstation, GE Healthcare, Chicago, Illinois, United States, Philips Medical System, Best, The Netherlands, or PaxeraUltima, Paxera Viewer version 5.0.9.6, PaxeraHealth, Newtone, MA, USA). Five neuroradiologists (Y.E.A., M.I.M., N.A.Z., E.M.H., and M.G.N. with 16, 15, 14, 14, and 12 years of experience in neuroimaging, respectively) independently evaluated the initial two consecutive MR images. Radiologists were aware of the patients' preoperative images, operative data, received treatment, and clinical states. The radiologists received three hours of detailed education on the BT-RADS before the start of the study, which included 20 practical cases chosen from a different time period than the study population. On the MRI images, the following features were assessed for each postoperative glioma: enhancing component (unchanged, decrease, increase <25%, or increase >25%), FLAIR component (unchanged, decrease, increase <25%, or increase >25%), mass effect (unchanged, decrease, or increase), and new lesion (no, yes (indeterminant), or yes (definite)). Finally, each radiologist independently assigned a BT-RADS category for each postoperative glioma using the BT-RADS developed by Weinberg et al. [13] to predict TP based on MRI features and the clinical state of the patients. The BT-RADS categories are described in Table 1.

**Table 1.** BT-RADS categories and description.

| Category | Description | Imaging Patterns | Management Recommendation |
|---|---|---|---|
| BT-RADS-0 | No score. | New baseline, incomplete study, or inability to categorize. | Continued follow-up, no change. |
| BT-RADS-1a | Improvement suspected due to a decrease in TP and/or treatment effects. Clinically stable or improved. | Reduction in enhancing component, FLAIR component, mass effect, or resolution of lesions compared with prior MRI. | Continued follow-up, no change. |
| BT-RADS-1b | Improvement potentially due to treatment effect. Clinically stable or improved. | Reduction in enhancing component, FLAIR component, mass effect, or resolution of lesions compared with prior MRI. | Continued follow-up, no change. |

**Table 1.** *Cont.*

| Category | Description | Imaging Patterns | Management Recommendation |
|---|---|---|---|
| BT-RADS-2 | No significant change. Clinically stable. | No substantial change in enhancing component, FLAIR component, mass effect, or new lesions compared with prior MRI. | Continued follow-up, no change. |
| BT-RADS-3a | Worsening may represent the treatment effect. Clinically stable. | Mild increase (<25%) in enhancing component, FLAIR component, or mass effect compared with prior MRI. | Decreased time interval of follow-up. |
| BT-RADS-3b | Indeterminate. Worsening may be a mix of TP and treatment effects. Clinically stable. | Moderate increase in enhancing component, FLAIR component, mass effect, or new lesion compared with prior MRI. | Decreased time interval of follow-up. |
| BT-RADS-3c | Worsening favors TP. Clinically worsening. | Significant increase (>25%) in enhancing component, FLAIR component, mass effect, or definite new lesion compared with prior MRI. | Change in management vs. decreased time interval of follow-up. |
| BT-RADS-4 | Worsening, highly suspicious for TP. Clinically worsening. | Substantial increase in enhancing component, FLAIR component, mass effect, and/or multiple new lesions compared with prior MRI. | Change in management. |

MRI = magnetic resonance imaging; TP = tumor progression; FLAIR = Fluid attenuated inversion recovery.

### 2.6. Reference Standards

The final diagnosis of TP was established based on several factors. The first factor was (a) clinical and imaging follow-up (n = 46): each patient had serial follow-up MRIs (at least two) after the initial two consecutive MRIs. TP was based on follow-up MRIs reviewed by a multidisciplinary brain tumor team, including neuroradiology, neuro-oncology, radiation oncology, and neurosurgery members. The team was blinded to the patients' previous imaging findings and clinical data. The neuroradiology team consisted of two neuroradiologists (S.A.A. and M.A.A.B. with over 20 years of experience in neuroimaging). They consensually assessed the follow-up MRI findings. Clinical follow-up was performed according to the NANO scale. TP was defined as a progressive increase in enhancing or FLAIR component, increased mass effect, or the appearance of a new lesion on follow-up MRIs. Non-progression was defined as the stability or improvement of lesions on follow-up MRIs for >6 months [18,19]. (b) The second factor was histopathological examination after repeating surgical resection or stereotactic biopsy (n = 27): biopsy-proven TP was required if imaging findings showed definite TP. An experienced neuropathologist blinded to the imaging findings reported the pathological results. More than 75% of tumor cells in tissue specimens indicated TP, whereas none or scanty tumor cells (<25%) indicated non-progression [20].

### 2.7. Reader Acceptance

The radiologists were asked to complete a brief survey after reviewing all MR images to assess BT-RADS acceptance. A 5-point ordinal scale was used to evaluate the response (one = strongly disagree; two = disagree; three = neither agree nor disagree; four = agree; five = strongly agree). The scores were calculated as the sum of each reader's points.

### 2.8. Statistical Analysis

The collected data were analyzed using SPSS version 26 (IBM, Armonk, NY, USA) or MedCalc version 15.8 (Mariakerke, Belgium). Continuous variables were described using means and standard deviations, whereas categorical variables were described using numbers and percentages. Using the receiver operating characteristic (ROC) curve, we established the best cutoff value of the BT-RADS that predicted TP and calculated its diagnostic accuracy based on imaging, clinical follow-up, and pathological results. The agreement between the radiologists regarding the final BT-RADS category and MR features was calculated using kappa (κ) statistics. A κ value of 0.00–0.20 indicated poor agreement; 0.21–0.40 indicated fair agreement; 0.41–0.60 indicated moderate agreement; 0.61–0.80 indicated good agreement; 0.81–1.00 indicated very good agreement. Statistical significance was determined by a $p$-value of 0.05.

## 3. Results

### 3.1. Patients and Glioma Characteristics

The final analysis of our study involved 73 patients (male, 67.1%; female, 32.9%; mean age, 43.2 ± 12.9; age range, 31–67 years) with 322 MRI examinations. The clinicopathological features of the patients and their gliomas are summarized in Table 2. The mean follow-up period of the study was 15.3 ± 4.8 months (range, 8–25 months). The mean interval to the first postoperative MRI scan was 4.3 ± 1.5 months (range, 1–10 months). The mean interval between the first and second postoperative MRI scans was 3.8± 0.9 months (2–7 months). Grade III gliomas were pathologically confirmed in 31 patients, whereas 42 patients had grade IV gliomas. Twenty-seven patients underwent histopathological examination after stereotactic biopsy or repeat surgery, and 42 underwent clinical and imaging follow-up. The final assessment of gliomas showed 35 patients with TP and 38 without TP. Figure 1 demonstrates the flowchart of the study.

**Table 2.** Clinicopathologic characteristics of patients and gliomas.

| Characteristic | Value |
|---|---|
| Total No. of patients | 73 |
| Age, (years), mean ± SD (range) | 43.2 ± 12.9 (31–67) |
| Sex | |
| Male | 49 (67.1) |
| Female | 24 (32.9) |
| Total No. of MRI | 322 |
| Follow-up period (months), mean ± SD (range) | 15.3± 4.8 (8–25) |
| Mean interval to the first postoperative MRI scan (months), mean ± SD (range) | 4.3± 1.5 (1–10) |
| Mean interval between first and second postoperative MRI scan (months), mean ± SD (range) | 3.8± 0.9 (2–7) |
| Maximum tumor diameter (mm), mean ± SD (range) | 57.5 ± 17.8 (28–90) |
| Primary surgery | |
| Total resection | 60 (82.2) |
| Subtotal resection | 13 (17.8) |
| Tumor grade | |
| Grade III | 31 (42.5) |
| Grade IV | 42 (57.5) |
| Treatment | |
| Cortisone intake | 28 (38.4) |
| Radiotherapy | 73 (100) |
| Chemotherapy | 62 (84.9) |
| Methods of final diagnosis | |
| Repeat surgical resection | 27 (37) |
| Clinical/imaging follow-up | 46 (63) |
| Final diagnosis | |
| TP | 35 (47.9) |
| No TP | 38 (52.1) |

Unless otherwise indicated, data are the number of patients with percentages in parentheses. MRI = magnetic resonance imaging; SD = standard deviation; TP = tumor progression.

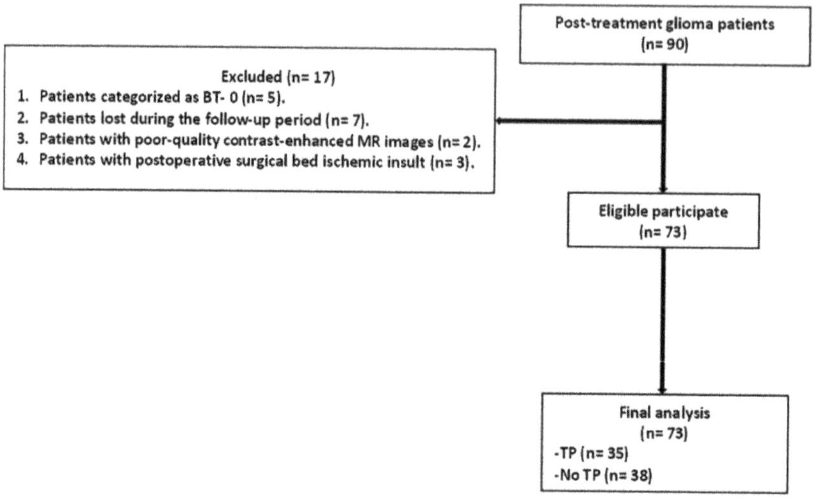


**Figure 1.** Flowchart of the study.

### 3.2. Distributions of BT-RADS Categories

Table 3 shows the frequency distributions of the assigned BT-RADS categories after the initial two postoperative MRIs, stratified by reader and final diagnosis (TP or no TP). Across the five readers, the number of gliomas assigned to each category ranged as follows: 7–11 for BT-RADS-1a, 9–12 for BT-RADS-1b, 12–14 for BT-RADS-2, 4–12 for BT-RADS-3a, 8–15 for BT-RADS-3b, 4–11 for BT-RADS-3c, and 7–17 for BT-RADS-4.

**Table 3.** Frequency distributions of BT-RADS categories of gliomas after the initial two postoperative MRI examinations stratified by readers and final diagnosis.

| Category | R 1 | | R 2 | | R 3 | | R 4 | | R 5 | |
|---|---|---|---|---|---|---|---|---|---|---|
| | TP | No TP | TP | No TP | TP | No TP | TP | No TP | TP | No TP |
| 1a | 1 (2.9) | 6 (15.8) | 0 (0) | 8 (21.1) | 2 (5.7) | 9 (23.7) | 1 (2.9) | 8 (21.1) | 0 (0) | 10 (26.3) |
| 1b | 3 (8.6) | 9 (23.7) | 1 (2.9) | 8 (21.1) | 1 (2.9) | 8 (21.1) | 1 (2.9) | 10 (26.3) | 2 (5.7) | 9 (23.7) |
| 2 | 2 (5.7) | 12 (31.6) | 3 (8.6) | 9 (23.7) | 3 (8.6) | 9 (23.7) | 4 (11.4) | 9 (23.7) | 2 (5.7) | 11 (28.9) |
| 3a | 5 (14.3) | 7 (18.4) | 3 (8.6) | 8 (21.1) | 4 (2.9) | 6 (15.8) | 3 (8.6) | 8 (21.1) | 1 (2.9) | 3 (7.9) |
| 3b | 8 (22.9) | 1 (2.6) | 11 (31.4) | 4 (10.5) | 8 (22.9) | 5 (13.2) | 7 (20) | 1 (2.6) | 11 (31.4) | 3 (7.9) |
| 3c | 5 (14.3) | 2 (5.3) | 10 (28.6) | 1 (2.6) | 8 (22.9) | 1 (2.6) | 6 (17.1) | 2 (5.3) | 3 (8.6) | 1 (2.6) |
| 4 | 11 (31.4) | 1 (2.6) | 7 (20) | 0 (0) | 9 (25.7) | 0 (0) | 13 (37.1) | 0 (0) | 16 (45.7) | 1 (2.6) |

Data are the number of gliomas with percentages in parentheses. BT-RADS = brain tumor reporting and data system; MRI = magnetic resonance imaging; R = reader; TP = tumor progression.

### 3.3. Incidence of TP across BT-RADS Categories

The incidence of pathologically or clinically confirmed TP increased with higher assigned BT-RADS categories (Table 3). According to the readers, the proportion of patients with TP ranged from 0–18.2% for BT-RADS-1a, 10–25% for BT-RADS-1b, 14.3–30.8% for BT-RADS-2, 25–41.7% for BT-RADS-3a, 61.5–88.9% for BT-RADS-3b, 75–90.9% for BT-RADS-3c, and 91.7–100% for BT-RADS-4.

### 3.4. Diagnostic Accuracy of BT-RADS

Table 4 presents the ROC analysis using different BT-RADS cutoff categories to predict TP status. ROC analysis demonstrated that the BT-RASDS classification had lower sensitivities with the worsening categories (BT-RADS-3 and 4), and the best cutoff value for predicting TP was the BT-RADS-3a category or higher. According to the reader, when >BT-RADS-3a was used as a predictor for TP, the sensitivity, specificity, accuracy, positive predictive value (PPV), and negative predictive value (NPV) of the BT-RADS were 68.6–85.7%, 84.2–92.1%, 78.1–86.3%, 80.6–89.7%, and 75.6–86.8%, respectively (Table 5).

**Table 4.** ROC analysis for TP prediction at BT-RADS category cutoff values.

| Category | Sensitivity | Specificity |
|---|---|---|
| ≥BT-RADS-1a | 100 | 0.0 |
| >BT-RADS-1a | 94.3–100 | 15.8–26.3 |
| >BT-RADS-1b | 88.6–97.1 | 39.5–50.0 |
| >BT-RADS-2 | 68.6–88.6 | 65.8–79.0 |
| >BT-RADS-3a * | 68.6–85.7 | 84.2–92.1 |
| >BT-RADS-3b | 45.7–54.3 | 92.1–97.4 |
| >BT-RADS-3c | 17.1–45.7 | 94.7–97.4 |
| >BT-RADS-4 | 0.0 | 100.0 |

Data are the range of percentages according to the reviewers. ROC = receiver operating characteristic; BT-RADS = Brain Tumor Reporting and Data System; TP = tumor progression; * = the optimal cutoff value that maximized the average of sensitivity and specificity.

**Table 5.** Diagnostic accuracy of BT-RADS for predicting TP stratified by readers.

| Parameters | R 1 | R 2 | R 3 | R 4 | R 5 |
|---|---|---|---|---|---|
| Cutoff | >BT-RADS-3a | >BT-RADS-3a | >BT-RADS-3a | >BT-RADS-3a | >BT-RADS-3a |
| True-positive findings (n) | 24 | 28 | 25 | 26 | 30 |
| False-negative findings (n) | 11 | 7 | 10 | 9 | 5 |
| False-positive findings (n) | 4 | 5 | 6 | 3 | 5 |
| True-negative findings (n) | 34 | 33 | 32 | 35 | 33 |
| Accuracy (%) | 79.5 (58/73) [68.4–88.0] | 83.6 (61/73) [73.1–91.2] | 78.1 (57/73) [66.9–86.9] | 83.6 (61/73) [73.1–91.2] | 86.3 (63/73) [76.3–93.2] |
| Sensitivity (%) | 68.6 (24/35) [50.7–83.2] | 80.0 (28/35) [63.1–91.6] | 71.4 (25/35) [53.7–85.4] | 74.3 (26/35) [56.7–87.5] | 85.7 (30/35) [69.7–95.2] |
| Specificity (%) | 89.5 (34/38) [75.2–97.1] | 86.8 (33/38) [71.9–95.6] | 84.2 (32/38) [68.8–94.0] | 92.1 (35/38) [78.6–98.3] | 86.8 (33/38) [71.9–95.6] |
| Positive predictive value (%) | 85.7 (24/28) [69.8–94.0] | 84.8 (28/33) [70.8–92.8] | 80.6 (25/31) [66.0–89.9] | 89.7 (26/29) [74.2–96.3] | 85.7 (30/35) [72.4–93.2] |
| Negative predictive value (%) | 75.6 (34/45) [65.2–83.6] | 82.5 (33/40) [70.6–90.2] | 76.2 (32/42) [65.1–84.6] | 79.6 (35/44) [68.7–87.3] | 86.8 (33/38) [74.4–93.8] |

Data in parentheses were used to calculate percentages. Data in brackets are 95% confidence intervals. BT-RADS = Brain Tumor Reporting and Data System; n = number; TP = tumor progression; R = reader.

### *3.5. Inter-Reader Reliability of BT-RADS*

The inter-reader agreement (IRA) for evaluating the key MRI features and assigning the final BT-RADS category is summarized in Table 6. Overall, the IRA was very good for identifying new lesions ($\kappa$ = 0.89) and good for evaluating the FLAIR component ($\kappa$ = 0.67), mass effect ($\kappa$ = 0.69), and determining the final BT-RADS classification ($\kappa$ = 0.75). The agreement was moderate for assessing the enhancing component ($\kappa$ = 0.54).

Table 6. Inter-reader agreement for MR imaging features and BT-RADS categorization.

| Readers | Features | R 2 | R 3 | R 4 | R 5 | Overall |
|---|---|---|---|---|---|---|
| R 1 | Enhanced component | 0.28 (0.15–0.40) | 0.24 (0.07–0.42) | 0.29 (0.13–0.44) | 0.36 (0.23–0.50) | |
| | FLAIR component | 0.60 (0.46–0.74) | 0.43 (0.27–0.60) | 0.56 (0.41–0.71) | 0.68 (0.56–0.80) | |
| | Mass effect | 0.62 (0.46–0.79) | 0.70 (0.55–0.84) | 0.64 (0.48–0.79) | 0.82 (0.70–0.95) | |
| | New lesion | 0.79 (0.59–0.99) | 1.00 (1.00–1.00) | 0.79 (0.51–1.00) | 1.00 (1.00–1.00) | |
| | BT-RADS | 0.57 (0.46–0.67) | 0.51 (0.38–0.64) | 0.54 (0.43–0.66) | 0.70 (0.58–0.81) | |
| R 2 | Enhanced component | | 0.33 (0.17–0.50) | 0.26 (0.08–0.44) | 0.65 (0.51–0.80) | |
| | FLAIR component | | 0.59 (0.44–0.73) | 0.65 (0.51–0.80) | 0.69 (0.57–0.82) | |
| | Mass effect | | 0.73 (0.58–0.88) | 0.79 (0.65–0.93) | 0.68 (0.52–0.84) | |
| | New lesion | | 0.79 (0.59–0.99) | 0.65 (0.38–0.93) | 0.79 (0.59–0.99) | |
| | BT-RADS | | 0.51 (0.41–0.62) | 0.60 (0.49–0.70) | 0.72 (0.63–0.80) | |
| R 3 | Enhanced component | | | 0.28 (0.11–0.45) | 0.44 (0.29–0.59) | |
| | FLAIR component | | | 0.46 (0.29–0.63) | 0.57 (0.42–0.71) | |
| | Mass effect | | | 0.71 (0.56–0.86) | 0.73 (0.58–0.87) | |
| | New lesion | | | 0.79 (0.51–1.00) | 1.00 (1.00–1.00) | |
| | BT-RADS | | | 0.54 (0.42–0.66) | 0.72 (0.62–0.81) | |
| R 4 | Enhanced component | | | | 0.46 (0.29–0.63) | |
| | FLAIR component | | | | 0.67 (0.53–0.81) | |
| | Mass effect | | | | 0.63 (0.47–0.80) | |
| | New lesion | | | | 0.79 (0.51–1.00) | |
| | BT-RADS | | | | 0.73 (0.64–0.83) | |
| Overall | Enhanced component | | | | | 0.45 (0.34–0.56) |
| | FLAIR component | | | | | 0.67 (0.58–0.76) |
| | Mass effect | | | | | 0.69 (0.60–0.77) |
| | New lesion | | | | | 0.89 (0.86–0.93) |
| | BT-RADS | | | | | 0.75 (0.68–0.82) |

Data are Kappa values. Data in parentheses are 95% confidence intervals. BT-RADS = Brain Tumor Reporting and Data System; R = reader; FLAIR = fluid-attenuation inversion recovery; MR = magnetic resonance. The overall inter-reader agreement was calculated using the intraclass correlation coefficient (ICC) with 95% confidence intervals. The inter-reader agreement between single readers was calculated using the weighted kappa statistics with 95% confidence intervals. The k values were interpreted as follows: 0.00–0.20 = poor agreement; 0.21–0.40 = fair agreement; 0.41–0.60 = moderate agreement; 0.61–0.80 = good agreement; and 0.81–1.00 = very good agreement.

### *3.6. Reader Acceptance of BT-RADS*

Table 7 shows the readers' responses to the survey statements about implementing the BT-RADS in clinical practice. The readers strongly agreed that "the application of BT-RADS should be encouraged" (score = 25) and that "reporting of post-treatment glioma imaging should follow a structured format" (score = 24). They also indicated that "BT-RADS needs further modification" (score = 24). There was a moderately high agreement with statements like "the structured reporting template of BT-RADS maintains consistency of reports" (score = 24) and "the application of BT-RADS constrains radiologists' and clinicians' communication" (score = 23). However, the readers gave lower scores to statements suggesting BT-RADS currently provides clarity, with "radiology reports following BT-RADS clarify significant findings" (score = 17. The lowest agreement was with the statement "BT-RADS is equivalent to other ACR reporting systems, as BI-RADS and LI-RADS" (score = 14).

Table 7. Reader acceptance of the BT-RADS classification system.

| Questions | Score |
|---|---|
| Reporting of post-treatment glioma imaging should follow a structured format | 24 |
| Radiology reports following BT-RADS clarify the significant findings | 17 |
| The structured reporting template of BT-RADS maintains the consistency of the reports | 24 |
| The application of BT-RADS constrains radiologists' and clinicians' communication | 23 |

**Table 7.** *Cont.*

| Questions | Score |
|---|---|
| The application of BT-RADS enables a more understandable and concise report | 21 |
| Much training is required before the application of BT-RADS | 21 |
| BT-RADS is an easily applicable system by both junior and senior radiologists | 16 |
| The application of BT-RAD needs highly experienced radiologists | 21 |
| Reporting using BT-RADS helps trainee education | 20 |
| Using a structured reporting template of BT-RADS saves time for radiologists | 18 |
| BT-RADS promotes confidence in the final categorization of post-glioma imaging | 16 |
| BT-RADS enables the determination of appropriate management strategy | 19 |
| Some post-glioma imaging was difficult to interpret by BT-RADS | 22 |
| BT-RADS is equivalent to other ACR reporting systems, such as BI-RADS and LI-RADS | 14 |
| The current BT-RADS system is satisfied | 15 |
| BT-RADS needs further modification | 24 |
| The application of BT-RADS should be encouraged | 25 |

The score represents the sum of points provided by each reviewer. BT-RADS = Brain Tumor Reporting and Data System; BI-RADS = Breast Imaging Reporting and Data System; LI-RADS = Liver Imaging Reporting and Data System; ACR = American College of Radiology. To assess reviewer acceptance of BT-RADS, all reviewers completed a short survey after achieving their review. Answers were provided on an ordinal scale of 5 points. The scores were calculated as the sum of each reader's points.

Figures 2–4 represent some cases from our study.

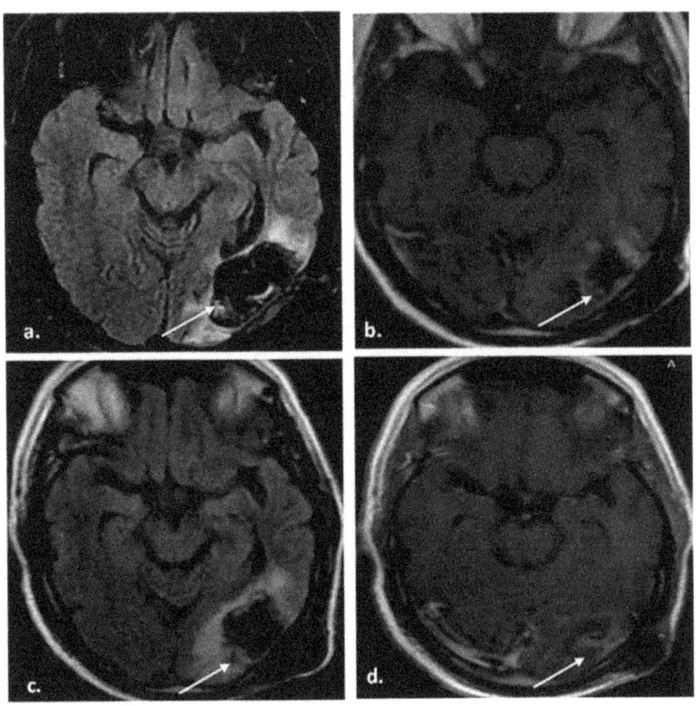


**Figure 2.** *Cont.*

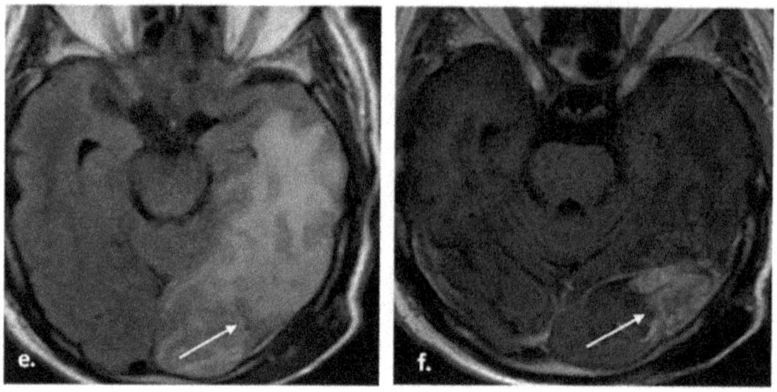


**Figure 2.** A 67-year-old male patient underwent total resection of pathologically proven HGG. The patient did not receive any antiangiogenic treatment. (**a**,**b**) MRI was performed 2 months after the operation while the patient was undergoing radiotherapy. (**a**) Axial FLAIR shows a CSF-like signal intensity resection cavity (arrow). (**b**) The axial postcontrast T1WI shows a mild peripheral enhancement of the cavity (arrow). (**c**,**d**) A follow-up MRI was performed 45 days after the completion of radiotherapy; the patient was clinically stable. (**c**) The axial FLAIR image shows an increased signal at the periphery of the resection cavity (arrow). (**d**) The axial post-contrast T1WI shows patchy central enhancement with less peripheral enhancement (arrow). The patient was categorized as BT-RADS-3a. (**e**,**f**) A follow-up MRI was performed 6.5 months after the completion of radiotherapy; the patient was clinically deteriorating. (**e**) The axial FLAIR shows a new intermediate signal-intensity lesion (arrow) with surrounding edema and mass effect. (**f**) The axial post-contrast T1WI image shows heterogeneous enhancement of the lesion (arrow). The multidisciplinary brain tumor team considered tumor progression, the patient underwent reoperation, and histopathology showed recurrence.

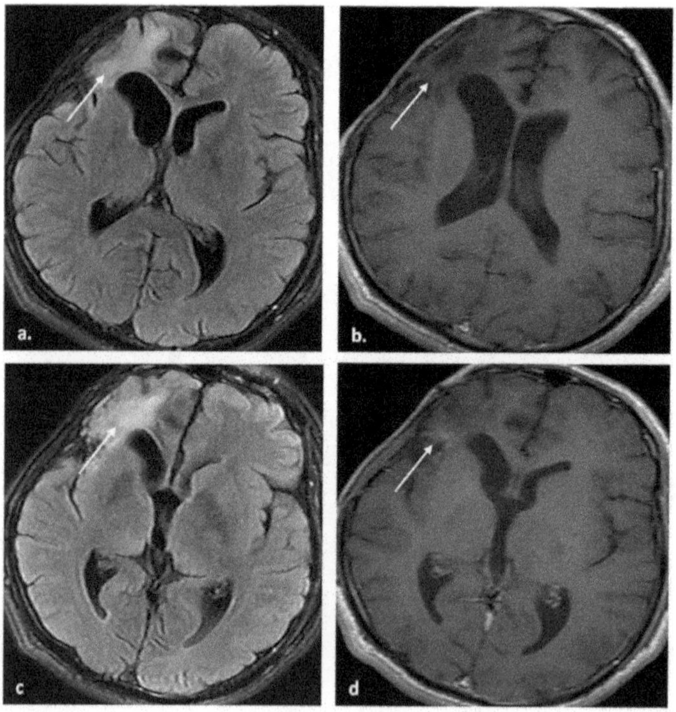


**Figure 3.** *Cont.*

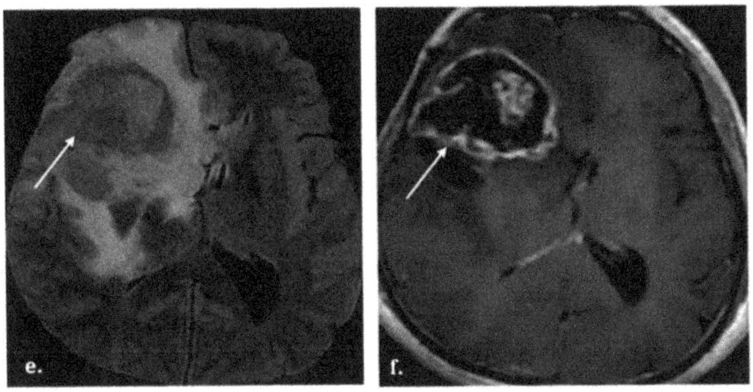

**Figure 3.** A 31-year-old male patient underwent total resection of HGG. The patient did not receive antiangiogenic therapy. (**a**,**b**) MRI was performed 9 months after completion of radiotherapy. (**a**) Axial FLAIR image shows right frontal high-signal intensity gliosis at the operative bed (arrow) with a negative mass effect on the ipsilateral ventricular system. (**b**) The axial post-contrast T1WI image shows no enhancement (arrow). (**c**,**d**) A follow-up MRI was performed 15 months after the completion of radiotherapy; the patient was clinically stable. (**c**) The axial FLAIR image shows stable high-signal intensity gliosis (arrow) with decreased negative mass effect on the ventricular system. (**d**) The axial post-contrast T1WI image shows small, enhanced foci (arrow). The patient was categorized as BT-RADS-3b. (**e**,**f**) A follow-up MRI was performed 22 months after the completion of radiotherapy; the patient was clinically deteriorating. (**e**) The axial FLAIR image shows a new intermediate signal intensity lesion (arrow) with a positive mass effect on the ipsilateral ventricular system. (**f**) The axial post-contrast T1WI image shows heterogeneous marginal enhancement of the lesion (arrow). The multidisciplinary brain tumor team considered tumor progression, the patient underwent reoperation, and histopathology showed recurrence.

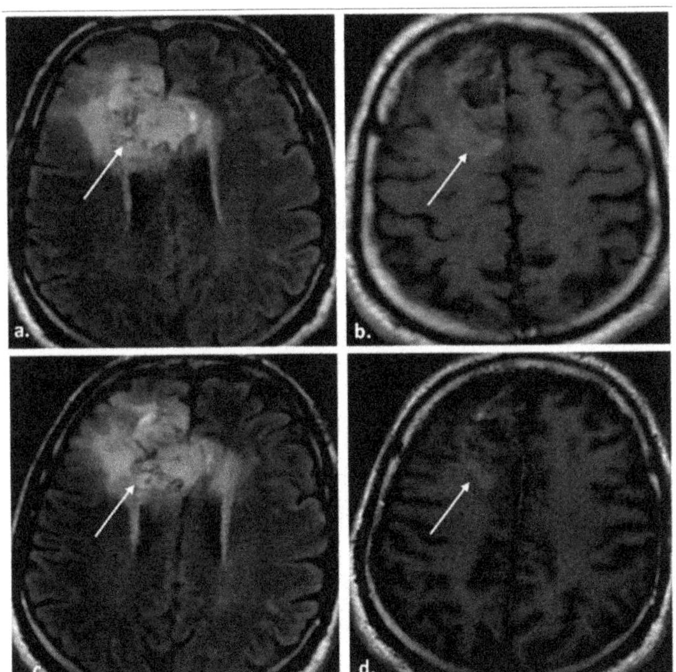

**Figure 4.** *Cont.*

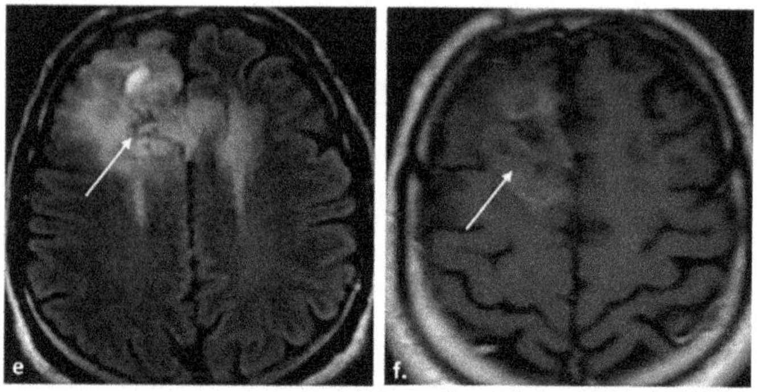


**Figure 4.** A 50-year-old male patient underwent total resection of HGG. The patient did not receive antiangiogenic therapy. (**a**,**b**) MRI was performed 4 months after the completion of radiotherapy. (**a**) The axial FLAIR image shows a right frontal high signal intensity area at the operative bed (arrow) extending to the genu of the corpus callosum to the opposite side with a mild positive mass effect on the related cortical sulci. (**b**) The axial post-contrast T1WI image shows no enhancement (arrow). (**c**,**d**) A follow-up MRI was performed 8 months after the completion of radiotherapy; the patient was clinically stable. (**c**) The axial FLAIR image shows a stable FLAIR component and mass effect (arrow). (**d**) The axial post-contrast T1WI image shows a stable enhancement pattern (arrow). The patient was categorized as BT-RADS-2. (**e**,**f**) A follow-up MRI was performed 12 months after the completion of radiotherapy; the patient was clinically stable. (**e**) The axial FLAIR image shows a stable FLAIR component and no mass effect (arrow). (**f**) The axial post-contrast T1WI image shows a stable enhancement pattern (arrow). The multidisciplinary brain tumor team considered the lesion to be non-progressive.

## 4. Discussion

Understanding the BT-RADS lexicon and how to apply it accurately can improve care in daily practice by reducing bias in imaging assessment, promoting the prediction of glioma progression, and avoiding unnecessary repeated operations or alterations in treatment. The present prospective multicenter study provides important novel insights into the diagnostic accuracy, reliability, and clinical acceptance of using the BT-RADS structured reporting system for postoperative monitoring of glioma patients. To our knowledge, this is one of the largest evaluations of the BT-RADS to date, with 73 patients and over 300 MRI examinations interpreted by 5 independent experienced neuroradiologists. A key novel finding is that the BT-RADS demonstrates good diagnostic accuracy for predicting TP when using a cutoff value of >BT-RADS 3a. Across readers, the sensitivity ranged from 68.6 to 85.7% and specificity from 84.2 to 92.1%, with an overall accuracy of 78.1 to 86.3%. These performance characteristics support the utility of the BT-RADS as a clinical tool to reliably identify patients requiring changes in management due to TP versus those who can continue routine monitoring. A recent study published by Kim et al. [21] validated the prognostic role of the BT-RADS and reported that the BT-RADS with the highest probability of worsening at the subsequent follow-up was BT-RADS-3b. A retrospective study by Yang et al. [22] reported that the BT-RADS with conventional MRI yielded a sensitivity of 88%, specificity of 55%, and accuracy of 74% for differentiating TP from non-progression in postoperative high-grade glioma (HGG) patients with BT-RADS-3 lesions. Another recent prospective study assessed the accuracy of BT-RADS-3 in detecting TP and showed 76.9% sensitivity, 64.3% specificity, and 70.4% accuracy [23].

The current study showed that the BT-RADS classification system has limited sensitivity for predicting TP in postoperative gliomas, specifically for worsening categories (68.6–85.7% for BT-RADS-3a, 45.7–54.3% for BT-RADS-3b, and 17.1–45.7% for BT-RADS-3c). This finding is not surprising, as the BT-RADS depends on conventional MRI, which previous studies have shown to be ineffective in distinguishing TP from non-progression [24–27].

A meta-analysis by van Dijken et al. [24] concluded that conventional MRI is unreliable for assessing treatment response in HGG, with a sensitivity of 68% and a specificity of 77%.

As a result of the limited sensitivity of the BT-RADS in predicting TP, particularly with the worsening categories (BT-RADS-3), some authors [22,23] have tried to enhance the diagnostic power of the BT-RADS with the incorporation of advanced MR imaging techniques. Yang et al. [22] found that adding DWI and perfusion-weighted imaging (PWI) to the BT-RADS can considerably enhance the diagnostic accuracy for distinguishing TP from non-progression in postoperative HGG patients with a sensitivity of 98%, a specificity of 85%, and an accuracy of 95%. Also, Metwally et al. [23] concluded that the sensitivity of BT-RADS-3 for predicting TP increased after adding DWI to the BT-RADS. A meta-analysis without the BT-RADS conducted by van Dijken et al. [24] proved that advanced MRI techniques are more accurate than conventional MRI in differentiating TP from non-progression. However, because of the cost burden, different acquisition techniques, and non-routinely requested PWI in radiological centers, we recommend adding DWI to the BT-RADS and restricting other advanced MR techniques, such as PWI, to doubted cases when there is still uncertainty of the BT-RADS category after using conventional MRI and DWI.

We assessed the incidence of TP and found that it increased with increasing a BT-RADS category. According to our readers, TP was noted in 25–41.7% of patients classified as BT-RADS-3a, 61.5–88.9% of patients classified as BT-3b, 75–90.9% of patients classified as BT-3c, and 91.7–100% of patients classified as BT-RADS-4. According to Yang et al. [22], the recurrence rate for BT-RADS-3a was 21.4%, for BT-RADS-3b was 61.5%, and for BT-RADS-3c was 78.4%.

We comprehensively evaluated the inter-reader reliability of the BT-RADS application across multiple key imaging features. The results revealed very good agreement for identifying new lesions ($\kappa = 0.89$) and good agreement for assessing FLAIR changes ($\kappa = 0.67$), mass effect ($\kappa = 0.69$), and determining the final BT-RADS categorization ($\kappa = 0.75$). This demonstrates that the BT-RADS criteria can be applied consistently by different radiologists, which is essential for generalizability across practices. These results are similar to those of Parillo et al. [28], who concluded that the BT-RADS had good inter-rater reliability with a high agreement rate among expert radiologists and radiologists in training. Cooper et al. [29] reported that the overall agreement rate between primary and secondary reviews was 82.2%, with perfect agreement for studies with improvement categories (BT-RADS-1a or BT-RADS-1b) and lower levels of agreement for studies with worsening categories (BT-RADS-3a and BT-RADS-4). Additionally, we reported IRA for MR features of gliomas. The overall IRA was moderate for the enhancing component ($\kappa = 0.54$), good for the FLAIR component ($\kappa = 0.67$) and mass effect ($\kappa = 0.69$), and very good for new lesions ($\kappa = 0.89$). However, to date, no studies have assessed the IRA of such features in the BT-RADS. Thus, our results are not directly comparable to the literature.

Implementing a BT-RADS at a large university hospital enhanced radiology report perceptions among referring physicians and radiologists by improving report consistency and communication and assisting decision-making and research [15,16]. We assessed our readers' acceptance of the routine use of the BT-RADS in reporting postoperative glioma imaging. A relatively lower score was given for "radiology reports following BT-RADS clarify significant findings". Our readers explained this by deficient data on glioma extension regarding the cross to the opposite side, corpus callosum involvement, transependymal spread, and the lack of diffusion criteria and metabolite assessment by MRS. Additionally, "BT-RADS is equivalent to other ACR reporting systems, as BI-RADS and LI-RADS" was given a lower score as the BT-RADS, unlike other RAD systems, is a classification system whose objective is to monitor patient surveillance rather than to provide a diagnosis. Moreover, the value of the BT-RADS cannot be based on a single MRI examination but on multiple follow-up examinations. However, "The application of BT-RADS should be encouraged" and "post-treatment glioma imaging should follow a structured format" received high scores.

While multidisciplinary teams remain crucial for determining diagnosis and treatment plans for brain tumor patients, the BT-RADS can be a valuable complementary tool in the clinical setting. Its potential applications include integrating multidisciplinary tumor board discussions and utilizing its structured reporting and standardized lexicon to facilitate clear communication among specialists. Additionally, BT-RADS' comprehensive assessment, incorporating advanced imaging techniques, can aid in evaluating treatment response, particularly in distinguishing true TP from treatment-related changes in high-grade glioma patients undergoing multimodal therapies. Adopting the BT-RADS can lead to more standardized and consistent imaging reports, benefiting large healthcare systems and patient transfers. Furthermore, incorporating the BT-RADS into training curricula for radiologists, neuro-oncologists, and other specialists can foster interdisciplinary collaboration and communication from an early stage. While not replacing multidisciplinary decision-making, the BT-RADS can enhance clinical workflows by improving communication, treatment response evaluation, and standardization across different healthcare settings.

RANO (Response Assessment in Neuro-Oncology) criteria were established in 2010 as guidelines for assessing treatment response in clinical trials for high-grade gliomas. RANO mainly focuses on measuring contrast-enhancing tumor portions on MRI [30]. While RANO is valuable for standardizing response evaluation in trials, it has limitations in clinical practice, requiring a more comprehensive assessment. The BT-RADS serves as a complementary tool by providing a structured reporting system with standardized terminology to describe overall brain tumor imaging findings beyond just contrast enhancement. The BT-RADS incorporates evaluations of tumor morphology, the extent of surrounding edema/mass effect, and patterns of enhancing and non-enhancing tumor components. It uses these factors to generate a scoring system that estimates the likelihood of tumor type and behavior [14]. Rather than replacing RANO's focus on treatment response metrics, the BT-RADS aims to comprehensively characterize and communicate the full range of imaging findings to facilitate optimal multidisciplinary treatment planning alongside criteria like RANO. Its structured lexicon promotes clear communication among the cancer care team [16].

Our findings clearly encourage the application of the BT-RADS in postoperative glioma imaging. However, the BT-RADS still needs some improvements to increase its diagnostic accuracy and become useful and comprehensive for all pertinent descriptors and definitions. The key to these improvements may include the incorporation of DWI into the BT-RADS, the use of other advanced MR techniques in doubted cases when there is still uncertainty of the BT-RADS category, a more descriptive analysis of TP regarding its relation to the corpus callosum and ependymal surface of the ventricular system, and the assessment of the infiltrative nature of the glioma in the surrounding apparently normal brain parenchyma, mainly by using advanced imaging modalities such as MRS and PWI.

Finally, this study makes several key novel contributions to the literature on BT-RADS validation and clinical implementation for postoperative glioma monitoring: (1) establishing the diagnostic accuracy benchmarks across multiple readers, (2) demonstrating good inter-reader reliability, and (3) providing insights from radiologists' perspective on clinical acceptance and potential areas for improvement. These findings support the increasing adoption of the BT-RADS while underscoring the need for continued refinements to maximize its clinical utility.

A notable strength of our study is the prospective design, which evaluates BT-RADS performance on actual clinical cases rather than retrospective assignments. Moreover, the multi-institutional nature increases the generalizability of the findings. Additionally, the study provides novel insights into radiologists' perspectives on the BT-RADS through the reader acceptance survey.

## 5. Limitations

This study was subjected to several limitations: (1) Sample size: Though adequate for preliminary exploration of the BT-RADS in the postoperative glioma monitoring, it may limit our findings' generalizability. (2) Prognostic evaluation: The role of the BT-RADS as a prognostic tool for overall survival in glioma patients was not assessed, marking a potential area for future research. (3) Expertise in image evaluation: The evaluation of all MR images was conducted by experienced radiologists, which might influence the perceived diagnostic accuracy of the BT-RADS. (4) Diagnosis confirmation: Not all diagnoses were confirmed via tissue pathology; more than half of the patients (63%) were confirmed by clinical and imaging follow-up, which could affect the validity of our findings. (5) Glioma subtypes: The study did not differentiate among various types of gliomas for sub-analyses due to the relatively small sample size. This limitation underscores the need for larger cohort studies to explore BT-RADS' performance across different glioma subtypes. (6) Imaging technology: The study exclusively used a 1.5 Tesla MRI for all examinations, which may not reflect the diagnostic potential of higher-field MRIs. (7) Clinician familiarity with the BT-RADS: The adoption and effectiveness of the BT-RADS may be hindered by its relatively new introduction and current unfamiliarity among many clinicians. Addressing these limitations will be crucial for enhancing the applicability and accuracy of the BT-RADS in clinical settings and for future research aimed at refining glioma management strategies.

## 6. Conclusions

The current study indicates that the BT-RADS has good diagnostic accuracy and reliability for predicting TP in postoperative gliomas. However, the BT-RADS-3 needs further improvements to increase its diagnostic accuracy.

**Author Contributions:** Conceptualization: Y.E.A., M.A.A.B., M.I.M., N.A.Z., M.G.N., E.M.H., S.K.A., A.A.M., S.A.A., R.B., M.A.E.-A.M.F., A.Z.A.A., M.M.A. and A.M.A.B.; Methodology: Y.E.A., M.A.A.B., M.I.M., N.A.Z., S.A.A., M.G.N., E.M.H. and S.K.A.; Software: A.A.M., R.B., A.Z.A.A., M.A.E.-A.M.F., M.M.A. and A.M.A.B.; Validation: M.A.A.B., M.I.M., N.A.Z., M.G.N., E.M.H., S.K.A. and A.A.M.; Formal analysis: Y.E.A., M.A.A.B., M.I.M., N.A.Z., M.G.N., S.A.A. and E.M.H.; Investigation: A.A.M., R.B., A.Z.A.A. and M.M.A.; Resources: A.A.M., R.B., A.Z.A.A., M.M.A. and A.M.A.B.; Writing—original draft preparation: Y.E.A., M.A.A.B., M.I.M., N.A.Z., M.G.N., E.M.H. and S.K.A.; Writing—review and editing: Y.E.A., M.A.A.B., M.I.M., N.A.Z., M.G.N., E.M.H., S.K.A., A.A.M., S.A.A., R.B., M.A.E.-A.M.F., A.Z.A.A., M.M.A. and A.M.A.B.; Visualization: Y.E.A., M.A.A.B., M.I.M., N.A.Z., M.G.N., E.M.H. and S.K.A.; Supervision: Y.E.A., M.A.A.B., M.I.M., N.A.Z., M.A.E.-A.M.F., M.G.N., E.M.H., S.K.A. and A.M.A.B.; Project administration: Y.E.A., S.K.A., E.M.H., S.K.A., A.A.M., R.B., A.Z.A.A., M.M.A. and A.M.A.B.; Funding acquisition: Y.E.A., M.A.A.B. and S.K.A. All authors have read and agreed to the published version of the manuscript.

**Funding:** This research was funded by the Deanship of Scientific Research, Najran University, Kingdom of Saudi Arabia (grant code: NU/DRP/MRC/12/39). The authors are thankful to the Deanship of Scientific Research at Najran University for funding this work under the General Research Funding program grant code (NU/DRP/MRC/12/39).

**Institutional Review Board Statement:** This study was approved by the Institutional Review Board of Zagazig University Hospitals in Egypt (Approval No. ZU-9857).

**Informed Consent Statement:** Informed consent was obtained from all the participants.

**Data Availability Statement:** The clinical and imaging datasets used and/or analyzed during the current study are available from the corresponding author upon reasonable request.

**Conflicts of Interest:** The authors declare no relevant conflicts of interest and no relationships with any companies whose products or services may be related to the subject matter of this article.

## References

1. Ostrom, Q.T.; Gittleman, H.; Truitt, G.; Boscia, A.; Kruchko, C.; Barnholtz-Sloan, J.S. CBTRUS Statistical Report: Primary Brain and Other Central Nervous System Tumors Diagnosed in the United States in 2011–2015. *Neuro Oncol.* **2018**, *20* (Suppl. S4), iv1–iv86. [CrossRef] [PubMed]
2. Nieder, C.; Mehta, M.P.; Jalali, R. Combined radio- and chemotherapy of brain tumours in adult patients. *Clin. Oncol. (R. Coll. Radiol.)* **2009**, *21*, 515–524. [CrossRef] [PubMed]

3. Mueller, S.; Chang, S. Pediatric brain tumors: Current treatment strategies and future therapeutic approaches. *Neurotherapeutics* **2009**, *6*, 570–586. [CrossRef] [PubMed]
4. Stupp, R.; Hegi, M.E.; Mason, W.P.; van den Bent, M.J.; Taphoorn, M.J.; Janzer, R.C.; Ludwin, S.K.; Allgeier, A.; Fisher, B.; Belanger, K.; et al. Effects of radiotherapy with concomitant and adjuvant temozolomide versus radiotherapy alone on survival in glioblastoma in a randomised phase III study: 5-year analysis of the EORTC-NCIC trial. *Lancet Oncol.* **2009**, *10*, 459–466. [CrossRef] [PubMed]
5. Quant, E.C.; Wen, P.Y. Response assessment in neuro-oncology. *Curr. Oncol. Rep.* **2011**, *13*, 50–56. [CrossRef] [PubMed]
6. Parvez, K.; Parvez, A.; Zadeh, G. The diagnosis and treatment of pseudoprogression, radiation necrosis and brain tumor recurrence. *Int. J. Mol. Sci.* **2014**, *15*, 11832–11846. [CrossRef] [PubMed]
7. Vos, M.J.; Uitdehaag, B.M.; Barkhof, F.; Heimans, J.J.; Bitz-van der Zwan, A.E.; Gons, R.A.; van Waesberghe, J.H.; Polman, C.H. Interobserver variability in the radiological assessment of response to chemotherapy in glioma. *Neurology* **2003**, *60*, 826–830. [CrossRef] [PubMed]
8. van den Bent, M.J.; Afra, D.; de Witte, O.; Ben Hassel, M.; Schraub, S.; Hoang-Xuan, K.; Malmström, P.O.; Hanzen, C.; Collette, L.; Piérart, M.; et al. Long-term efficacy of early versus delayed radiotherapy for low-grade astrocytoma and oligodendroglioma in adults: The EORTC 22845 randomised trial. *Lancet* **2005**, *366*, 985–990, Erratum in *Lancet* **2006**, *367*, 1818. [CrossRef] [PubMed]
9. Huang, R.Y.; Wen, P.Y. Response Assessment in Neuro-Oncology Criteria and Clinical Endpoints. *Magn. Reson. Imaging Clin. N. Am.* **2016**, *24*, 705–718. [CrossRef]
10. Macdonald, D.R.; Cascino, T.L.; Schold, S.C., Jr.; Cairncross, J.G. Response criteria for phase II studies of supratentorial malignant glioma. *J. Clin. Oncol.* **1990**, *8*, 1277–1280. [CrossRef]
11. Wen, P.Y.; Macdonald, D.R.; Reardon, D.A.; Cloughesy, T.F.; Sorensen, A.G.; Galanis, E.; Degroot, J.; Wick, W.; Gilbert, M.R.; Lassman, A.B.; et al. Updated response assessment criteria for high-grade gliomas: Response assessment in neuro-oncology working group. *J. Clin. Oncol.* **2010**, *28*, 1963–1972. [CrossRef] [PubMed]
12. Rao, B.; Ikuta, I.; Mahajan, A.; Torskaya, V.; Patsalides, A.; Weinberg, B.D. Brain Tumor Reporting and Data System: A Pictorial Review. *Neurographics* **2021**, *11*, 175–185. [CrossRef]
13. Weinberg, B.D. BT-RADS Brain Tumor Reporting & Data System. 2023. Available online: https://btrads.com/resources/ (accessed on 11 January 2023).
14. Weinberg, B.D.; Gore, A.; Shu, H.G.; Olson, J.J.; Mu, Y.; Lindell, P.K.; Cogswell, P.M.; Wawrzynczak, C.J.; Johnson, J.D.; Wang, K.J.; et al. Management-Based Structured Reporting of Posttreatment Glioma Response with the Brain Tumor Reporting and Data System. *J. Am. Coll. Radiol.* **2018**, *15*, 767–771. [CrossRef] [PubMed]
15. Gore, A.; Hoch, M.J.; Shu, H.G.; Lindell, P.K.; Weinberg, B.D. Institutional Implementation of a Structured Reporting System: Our Experience with the Brain Tumor Reporting and Data System. *Acad. Radiol.* **2019**, *26*, 974–980. [CrossRef] [PubMed]
16. Zhang, J.Y.; Weinberg, B.D.; Hu, R.; Gore, A.; Shu, H.G.; Olson, J.J.; Lindell, P.K.; Wawrzynczak, C.J.; Cogswell, P.M.; Wang, K.J.; et al. Quantitative Improvement in Brain Tumor MRI Through Structured Reporting (BT-RADS). *Acad. Radiol.* **2020**, *27*, 780–784. [CrossRef]
17. Nayak, L.; DeAngelis, L.M.; Brandes, A.A.; Peereboom, D.M.; Galanis, E.; Lin, N.U.; Soffietti, R.; Short, S.C.; Deangelis, L.M.; Chamberlain, M.C.; et al. The Neurologic Assessment in Neuro-Oncology (NANO) scale: A tool to assess neurologic function for integration into the Response Assessment in Neuro-Oncology (RANO) criteria. *Neuro Oncol.* **2017**, *19*, 625–635. [CrossRef] [PubMed]
18. Ma, B.; Blakeley, J.O.; Hong, X.; Yu, Y.; Zou, J.; Zhang, W.; Zhang, M.; van Zijl, P.C.; Mori, S.; Jiang, S.; et al. Applying amide proton transfer-weighted MRI to distinguish pseudoprogression from true progression in malignant gliomas. *J. Magn. Reson. Imaging* **2016**, *44*, 456–462. [CrossRef] [PubMed]
19. Tran, D.K.; Jensen, R.L. Treatment-related brain tumor imaging changes: So-called "pseudoprogression" vs. tumor progression: Review and future research opportunities. *Surg. Neurol. Int.* **2013**, *4* (Suppl. S3), S129–S135. [CrossRef] [PubMed]
20. Wang, S.; Martinez-Lage, M.; Sakai, Y.; Chawla, S.; Woo, D.C.V.; Alonso-Basanta, M.; Shadowitz, S.; Carralero, J.D.; Desai, A.; Prah, M.; et al. Differentiating Tumor Progression from Pseudoprogression in Patients with Glioblastomas Using Diffusion Tensor Imaging and Dynamic Susceptibility Contrast MRI. *AJNR Am. J. Neuroradiol.* **2016**, *37*, 28–36. [CrossRef]
21. Kim, S.; Hoch, M.J.; Peng, L.; Shao, J.; Go, J.L.; Akers, A.; Moon, J.J.; Cha, Y.J.; Choi, S.W.; Ivanidze, J.; et al. A brain tumor reporting and data system to optimize imaging surveillance and prognostication in high-grade gliomas. *J. Neuroimaging* **2022**, *32*, 1185–1192. [CrossRef]
22. Yang, Y.; Yang, Y.; Wu, X.; Liu, J.; Lin, N.; Xu, Y.; Cheng, L.; Jiang, Y.; Wang, T.; Lv, X.; et al. Adding DSC PWI and DWI to BT-RADS can help identify postoperative recurrence in patients with high-grade gliomas. *J. Neurooncol.* **2020**, *146*, 363–371. [CrossRef] [PubMed]
23. Metwally, M.I.; Hafez, F.F.; Ibrahim, S.A.; Amin, D.H.; EL-Saied, W.E.F.F.; Amer, A.M. The value of adding DWI and FLAIR signal changes in the resection cavity on the diagnostic performance of BT-RADS category 3 for tumor progression prediction in post-treated glioma patients: A prospective pilot study. *Egypt. J. Radiol. Nucl. Med.* **2023**, *54*, 52. [CrossRef]
24. van Dijken, B.R.J.; van Laar, P.J.; Holtman, G.A.; van der Hoorn, A. Diagnostic accuracy of magnetic resonance imaging techniques for treatment response evaluation in patients with high-grade glioma, a systematic review and meta-analysis. *Eur Radiol.* **2017**, *27*, 4129–4144. [CrossRef] [PubMed]

25. Da Cruz, L.H.; Rodriguez, I.; Domingues, R.C.; Gasparetto, E.L.; Sorensen, A.G. Pseudoprogression and pseudoresponse: Imaging challenges in the assessment of posttreatment glioma. *AJNR Am. J. Neuroradiol.* **2011**, *32*, 1978–1985. [CrossRef] [PubMed]
26. Delgado-López, P.D.; Riñones-Mena, E.; Corrales-García, E.M. Treatment-related changes in glioblastoma: A review on the controversies in response assessment criteria and the concepts of true progression, pseudoprogression, pseudoresponse and radionecrosis. *Clin. Transl. Oncol.* **2018**, *20*, 939–953. [CrossRef] [PubMed]
27. Zikou, A.; Sioka, C.; Alexiou, G.A.; Fotopoulos, A.; Voulgaris, S.; Argyropoulou, M.I. Radiation Necrosis, Pseudoprogression, Pseudoresponse, and Tumor Recurrence: Imaging Challenges for the Evaluation of Treated Gliomas. *Contrast Media Mol. Imaging* **2018**, *2018*, 6828396. [CrossRef] [PubMed]
28. Parillo, M.; Mallio, C.A.; Pileri, M.; Salicone, E.; Bandi, N.; Mazzanti, G.; Perocchio, F.; Rabni, M.; Viale, M.; Laino, M.; et al. Interrater reliability of Brain Tumor Reporting and Data System (BT-RADS) in the follow up of adult primary brain tumors: A single institution experience in Italy. *Quant. Imaging Med. Surg.* **2023**, *13*, 7423–7431. [CrossRef] [PubMed]
29. Cooper, M.; Hoch, M.; Abidi, S.; Mathura, G.; Almero, J.; Nasr, K.; Pantel, P.; Reddy, V.; Naemi, A.; Davis, M. Brain tumor mri structured reporting allows calculation of interrater agreement of patients reviewed at tumor board. *Neuro Oncol.* **2019**, *21* (Suppl. S6), vi163. [CrossRef]
30. Wen, P.Y.; van den Bent, M.; Youssef, G.; Cloughesy, T.F.; Ellingson, B.M.; Weller, M.; Galanis, E.; Barboriak, D.P.; de Groot, J.; Gilbert, M.R.; et al. RANO 2.0: Update to the Response Assessment in Neuro-Oncology Criteria for High- and Low-Grade Gliomas in Adults. *J. Clin. Oncol.* **2023**, *41*, 5187–5199. [CrossRef]

biomedicines

MDPI

*Article*

# Brain Tumor MRI Classification Using a Novel Deep Residual and Regional CNN

**Mirza Mumtaz Zahoor [1], Saddam Hussain Khan [2], Tahani Jaser Alahmadi [3,*], Tariq Alsahfi [4], Alanoud S. Al Mazroa [3], Hesham A. Sakr [5], Saeed Alqahtani [6], Abdullah Albanyan [7] and Bader Khalid Alshemaimri [8]**

1 Faculty of Computer Sciences, Ibadat International University, Islamabad 44000, Pakistan; mumtazzahoor5@gmail.com
2 Department of Computer System Engineering, University of Engineering and Applied Science (UEAS), Swat 19060, Pakistan; saddamhkhan@ueas.edu.pk
3 Department of Information Systems, College of Computer and Information Sciences, Princess Nourah bint Abdulrahman University, P.O. Box 84428, Riyadh 11671, Saudi Arabia; asalmazroa@pnu.edu.sa
4 Department of Information Systems and Technology, College of Computer Science and Engineering, University of Jeddah, Jeddah 21959, Saudi Arabia; tmalsahfi@uj.edu.sa
5 Nile Higher Institute for Engineering and Technology, Mansoura 35511, Dakahlia, Egypt; heshamsakr535@nilehi.edu.eg
6 Radiological Sciences Department, College of Applied Medical Sciences, Najran University, Najran 61441, Saudi Arabia; salqahtani@nu.edu.sa
7 College of Computer Engineering and Sciences, Prince Sattam bin Abdulaziz University, Al-Kharj 16278, Saudi Arabia; a.albanyan@psau.edu.sa
8 Software Engineering Department, King Saud University, Riyadh 11671, Saudi Arabia; balshemaimri@ksu.edu.sa
* Correspondence: tjalahmadi@pnu.edu.sa

**Abstract:** Brain tumor classification is essential for clinical diagnosis and treatment planning. Deep learning models have shown great promise in this task, but they are often challenged by the complex and diverse nature of brain tumors. To address this challenge, we propose a novel deep residual and region-based convolutional neural network (CNN) architecture, called Res-BRNet, for brain tumor classification using magnetic resonance imaging (MRI) scans. Res-BRNet employs a systematic combination of regional and boundary-based operations within modified spatial and residual blocks. The spatial blocks extract homogeneity, heterogeneity, and boundary-related features of brain tumors, while the residual blocks significantly capture local and global texture variations. We evaluated the performance of Res-BRNet on a challenging dataset collected from Kaggle repositories, Br35H, and figshare, containing various tumor categories, including meningioma, glioma, pituitary, and healthy images. Res-BRNet outperformed standard CNN models, achieving excellent accuracy (98.22%), sensitivity (0.9811), F1-score (0.9841), and precision (0.9822). Our results suggest that Res-BRNet is a promising tool for brain tumor classification, with the potential to improve the accuracy and efficiency of clinical diagnosis and treatment planning.

**Keywords:** brain tumor classification; deep learning; convolutional neural networks; magnetic resonance imaging

## 1. Introduction

The human brain is the most complicated and imperative organ in the body, governing the neurological system. The most deadly brain tumor is caused by erratic and out-of-control cell growth in the brain [1]. Patient survival depends on the type of glioma type; low-grade gliomas have survival rates of 5 years as high as 80%, whereas the survival rates of 5 years are under 5% for high-grade gliomas [2]. Timely brain tumor recognition and categorization is an imperative research topic in the clinical imaging domain, and it assists in choosing the most suitable treatment plan for a patient's lifesaving.

Several screening methods, either invasive or non-invasive, are employed to identify tumors in the human brain [3]. Magnetic resonance imaging (MRI) is a preferable, less harmful scanning modality since it provides rich information about the location of brain tumors, shape, and size in medical images (MI), and is generally considered quicker, cheaper, and safer [4]. The manual assessment of brain MRIs is challenging for radiologists to identify and categorize brain tumors from MIs. A computer-aided diagnosis (CADx) is required to reduce the burden and assist radiologists or doctors with an MI assessment.

Many research areas are being explored in medical image analysis. It includes medical imaging domains like identification, detection, and segmentation [5–11]. Traditional ML approaches comprise numerous steps, pre-processing, feature extraction and selection, and classification. More discriminative feature acquisition is essential, as classification accuracy relies on obtained features [12,13].

In conclusion, conventional ML techniques have two key challenges in the feature extraction step: one, that it solely emphasizes low- or high-level attributes. Secondly, standard ML techniques rely on hand-crafted features that require significant prior knowledge, such as the position of the tumor in a medical scan. However, there is a considerable risk of human error. Designing an effective system to incorporate high- and low-level features with no human intervention is crucial. As brain tumor datasets are being expanded, there is a need for technological improvements in feature extraction focusing on confined and imbalanced MR imaging datasets of brain abnormalities and other irregularities of the human organs [14–16].

In recent times, deep learning (DL) methods have frequently been employed for brain MRI categorization, including patients with disabilities [17]. While feature mining and classification were integrated into self-learning, deep learning methods do not necessitate a manual process for feature extraction. The DL approach requires a dataset, and minimal pre-processing is required for selecting salient features in a self-learning way [18]. MR imaging categorization faces a significant challenge in diminishing the semantic space among high-level spatial details observed by a human assessor and low level acquired using the imagery mechanism. One of the well-known neural network models, convolutional neural networks (CNNs), specially designed for images, is utilized for feature extraction to acquire the important characteristics to categorize and minimize the semantic gap [19].

Recently, in many studies, CNNs have been widely employed to classify brain MRIs, and are validated on a different dataset of brain tumors [20–26]. A deep CNN-based model was proposed in [27] for brain MRI image categorization into distinct classes. The authors used brain MRI images from a publicly available dataset to prevent model ambiguity. The suggested model has a classification accuracy of 91.4%. Deepak and Ameer [28] employed a pre-train deep CNN, GoogLeNet, to extract key attributes using brain MR images and classify tumors into three classes with 98% accuracy. Ahmet and Muhammad [29] categorized brain MR images using various CNN models and attained satisfactory accuracy. They modify a pre-trained ResNet-50 DCNN by excluding the final five layers and introducing an additional eight layers. The model achieved the highest among all pre-trained models, with an accuracy of 97.2%. Sultan et al. [30] suggested a CNN-based deep learning model utilizing two publicly accessible datasets which have 3064 (glioma, meningioma, and pituitary tumors) and 516 (Grade II, Grade III, and Grade IV) brain medical scans. The proposed method has the best accuracy of 96.13% and 98.7%. Khwaldeh et al. [31] used several CNNs to classify brain MRI images and achieved good results. Using a modified pre-trained Alexnet CNN, they achieved a higher accuracy of 97.2%. Khan, M.A. et al. [32] developed a multi-model-based technique to differentiate brain tumors with DL. The presented system includes many stages. They used partial least-squares (PLS) to concatenate the features and ELM for classification. Their methodology stated improvements of 97.8%, 96.9%, and 92.5% on BraTs-2015, BraTs-2017, and BraTs-2018, respectively. Özyurt et al. [33] presented a technique for detecting brain tumors. They began with MRI tumor image segmentation with the NS-EMFSE algorithm. They obtained features from the segmented image using

AlexNet, and then using the SVM, they detected and classified brain tumor images as benign or malignant with a 95.62% accuracy.

However, most of these models are assessed on small-scale datasets due to the inaccessibility of the data repositories. Likewise, the majority of earlier research was based on pre-trained CNN models, which were developed generally for a dataset of natural images. Pretrained models are customized for the brain tumor task without designing them to distinguish brain tumor patterns; thus, it limits the use of pre-trained CNN models for brain tumor diagnosis.

In this study, a new deep CNN-based brain tumor classification scheme is developed for MRI image categorization. A novel CNN architecture, Res-BRNet, is suggested for brain tumor classification. Performance assessment is performed using standard measures like sensitivity, precision, F1-score, accuracy, and AUC-PR/ROC. Moreover, we have gathered a large dataset by collecting the brain MRI images of three tumor types and normal brain images from publicly accessible sources. The prediction ability of the developed approach is assessed on the test dataset and assessed with a comparison of numerous existing DCNNs, and the proposed technique concept is also compared with standard existing approaches. The proposed work has the following contributions:

- A novel deep residual and regional CNN architecture, Res-BRNet, has been developed for brain tumors classification.
- The proposed Res-BRNet integrates spatial and residual blocks to learn complex tumor patterns from brain MRIs, and it enhances the performance of the developed model for brain tumor classification.
- The developed Res-BRNet employed regional and boundary-based operations in a systematic order within the customized spatial and residual blocks to exploit spatial correlation information and textural variations from brain tumor MRIs.
- The systematic integration of residual and spatial blocks within the proposed Res-BRNet CNN improves the discriminative capability and generalization. Moreover, spatial blocks extract homogeneity and boundary-defined features at the abstract level. Furthermore, residual blocks at the target level effectively learn local and global texture variations of different brain tumors.

The rest of the manuscript is organized as follows: Section 2: incorporates the proposed methodology. The results and discussion are described in Section 3, and Section 4 concludes the entire paper.

## 2. Materials and Methods

In this work, a new deep residual and regional CNN architecture is designed for automated brain tumor classification from MRI images. The discriminating ability of the proposed classification method is empirically assessed using several standard performance measures, and the results are evaluated by comparing them with existing DCNNs. A better generalization is achieved by augmenting the training samples in the experimental setup. A brief and detailed description of the developed brain tumor classification technique is shown in Figure 1A,B.

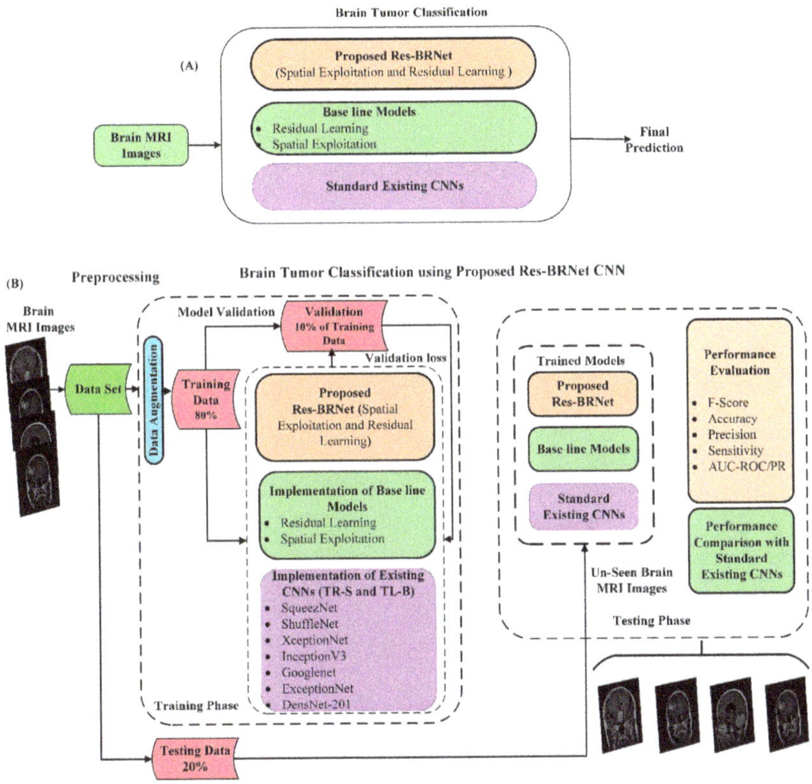


**Figure 1.** (**A**) Brief design of proposed brain tumor MRI image classification technique. (**B**) The detailed design of the proposed brain tumor MRI image classification technique.

### *2.1. Dataset*

In this work, we have collected a dataset containing the 2D MRI images of healthy individuals and three diverse types of brain tumors. MRI scans of four classes are gathered from open-source Kaggle repositories [34], Br35H [35], and figshare [36], collected from 2005 to 2010 by Nanfang Hospital, Guangzhou, and General Hospital, Tianjin Medical University, China. For this experimental setup, we collected 2044 brain normal, 2352 glioma_tumor, 1645 meningioma_tumor, and 1831 pituitary_tumor MRI images from these repositories; hence, in nature, the acquired dataset is unbalanced, as shown in Table 1. Each image was resized according to the input size of DCNNs. Some of the four classes' images are displayed in Figure 2.

**Table 1.** Details of the collected dataset having MRI images of normal and different types of tumors.

| | Glioma | Meningioma | Pituitary | Normal | Total |
|---|---|---|---|---|---|
| **Training (80%)** | 1882 | 1316 | 1465 | 1635 | 6298 |
| **Validation (10% of training)** | 188 | 132 | 146 | 163 | 629 |
| **Testing (20%)** | 470 | 329 | 366 | 409 | 1574 |
| **Total (100%)** | 2352 | 1645 | 1831 | 2044 | 7872 |

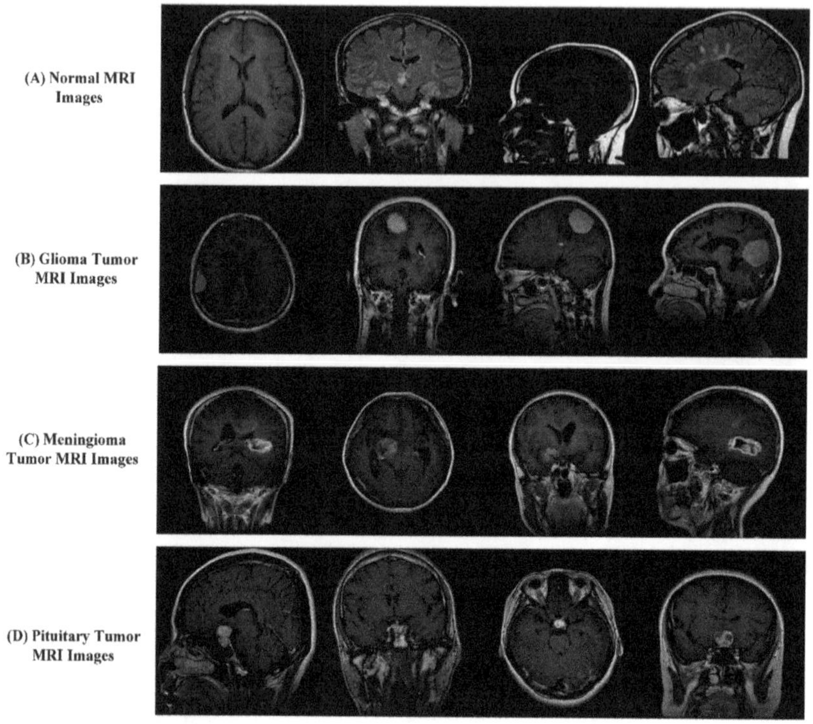


**Figure 2.** Example MRI images of normal and different types of tumors.

### *2.2. Data Augmentation*

On a small volume of data, deep learning models are generally overfitting. Thus, a significant amount of data are usually required to train deep CNNs and provide better generalizability. Data augmentation is generally employed for increasing the original dataset's samples [5,37,38]. In this experiment, random rotation (02013360 degrees), scaling (0.5–1), sharing ($\pm$ 0.05), and image reflecting ($\pm$1 range) are used to augment the dataset. These augmentation techniques are used to strengthen the model's generalization.

### *2.3. Performance Metrics*

The efficiency of the developed model was assessed using several standard evaluation measures. These measures include precision [39], sensitivity [40], accuracy [41], F1-score [42], PR, and ROC curves [43]. TP is defined as truly positive predictions, TN as truly negative predictions, FP as incorrectly positive predictions, and FN for incorrectly negative predictions. In Equation (1), accuracy is defined, calculating the total number of accurate selections. Accordingly, Sensitivity is in Equation (2), precision is denoted in Equation (3), and the F1-score is defined in Equation (4). In Equation (5), the standard error (S.E.) for the F1-score is calculated at a 95% confidence interval (CI), where z = 1.96 [44]. The CI is used as a statistical test to evaluate the uncertainty of the classification CNNs.

$$\text{Accuracy} = \left(\frac{\text{TP} + \text{TN}}{\text{TN} + \text{TP} + \text{FN} + \text{FP}}\right) \times 100 \tag{1}$$

$$\text{Sensitivity} = \frac{\text{TP}}{\text{TP} + \text{FN}} \tag{2}$$

$$\text{Precision} = \frac{\text{TN}}{\text{TN} + \text{FP}} \tag{3}$$

$$\text{F1} - \text{Score} = \frac{2 \times (\text{P} \times \text{R})}{\text{P} + \text{R}} \quad (4)$$

$$\text{CI} = z\sqrt{\frac{error(1 - error)}{total\ instances}} \quad (5)$$

### 2.4. The Developed Deep Res-BRNet-Based Categorization

In this work, we exploit the learning capability of a deep CNN to acquire the tumor's distinctive patterns in brain MRI scans. The strong potential of a deep CNN for learning specific features and patterns from images inspires us to employ them for classification and recognition tasks. Because of their effective learning capability, CNNs are largely employed for feature extraction and classification. In this proposed work, we designed a novel residual and regional CNN architecture using boundary- and region-based operations to classify tumor-specific abnormalities in brain MRI images and named it Res-BRNet. The proposed model is trained in an end-to-end way to learn the tumor-related patterns from MRI scans. The last fully connected layers, followed by softmax-based operation of the proposed deep CNNs, are used for the final classifications. The details of Res-BRNet are depicted in the section given below.

### 2.5. Structural Details of the Developed Res-BRNet

The architecture-level details of the developed Res-BRNet are inspired by standard image processing techniques [45,46]. It is developed to explore the hidden insights in MRI images. In this context, the region- and boundary-based operators with convolution operators are optimized in the proposed architecture to attain the brain tumor patterns excellently. In this work, we exploited spatial and residual blocks [47–49] as baselines to justify the advantages of region uniformity and boundary-related features for obtaining the tumor patterns using CNNs.

As illustrated in Figure 3 in a spatial block, input x is fed into the operation block, and all operators are applied sequentially on input, and at the output operation block, we get T (x) = $f_{conv.}$(x), as sown in Equation (6). The difference between a feed-forward spatial block and residual learning is that residual blocks skip connections from input x to the output of the encoding block and add up with the output of the encoding block $f_{conv.}$(x). At the output of the residual block, we get T (x) = $f_{conv.}$(x) +x, as shown in Equations (7) and (8). As compared to the spatial block, residual learning facilitates the model to capture minor textural and contrast variations and also facilitates overcoming the vanishing gradient problem, as well as improves the learned feature maps and model's convergence.

$$T(x) = f_{conv.}(x) \quad (6)$$

$$T(x) = f_{conv.}(x) + x \quad (7)$$

$$f_{conv.}(x) = T(x) - x \quad (8)$$

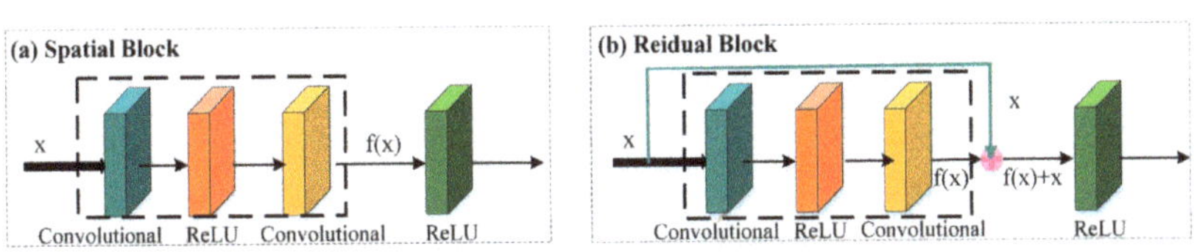


**Figure 3.** The difference in the process of (**a**) plain and (**b**) residual blocks.

The proposed Res-BRNet comprises three spatial blocks at the start, and four residual blocks are used after them. Every spatial block contains a single convolution layer (Equation (8)), batch normalization, and ReLU. The convolution layer exploits tumor-relating patterns, while ReLU performs as an activation function. To learn the region homogeneousness and boundary-related attributes of brain tumors, a max- or average-pooling operation is applied at the end of each spatial block, as shown in Equations (10) and (11).

Figure 4 illustrates the architecture of the developed Res-BRNet. The fully connected (FC) layers stated in Equation (12) are applied in the designed architecture to attain particular features for classification. Dropout layers are used with FC layers to minimize the risk of overfitting.

$$Z_{m,n} = \sum_{u=1}^{r}\sum_{v=1}^{s} Z_{m+u-1,\ n+v-1}\, k_{a,b} \tag{9}$$

$$Z^{Avg}_{m,n} = \frac{1}{T^2}\sum_{u=1}^{t}\sum_{v=1}^{t} Z_{m+u-1,\ n+v-1} \tag{10}$$

$$Z^{Max}_{m,n} = Max_{u=1\ldots t, v=1\ldots t} Z_{m+u-1,\ n+v-1} \tag{11}$$

$$Q = \sum_{b}^{B}\sum_{c}^{C} W_d Z_c \tag{12}$$

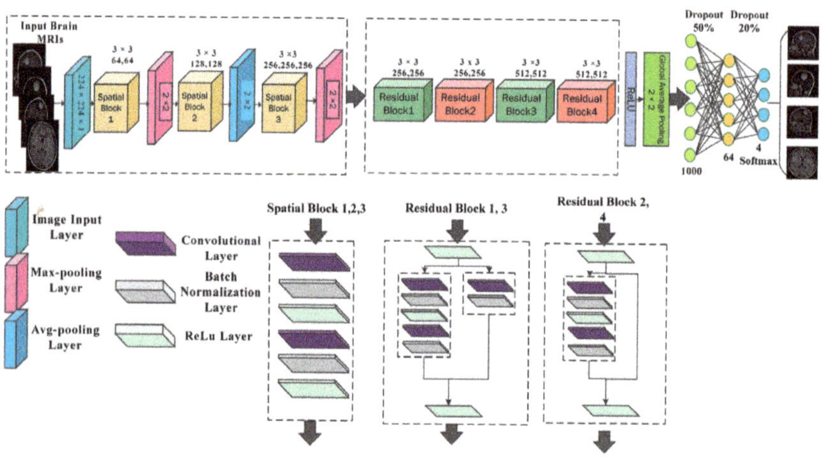

**Figure 4.** Blockwise details of the proposed Res-BRNet.

*Z* illustrates the source feature map of size × *N*, and the filter with size r × s is defined by k in the convolutional operator used in Equation (9). The output map of features is shown by *Z*. m and n have begun from 1 to $(M - r + 1)$ and $(N - s + 1)$, accordingly. As shown in Equations (9)–(11), we regulate the Zavg and Zmax methods, denoted by *Z* Avg and *Z* Max, similarly. In Equations (10) and (11), t indicates the average and max window dimensions. In Equation (12), the dense layer outcome is stated by Q, which employs global operation on $Z_c$. The FC-layer neurons are presented by $W_d$ and preserve essential features for the analysis. We used a kernel size of 3 × 3, padding of 0, 0, and ReLU as the activation function for Res-BRNet. Table 2 shows the used symbols and their description.

**Table 2.** List of symbols used in this work.

| Symbol | Description |
|---|---|
| $T(x)$ | Output of spatial block |
| $T(x)$ | Output of residual block |
| $f_{conv.}(x)$ | Output of convolutional block |
| $Z_{m,n}$ | Source feature map of size $M \times N$ |
| k | CNN's filter with size r × s |

**Table 2.** Cont.

| Symbol | Description |
|---|---|
| Z | Output feature map |
| $Z_{m,n}^{Avg}$ | Output of average pooling |
| $Z_{m,n}^{Max}$ | Output of max pooling |
| Q | Output of dense layer |

### 2.6. Benefits of the Proposed Res-BRNet for Image Content Analysis

Brain MRI scans reveal complex patterns with different intensity levels in distinct regions. Regional smoothness, textural differences, and edges make the basic structure of these patterns. In this study, the developed model is significantly improved by combining the convolutional operator (Equation (9)), enhancing the region homogeneity and boundary-based operations (Equations (10) and (11)), respectively, to differentiate the healthy instances from the tumor-affected MRI scans. In contrast to the developed model, the majority of existing CNN designs employ different convolutional combinations with simply one type of pooling layer to capture invariant features [47,48,50–52]. The proposed Res-BRNet employed regional and boundary-based operations in a systematic order within the modified spatial and residual blocks. The systematic use of boundary and regional operations within spatial blocks extract the brain tumor's homogeneity and heterogeneity patterns, and boundary-related features. Additionally, the residual blocks significantly capture local and global texture variations of brain tumors. The following are the significance of applying the proposed idea in CNN:

- The developed residual and regional CNN architecture aimed to dynamically exploit image smoothness and sharpness, and it may efficiently optimize the level of smoothness and sharpening in harmony with the spatial features of an image.
- Implementing the spatial block with residual learning improves the overall detection ability of the model by acquiring textural features along with spatial correlation from MRI images.
- The systematic use of boundary and regional operations within spatial blocks helps enhance the region homogeneity of various regions. Using average pooling, the region operator helps smooth the regional variations and eliminates noise caused by distortions captured during MRI imaging. On the other hand, Res-BRNet is aided by boundary operators to acquire discriminative local features with the max pooling operation.
- Residual blocks aid the model in capturing textural and minor contrast variations and overcoming the vanishing gradient problem, which is generally produced in very deep architectures.
- Down-sampling is also performed during pooling operations, which increases the model's robustness to small changes in the input image.

### 2.7. Employment of Existing CNNs

Competitive assessment is performed by implementing several existing deep CNN models, including SqueezeNet, ShuffleNet, VGG-16, Xception, ResNet-18, GoogleNet, Inception-V3, and DenseNet-201 [47,48,50,52–57]. Several researchers applied these CNNs to classify MRI images and they have been widely used for many image recognition tasks. Although these models' block architecture and design changed, they all employed a single-pooling operation along the network or changed this for a stridden convolution operation to control complexity. To fine-tune these CNNs for brain tumor classification, we added FC and a classification layer, and employed them in an end-to-end manner.

### 2.8. Implementation Details

A brain MRI dataset was split into two sets, the 80% train set and 20% test set using the hold-out method. Furthermore, the train set was divided into train and validation data to select optimized parameters. 'RMSprop' [58] was employed for optimization with a 'SquaredGradientDecayFactor' of 0.95 throughout the training of CNNs. The learning rate was initially set to 0.0001 with the "LearnRateDropFactor" to be 0.4 and 40 epochs. A small-batch-based technique is used to train models on a batch size of 16 for every epoch. As an activation function, softmax was used, and cross-entropy loss has been reduced for all of the deep CNN optimizations. We have trained deep CNN models using hyper-parameters that have been experimentally optimized by holdout cross-validation techniques, including learning rate, batch size, and number of epochs. These models were executed on MATLAB-based simulations, utilizing a hardware setup consisting of a 2.90-GHz Dell Core i7-7500 CPU and a Nvidia® GTX 1060 Tesla graphics card with CUDA support, Islamabad, Pakistan.

## 3. Results and Discussion

This study suggests a deep CNN-based system for identifying brain tumor patients using MRI images. We performed two different experiments for empirical evaluation of the developed technique. We initially explored the impacts of using simultaneously average and max pooling in spatial blocks of Res-BRNet. Secondly, a general assessment of brain tumor classification is carried out by comparing performances with well-known existing deep CNN models.

### 3.1. *Efficiency Analysis of the Proposed Res-BRNet*

In a comprehensive experimental investigation, the proficiency of the developed Res-BRNet was assessed with well-known CNNs on unseen test data using the Accuracy, F1-score, Sensitivity, Precision, ROC, and PR-AUC. In contrast to accuracy, the F1-score tends to give more weight to precision and sensitivity. The proposed Res-BRNet model correctly classified 1553 samples of three brain tumors and normal instances. Likewise, the proposed Res-BRNet performs similarly by correctly identifying 463 gliomas, 321 meningiomas, 365 pituitary, and 404 normal individuals correspondingly. It is observed that a change in the region and boundary arrangements, as illustrated in Figure 3, improves the overall performance. Figure 5 displays some of the brain MRI images that Res-BRNet misclassifies. The input images have minimum intensity = 0, maximum intensity = 255, mean intensity = 128.5, and standard deviation of 64.3. The training loss and accuracy chart for Res-BRNet are presented in Figure 6. The developed CNN converges smoothly and quickly to achieve its optimal value, as seen in Figure 6. Low contrast, irregular sample patterns, and varying illumination variations are the probable reasons for misclassification. The generalization and robustness enhancement of test samples are achieved by using several data augmentation strategies while developing CNNs.

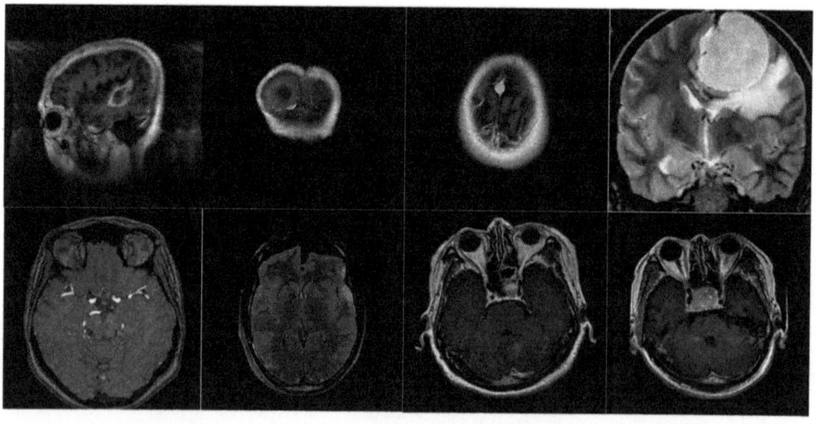

**Figure 5.** Normal and three tumor images misclassified by Res-BRNet.

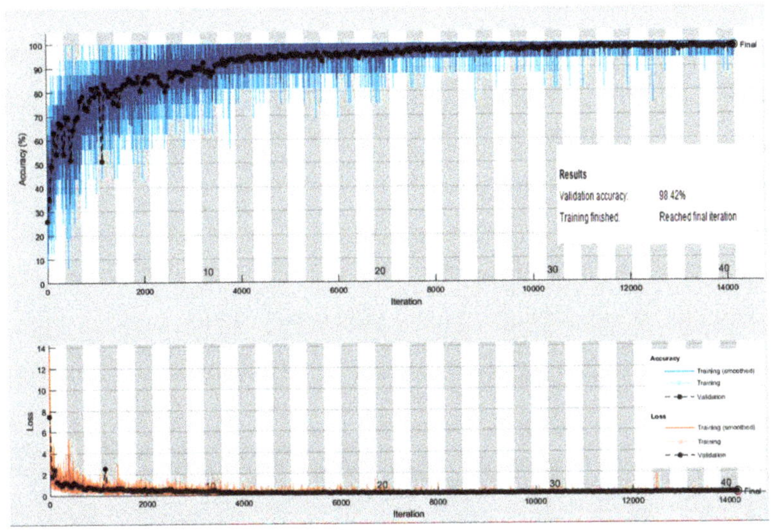


**Figure 6.** Plots of training the proposed Res-BRNet.

#### 3.1.1. Analysis of Performance with Baseline Methods

The significance of the anticipated idea is assessed by evaluating the performance, especially in comparison to residual learning and spatial exploitation-based architectures. Both baseline architectures, VGG-16 and ResNet-18, are almost as deep as Res-BRNet. Spatial block-based architectures exploit one type of down-sampling operation, and residual blocks use stridden convolution instead of pooling down.

Initially, both residual learning and spatial exploitation combinedly improve the classification ability of the proposed Res-BRNet, with an F1-score of 0.9385 and accuracy of 96.79%. Furthermore, employing both pooling operators (region and edge-based operators) in Res-BRNet improves the overall performance of the developed model, as shown in Table 3. Thus, according to performance comparison, the Res-BRNet consisting of region and edge-based operator shows exceptional performance compared to residual blocks and spatial block-based architectures in terms of F1-score (0.9641) and accuracy of (98.22%). The proposed Res-BRNet model significantly reduced the false positives and increased the true positives, leading to enhanced precision.

**Table 3.** Performance evaluation of the developed Res-BRNet with baseline architectures using the test data.

| Model | Performance Comparison with Baseline Architecture CNNs | | | |
|---|---|---|---|---|
| | Accuracy % | Sensitivity | Precision | F1-Score |
| VGG-16 (SB + DA *) | 93.32 | 0.9285 | 0.8231 | 0.8719 |
| TL_VGG-16 (SB+ DA *) | 94.66 | 0.9426 | 0.8569 | 0.8961 |
| ResNet-18 (SB+ DA *) | 95.67 | 0.9566 | 0.8788 | 0.9158 |
| TL_ResNet-18 (RB+ DA *) | 96.44 | 0.9641 | 0.8998 | 0.9303 |
| **Res-BRNet (RB+SB) (without region and edge operator + DA *)** | **96.79** | **0.9661** | **0.9467** | **0.9385** |
| **Res-BRNet (RB+SB) (region and edge operator, and without DA *)** | **97.59** | **0.9712** | **0.9723** | **0.9446** |
| **Proposed Res-BRNet (RB+SB) (region and edge operator, and DA *)** | **98.22** | **0.9811** | **0.9822** | **0.9641** |

Data augmentation (DA) *, spatial block (SB), residual block (RB).

### 3.1.2. Analysis of Performance with Reported Techniques

The performance of the proposed work has also been compared to previously reported studies and exhibited in Table 4. Irmak [59] introduced two CNN models: (i) tumor detection (tumor or no tumor) and (ii) tumor type classification. The reported testing accuracy for tumor type classification was 92.66%. E. M. Senan et al. [60] performed different experiments for brain tumor analysis by combining deep learning and conventional ML methods and achieved the best performance, with 95.10% accuracy, 95.25% sensitivity, and 98.50% specificity with AlexNet + SVM. M. F. Alanazi et al. [61] developed a TL-based model to detect the subclass of the tumor. The developed transfer-learned model exhibited an accuracy of 95.75% for the brain MRI images. Another study by Kang et al. [62] utilized deep features to train an SVM model. However, the training–vector feature size was large and required a significant amount of computational time for training.

**Table 4.** Performance evaluation of the developed Res-BRNet with reported techniques using the test data.

| Model | Performance Comparison with Reported Techniques | | | | | |
|---|---|---|---|---|---|---|
| | Dataset | Class | Accuracy % | Sensitivity | Precision | F1-Score ± CI |
| Custom-made CNN [59] | Figshare | 5 | 92.66 | - | - | - |
| AlexNet + SVM [60] | MRI dataset | 4 | 95.10 | 95.25 | - | - |
| Isolated-CNN [61] | Kaggle | 4 | 92.67 | - | - | - |
| (DenseNet-169 + Shufflenet + MnasNet) feature [62] | BT-large-4c | 4 | 93.72 | - | - | - |
| **Proposed Res-BRNet** | **Figshare + Kaggle** | **4** | **98.22** | **0.9811** | **0.9822** | **0.9641 ± 0.0359** |

The proposed framework achieved a classification accuracy of 93.72% when tested on an unseen dataset. Table 4 and Figures 7 and 8 show that the suggested Res-BRNet (with region and edge operator, and augmentation) considerably enhances the detection ability for all three brain tumors, as well as for normal MRI images compared to baseline residual learning and spatial exploitation architectures, without employing region and edge operators and data augmentation.

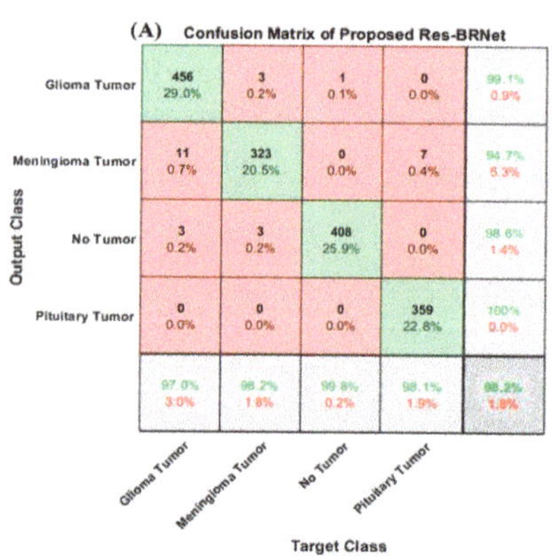


**Figure 7.** *Cont.*

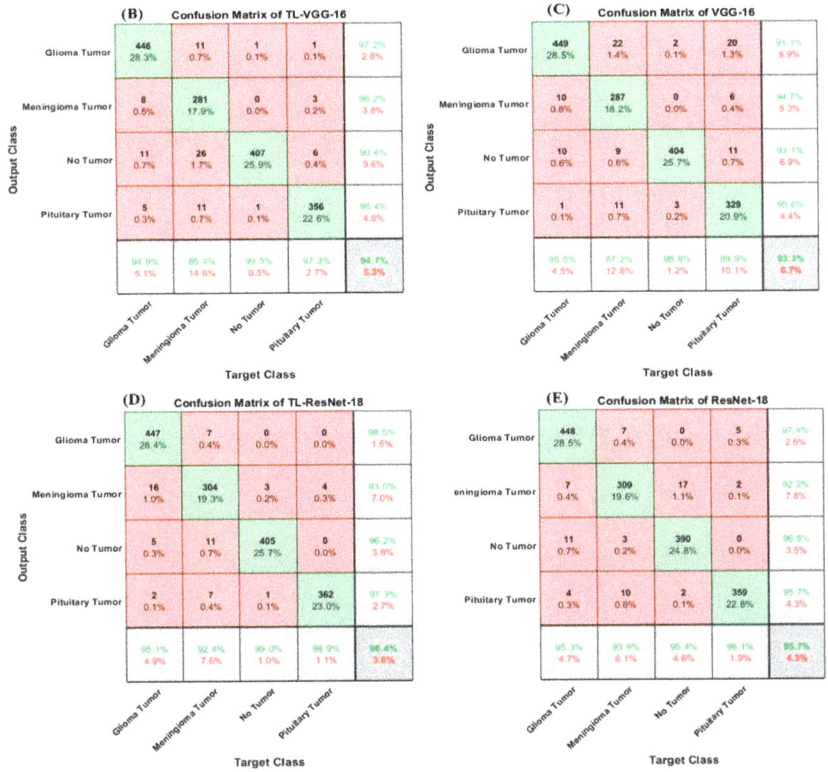


Figure 7. Confusion matrix-based performance assessment of the developed Res-BRNet and baseline architectures for different brain tumors.

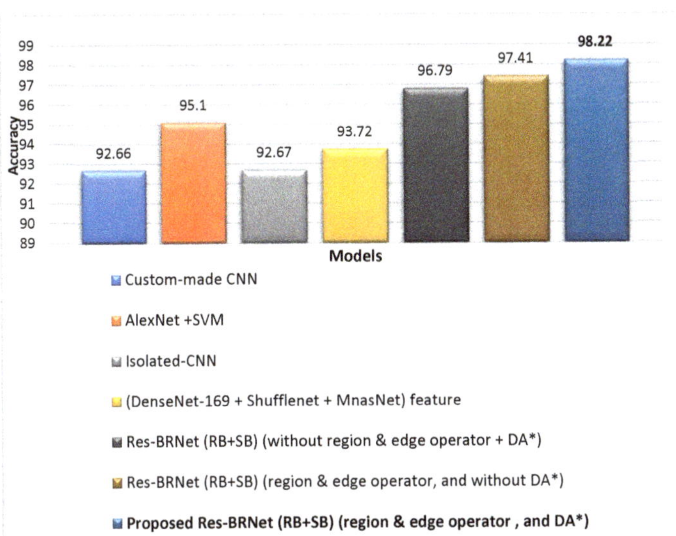


Figure 8. Performance assessment of the developed Res-BRNet with state-of-the-art CNN architectures. (* data augmentation (DA), spatial block (SB), residual block (RB)).

3.1.3. Performance Assessment with Existing CNNs

The effectiveness of the developed Res-BRNet is compared with custom-made training-from-scratch and transfer learning-based (TL-based) existing CNNs, namely SqueezeNet, ShuffleNet, VGG-16, Xception, ResNet-18, GoogleNet, InceptionV3, and DenseNet-201. Tables 5 and 6 illustrate that the proposed models' performance analysis indicates that Res-BRNet is more efficient at identifying the patterns specific to the brain tumors in MRI scans with standard measures of accuracy and the F1-score. This performance of the proposed model improved by using average_ and max_pooling operations systematically in the designed CNN. Overall, the developed Res-BRNet achieved a performance gain in terms of accuracy (TR-SC to proposed Res-BRNet (1.2–11.06%)), (TL-B to proposed Res-BRNet (0.45–7.31)), as shown in Table 7. In general, the model is encouraged to learn highly discriminative features and fine-grained information from the raw MR image by the use of these opposing pooling operations.

**Table 5.** Performance analysis of the existing standard custom CNNs and the proposed Res-BRNet on the testing data with the confidence interval (CI).

| Model | Performance Comparison with Custom-Made CNNs | | | |
|---|---|---|---|---|
| | **Accuracy %** | **Sensitivity** | **Precision** | **F1-Score ± CI** |
| SqueezeNet | 87.16 | 0.8691 | 0.6946 | 0.7671 ± 0.2430 |
| ShuffleNet | 89.45 | 0.8923 | 0.7411 | 0.8047 ± 0.1953 |
| VGG-16 | 93.32 | 0.9285 | 0.8231 | 0.8719 ± 0.1281 |
| Xception | 95.36 | 0.9531 | 0.8721 | 0.9101 ± 0.0899 |
| ResNet-18 | 95.67 | 0.9566 | 0.8788 | 0.9158 ± 0.0842 |
| GoogleNet | 95.87 | 0.9593 | 0.8851 | 0.9196 ± 0.0804 |
| Inception-V3 | 96.56 | 0.9676 | 0.9015 | 0.9331 ± 0.0669 |
| DenseNet-201 | 97.01 | 0.9668 | 0.9175 | 0.9406 ± 0.0594 |
| **Proposed Res-BRNet (with region and edge operator and augmentation)** | **98.22** | **0.9811** | **0.9822** | 0.9641 ± 0.0359 |

**Table 6.** Performance analysis of the existing standard TL-based CNNs and the proposed Res-BRNet on the testing data with the confidence interval (CI).

| Model | Performance Comparison with TL-Based CNNs | | | |
|---|---|---|---|---|
| | **Accuracy %** | **Sensitivity** | **Precision** | **F1-Score ± CI** |
| TL_SqueezeNet | 90.91 | 0.9108 | 0.7685 | 0.8315 ± 0.1685 |
| TL_ShuffleNet | 92.31 | 0.9155 | 0.8056 | 0.8521 ± 0.1479 |
| TL_VGG-16 | 94.66 | 0.9426 | 0.8569 | 0.8961 ± 0.1039 |
| TL_Xception | 96.37 | 0.9611 | 0.8996 | 0.9285 ± 0.0715 |
| TL_ResNet-18 | 96.44 | 0.9641 | 0.8998 | 0.9303 ± 0.0697 |
| TL_GoogleNet | 96.37 | 0.9641 | 0.8985 | 0.9291 ± 0.0709 |
| TL_Inception-V3 | 97.26 | 0.9711 | 0.9225 | 0.9459 ± 0.0541 |
| TL_DenseNet-201 | 97.77 | 0.9778 | 0.9349 | 0.9557 ± 0.0443 |
| **Proposed Res-BRNet (with region and edge operator, and augmentation)** | **98.22** | **0.9811** | **0.9822** | 0.9641 ± 0.0359 |

**Table 7.** A performance gain of the developed Res-BRNet compared to TR-SC and TL-based models.

| Improvement | Accuracy % | Sensitivity % | Precision % | F1-Score % |
|---|---|---|---|---|
| TR-SC to TL-B | 0.76–3.75 | 1.1–4.45 | 1.74–7.39 | 1.51–6.44 |
| TR-SC to Proposed Res-BRNet | 1.2–11.06 | 1.43–11.02 | 6.47–28.76 | 2.35–19.07 |
| TL-B to Proposed Res-BRNet | 0.45–7.31 | 1.1–4.17 | 1.74–7.39 | 1.51–6.44 |

### 3.1.4. Feature Space-Based Performance Analysis

To understand decision-making behavior, the proposed Res-BRNet and best-performing baseline architectures, ResNet-18 and VGG-16, are evaluated to examine their learned feature spaces. Characteristics of the feature space responsible for the discrimination capability of a classifier. Features with classes distinguishably improve the model's learning and lower the variance on distinct samples. T-distributed Stochastic Neighbor Embedding (t-SNE) [63] is an algorithm that is well-suited to visualize by embedding high-dimensional points in low dimensions based on similarities between points. Figure 9 illustrates the 2-D t-SNE plots for the proposed Res-BRNet, baseline models ResNet-18 and VGG-16, best-performing TL-B Densnet-201, and worst-performing TL-SqueezeNet using testing data. Data visualization shows that the feature space diversity is significantly improved by using both the boundary and regional operations, and it improves the model's performance.

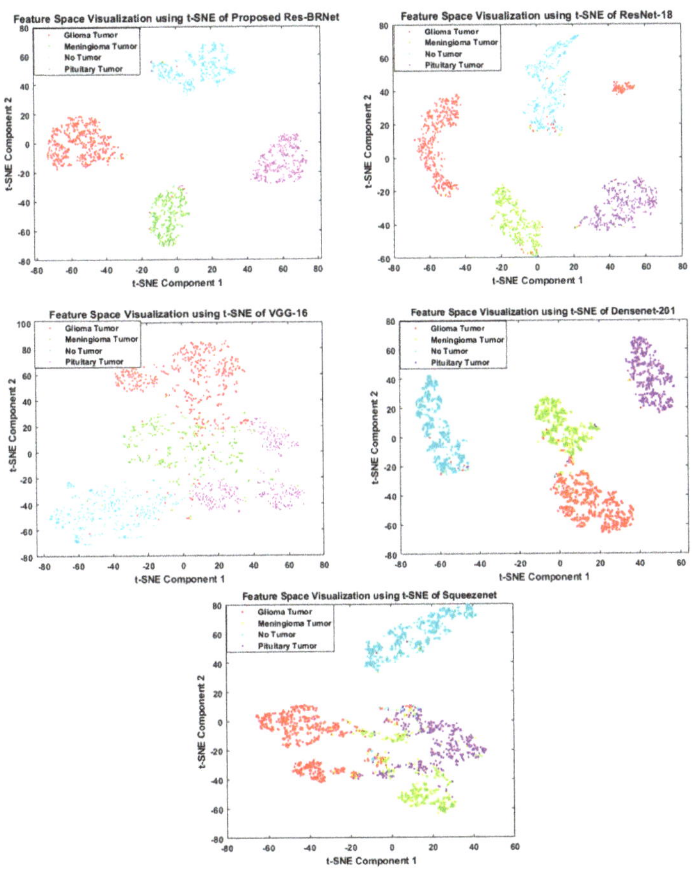


**Figure 9.** Feature space-based performance analysis of the developed Res-BRNet with baseline architectures (VGG-16, ResNet-18), best-performing TL-B model (DenseNet-201), and worst-performing TL-B model (SqueezeNet).

#### 3.1.5. ROC and PR-AUC-Based Analysis

The ROC curve is essential to achieve the optimal analytic threshold for the classifier. ROC curve graphically displays the classifier distinction capability at possible threshold values. As shown in Figure 10, the proposed Res-BRNet achieved an AUC of (ROC_AUC of 0.9921 and PR_AUC of 0.9702).

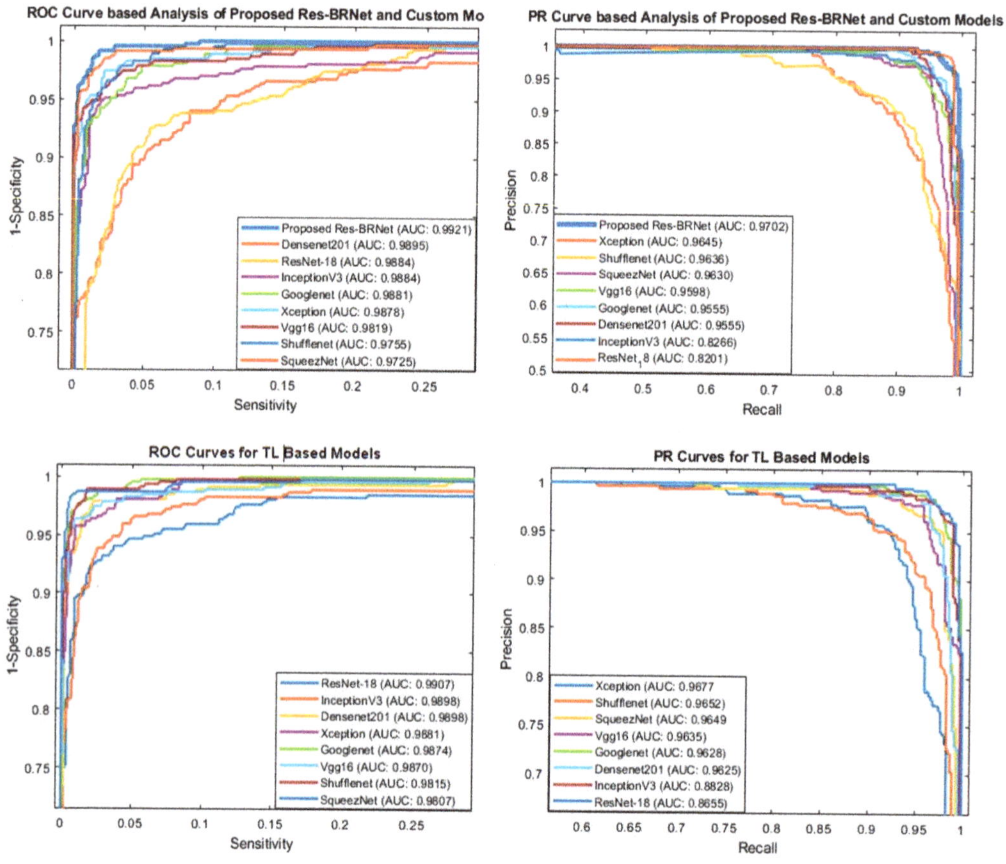


**Figure 10.** Detection rate analysis of the developed Res-BRNet with existing CNNs.

The brain MRI dataset includes patients with tumors. ROC and PR-AUC quantitative analysis proves that the suggested method enhances sensitivity by having the lowest false positive rate. This shows that the presented approach for classifying brain tumors has a lot of potential to be used in the analysis of brain tumors.

#### 3.1.6. Screening Effectiveness of the Proposed Technique

The precision and detection rates (sensitivity) are also critical metrics to evaluate a medical diagnostic system's efficiency, where the costs of false positives and false negatives can be high. The brain tumor detection system needs to have a good detection performance. As seen in Figure 10 and Tables 5 and 6, the detection rate and precision of the proposed approach are evaluated for brain MR images. As shown in the quantitative study (Figure 11), the Res-BRNet (Sensitivity: 0.9811, Precision: 0.9822) increases the prediction system's accuracy and has a high prediction rate. Consequently, it is expected to help radiologists with good accuracy and may be utilized to enhance efficiency by decreasing the burden on medical professionals.

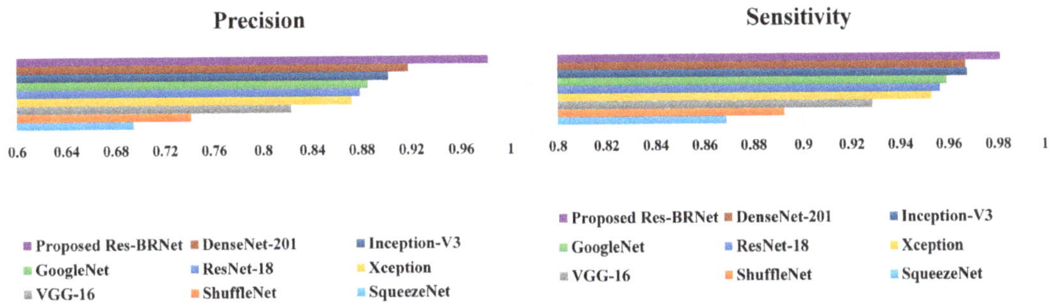


**Figure 11.** Performance analysis of the developed Res-BRNet with existing CNNs.

## 4. Conclusions

Brain tumor diagnosis at an early stage is crucial to cure the patient. Therefore, in this work, a new customized deep CNN model is developed to classify the brain MRI scans of meningioma, glioma, and pituitary tumor patients from healthy entities. The proposed model benefits from data augmentation and learning discriminative features using regional and boundary operators in the developed Res-BRNet. Moreover, the developed Res-BRNet employs spatial and residual ideas to acquire feature maps with diverse rich information, improving the capability to learn homogeneity, textural variation, and tumor's structural patterns. The performance exploration of the developed is analyzed with the existing deep CNN models. The experiment results show that the proposed Res-BRNet outperforms existing CNN architectures, indicating accuracy and F1-score improvement. The developed method classifies brain tumors with an accuracy of 98.22%, an F1-Score of 0.9641, and sensitivity and precision of 0.9811 and 0.9822, respectively. The proposed approach will likely facilitate healthcare professionals in making diagnoses of brain tumors. Additionally, it motivates us to explore different forms of abnormalities in brain MRI and other medical images. Exploring novel techniques and methodologies to further optimize and validate the Res-BRNet model for clinical use may be considered in future.

**Author Contributions:** Conceptualization, M.M.Z. and S.H.K.; Methodology, M.M.Z. and S.H.K.; Software, M.M.Z.; Validation, T.A.; Formal analysis, A.S.A.M., S.A., A.A. and B.K.A.; Investigation, S.H.K.; Resources, A.S.A.M. and A.A.; Data curation, T.J.A. and H.A.S.; Writing—original draft, M.M.Z.; Writing—review & editing, S.H.K. and H.A.S.; Visualization, S.H.K., T.A., S.A. and B.K.A.; Supervision, M.M.Z. and S.H.K.; Project administration, S.H.K.; Funding acquisition, T.J.A. All authors have read and agreed to the published version of the manuscript.

**Funding:** The funding of the Princess Nourah bint Abdulrahman University Researchers Supporting project number (PNURSP2024R513), Princess Nourah bint Abdulrahman University, P.O. Box 84428, Riyadh 11671, Saudi Arabia.

**Informed Consent Statement:** Not applicable.

**Data Availability Statement:** The code of this paper is available at https://github.com/MumtazZahoor1/Res-BRNet-.git (accessed on 7 April 2024). The brain tumor datasets collected during the current study are available in the standard open access Kaggle [34], Br35H [35], and Figshare [36] repositories, and are verified by medical experts. Moreover, the datasets are available in publicly accessible repositories, which are described in Section 2.1. Correspondence and requests for materials should be addressed to S.H.K.

**Acknowledgments:** The funding of the Princess Nourah bint Abdulrahman University Researchers Supporting project number (PNURSP2024R513), Princess Nourah bint Abdulrahman University, P.O. Box 84428, Riyadh 11671, Saudi Arabia. Moreover, we thank the Department of Faculty of Computer Sciences, Ibadat International University, Islamabad, and Computer Systems Engi-neering, University of Engineering and Applied Sciences (UEAS), Swat, for providing the necessary resources.

**Conflicts of Interest:** The authors declare that they have no known competing financial interests or personal relationships that could have appeared to influence the work reported in this paper.

## References

1. Behin, A.; Hoang-Xuan, K.; Carpentier, A.F.; Delattre, J.-Y. Primary brain tumours in adults. *Lancet* **2003**, *361*, 323–331. [CrossRef] [PubMed]
2. Miller, K.D.; Ostrom, Q.T.; Kruchko, C.; Patil, N.; Tihan, T.; Cioffi, G.; Fuchs, H.E.; Waite, K.A.; Jemal, A.; Siegel, R.L.; et al. Brain and other central nervous system tumor statistics. *CA. Cancer J. Clin.* **2021**, *71*, 381–406. [CrossRef] [PubMed]
3. El-Dahshan, E.-S.A.; Mohsen, H.M.; Revett, K.; Salem, A.-B.M. Computer-aided diagnosis of human brain tumor through MRI: A survey and a new algorithm. *Expert Syst. Appl.* **2014**, *41*, 5526–5545. [CrossRef]
4. Iftekharuddin, K.M.; Zheng, J.; Islam, M.A.; Ogg, R.J. Fractal-based brain tumor detection in multimodal MRI. *Appl. Math. Comput.* **2009**, *207*, 23–41. [CrossRef]
5. Zahoor, M.M.; Qureshi, S.A.; Bibi, S.; Khan, S.H.; Khan, A.; Ghafoor, U.; Bhutta, M.R. A New Deep Hybrid Boosted and Ensemble Learning-Based Brain Tumor Analysis Using MRI. *Sensors* **2022**, *22*, 2726. [CrossRef] [PubMed]
6. Khan, A.; Khan, S.H.; Saif, M.; Batool, A.; Sohail, A.; Khan, M.W. A Survey of Deep Learning Techniques for the Analysis of COVID-19 and their usability for Detecting Omicron. *arXiv* **2022**, arXiv:2202.06372. [CrossRef]
7. Khan, S.H.; Sohail, A.; Khan, A.; Lee, Y.S. Classification and region analysis of COVID-19 infection using lung CT images and deep convolutional neural networks. *arXiv* **2020**, arXiv:2009.08864.
8. Asam, M.; Hussain, S.J.; Mohatram, M.; Khan, S.H.; Jamal, T.; Zafar, A.; Khan, A.; Ali, M.U.; Zahoora, U. Detection of Exceptional Malware Variants Using Deep Boosted Feature Spaces and Machine Learning. *Appl. Sci.* **2021**, *11*, 10464. [CrossRef]
9. Zahoor, M.M.; Qureshi, S.A.; Khan, A.; Rehman, A.U.; Rafique, M. A novel dual-channel brain tumor detection system for MR images using dynamic and static features with conventional machine learning techniques. *Waves Random Complex Media* **2022**, 1–20. [CrossRef]
10. Khan, S.H.; Sohail, A.; Khan, A.; Lee, Y.-S. COVID-19 detection in chest X-ray images using a new channel boosted CNN. *arXiv* **2020**, arXiv:2012.05073. [CrossRef]
11. Rauf, Z.; Sohail, A.; Khan, S.H.; Khan, A.; Gwak, J.; Maqbool, M. Attention-guided multi-scale deep object detection framework for lymphocyte analysis in IHC histological images. *Microscopy* **2023**, *72*, 27–42. [CrossRef] [PubMed]
12. Zahoora, U.; Khan, A.; Rajarajan, M.; Khan, S.H.; Asam, M.; Jamal, T. Ransomware detection using deep learning based unsupervised feature extraction and a cost sensitive Pareto Ensemble classifier. *Sci. Rep.* **2022**, *12*, 15647. [CrossRef] [PubMed]
13. Khan, S.H.; Iqbal, J.; Hassnain, S.A.; Owais, M.; Mostafa, S.M.; Hadjouni, M.; Mahmoud, A. COVID-19 detection and analysis from lung CT images using novel channel boosted CNNs. *Expert Syst. Appl.* **2023**, *229*, 120477. [CrossRef]
14. Alqahtani, A.; Zahoor, M.M.; Nasrullah, R.; Fareed, A.; Cheema, A.A.; Shahrose, A.; Irfan, M.; Alqhatani, A.; Alsulami, A.A.; Zaffar, M.; et al. Computer Aided COVID-19 Diagnosis in Pandemic Era Using CNN in Chest X-ray Images. *Life* **2022**, *12*, 1709. [CrossRef] [PubMed]
15. Akkus, Z.; Galimzianova, A.; Hoogi, A.; Rubin, D.L.; Erickson, B.J. Deep learning for brain MRI segmentation: State of the art and future directions. *J. Digit. Imaging* **2017**, *30*, 449–459. [CrossRef]
16. Khan, S.H.; Khan, A.; Lee, Y.S.; Hassan, M.; Jeong, W.K. Segmentation of shoulder muscle MRI using a new Region and Edge based Deep Auto-Encoder. *Multimed. Tools Appl.* **2022**, *82*, 14963–14984. [CrossRef]
17. Domingues, I.; Pereira, G.; Martins, P.; Duarte, H.; Santos, J.; Abreu, P.H. Using deep learning techniques in medical imaging: A systematic review of applications on CT and PET. *Artif. Intell. Rev.* **2020**, *53*, 4093–4160. [CrossRef]
18. Goodfellow, I.; Bengio, Y.; Courville, A. *Deep Learning—Ian Goodfellow, Yoshua Bengio, Aaron Courville—Google Books*; MIT Press: Cambridge, MA, USA, 2016; p. 800.
19. Asam, M.; Khan, S.H.; Akbar, A.; Bibi, S.; Jamal, T.; Khan, A.; Ghafoor, U.; Bhutta, M.R. IoT malware detection architecture using a novel channel boosted and squeezed CNN. *Sci. Rep.* **2022**, *12*, 15498. [CrossRef]
20. Gómez-Flores, W.; Pereira, W.C.d.A. A comparative study of pre-trained convolutional neural networks for semantic segmentation of breast tumors in ultrasound. *Comput. Biol. Med.* **2020**, *126*, 2419–2428. [CrossRef]
21. Rawat, W.; Wang, Z. Deep convolutional neural networks for image classification: A comprehensive review. *Neural Comput.* **2017**, *29*, 2352–2449. [CrossRef]
22. Arabahmadi, M.; Farahbakhsh, R.; Rezazadeh, J. Deep Learning for Smart Healthcare—A Survey on Brain Tumor Detection from Medical Imaging. *Sensors* **2022**, *22*, 1960. [CrossRef] [PubMed]
23. Shirazi, A.Z.; Fornaciari, E.; McDonnell, M.D.; Yaghoobi, M.; Cevallos, Y.; Tello-Oquendo, L.; Inca, D.; Gomez, G.A. The Application of Deep Convolutional Neural Networks to Brain Cancer Images: A Survey. *J. Pers. Med.* **2020**, *10*, 224. [CrossRef] [PubMed]
24. Xie, Y.; Zaccagna, F.; Rundo, L.; Testa, C.; Agati, R.; Lodi, R.; Manners, D.N.; Tonon, C. Convolutional Neural Network Techniques for Brain Tumor Classification (from 2015 to 2022): Review, Challenges, and Future Perspectives. *Diagnostics* **2022**, *12*, 1850. [CrossRef] [PubMed]
25. Hoang, Q.T.; Yong, K.-T.; Liu, X.; Mahony, D.; Chaitarvornkit, A.; Cohen, A.; Grootswagers, T. Detecting mild traumatic brain injury for athletes using SSVEP classification: A case study. *Biomed. Signal Process. Control* **2023**, *86*, 105274. [CrossRef]

26. Abdel-Nabi, H.; Ali, M.Z.; Awajan, A. A multi-scale 3-stacked-layer coned U-net framework for tumor segmentation in whole slide images. *Biomed. Signal Process. Control* **2023**, *86*, 105273. [CrossRef]
27. Paul, J.S.; Plassard, A.J.; Landman, B.A.; Fabbri, D. Deep learning for brain tumor classification. *Med. Imaging 2017 Biomed. Appl. Mol. Struct. Funct. Imaging* **2017**, *10137*, 1013710. [CrossRef]
28. Deepak, S.; Ameer, P. Brain tumor classification using deep CNN features via transfer learning. *Comput. Biol. Med.* **2019**, *111*, 103345. [CrossRef] [PubMed]
29. Çinar, A.; Yildirim, M. Detection of tumors on brain MRI images using the hybrid convolutional neural network architecture. *Med. Hypotheses* **2020**, *139*, 109684. [CrossRef]
30. Sultan, H.H.; Salem, N.M.; Al-Atabany, W. Multi-Classification of Brain Tumor Images Using Deep Neural Network. *IEEE Access* **2019**, *7*, 69215–69225. [CrossRef]
31. Khawaldeh, S.; Pervaiz, U.; Rafiq, A.; Alkhawaldeh, R.S. Noninvasive grading of glioma tumor using magnetic resonance imaging with convolutional neural networks. *Appl. Sci.* **2018**, *8*, 27. [CrossRef]
32. Khan, M.A.; Ashraf, I.; Alhaisoni, M.; Damaševičius, R.; Scherer, R.; Rehman, A.; Bukhari, S.A.C. Multimodal Brain Tumor Classification Using Deep Learning and Robust Feature Selection: A Machine Learning Application for Radiologists. *Diagnostics* **2020**, *10*, 565. [CrossRef] [PubMed]
33. Özyurt, F.; Sert, E.; Avci, E.; Dogantekin, E. Brain tumor detection based on Convolutional Neural Network with neutrosophic expert maximum fuzzy sure entropy. *Measurement* **2019**, *147*, 106830. [CrossRef]
34. Br35H: Brain Tumor Detection 2020. Kaggle. Available online: https://www.kaggle.com/datasets/ahmedhamada0/brain-tumor-detection (accessed on 8 September 2022).
35. Brain Tumor Classification (MRI). Kaggle. Available online: https://www.kaggle.com/datasets/sartajbhuvaji/brain-tumor-classification-mri (accessed on 8 September 2022).
36. Jun, C. Brain Tumor Dataset. 2017. Available online: https://figshare.com/articles/brain_tumor_dataset/1512427 (accessed on 2 February 2022).
37. Shorten, C.; Khoshgoftaar, T.M. A survey on Image Data Augmentation for Deep Learning. *J. Big Data* **2019**, *6*, 60. [CrossRef]
38. Khan, A.; Sohail, A.; Zahoora, U.; Qureshi, A.S. A survey of the recent architectures of deep convolutional neural networks. *Artif. Intell. Rev.* **2020**, *53*, 5455–5516. [CrossRef]
39. Buckland, M.; Gey, F. The relationship between recall and precision. *J. Am. Soc. Inf. Sci.* **1994**, *45*, 12–19. [CrossRef]
40. Davis, J.; Goadrich, M. The relationship between Precision-Recall and ROC curves. In Proceedings of the 23rd International Conference on Machine Learning, ACM, Pittsburgh, PA, USA, 25–29 June 2006; pp. 233–240.
41. Diebold, F.X.; Mariano, R.S. Comparing predictive accuracy. *J. Bus. Econ. Stat.* **2002**, *20*, 134–144. [CrossRef]
42. Sokolova, M.; Japkowicz, N.; Szpakowicz, S. *Beyond Accuracy, F-Score and ROC: A Family of Discriminant Measures for Performance Evaluation*; Springer: Berlin/Heidelberg, Germany, 2006; pp. 1015–1021.
43. Cortes, C.; Mohri, M. Confidence intervals for the area under the ROC Curve. In *Advances in Neural Information Processing Systems 17*; MIT: Cambridge, MA, USA, 2005.
44. DiCiccio, T.J.; Efron, B. Bootstrap confidence intervals. *Stat. Sci.* **1996**, *11*, 189–228. [CrossRef]
45. Hussain, S.; Khan, A. Coronavirus Disease Analysis using Chest X-ray Images and a Novel Deep Convolutional Neural Network. *Photodiagnosis Photodyn. Ther.* **2020**, *35*, 102473.
46. Mallick, A.; Roy, S.; Chaudhuri, S.S.; Roy, S. Optimization of Laplace of Gaussian (LoG) filter for enhanced edge detection: A new approach. In Proceedings of the International Conference on Control, Instrumentation, Energy and Communication, CIEC 2014, Calcutta, India, 31 January–2 February 2014; pp. 658–661. [CrossRef]
47. He, K.; Zhang, X.; Ren, S.; Sun, J. Deep Residual Learning for Image Recognition. *Proc. IEEE Comput. Soc. Conf. Comput. Vis. Pattern Recognit.* **2015**, *2016*, 770–778. [CrossRef]
48. Simonyan, K.; Zisserman, A. Very deep convolutional networks for large-scale image recognition. *arXiv* **2014**, arXiv:1409.1556.
49. Khan, S.H.; Shah, N.S.; Nuzhat, R.; Majid, A.; Alquhayz, H.; Khan, A. Malaria parasite classification framework using a novel channel squeezed and boosted CNN. *Reprod. Syst. Sex. Disord.* **2022**, *71*, 271–282. [CrossRef] [PubMed]
50. Huang, G.; Liu, Z.; Van Der Maaten, L.; Weinberger, K.Q. Densely connected convolutional networks. In Proceedings of the 30th IEEE Conference on Computer Vision and Pattern Recognition, CVPR 2017, Honolulu, HI, USA, 21–26 July 2017; pp. 2261–2269. [CrossRef]
51. Zagoruyko, S.; Komodakis, N. Wide Residual Networks. *Proc. Br. Mach. Vis. Conf.* **2016**, *87*, 87.1–87.12.
52. Iandola, F.N.; Moskewicz, M.W.; Ashraf, K.; Han, S.; Dally, W.J.; Keutzer, K. SqueezeNet. *arXiv* **2016**, arXiv:1602.07360.
53. Zhang, X.; Zhou, X.; Lin, M.; Sun, J. ShuffleNet: An Extremely Efficient Convolutional Neural Network for Mobile Devices. In Proceedings of the IEEE Computer Society Conference on Computer Vision and Pattern Recognition, Salt Lake City, UT, USA, 18–23 June 2018. [CrossRef]
54. Chollet, F. Xception: Deep learning with depthwise separable convolutions. In Proceedings of the 2017 IEEE Conference on Computer Vision and Pattern Recognition (CVPR), Honolulu, HI, USA, 21–26 July 2017; pp. 1251–1258.
55. Szegedy, C.; Liu, W.; Jia, Y.; Sermanet, P.; Reed, S.; Anguelov, D.; Erhan, D.; Vanhoucke, V.; Rabinovich, A.; Liu, W.; et al. Going deeper with convolutions. In Proceedings of the 2015 IEEE Conference on Computer Vision and Pattern Recognition (CVPR), Boston, MA, USA, 7–12 June 2015; pp. 1–9.

56. Yadav, S.S.; Jadhav, S.M. Deep convolutional neural network based medical image classification for disease diagnosis. *J. Big Data* **2019**, *6*, 113. [CrossRef]
57. Szegedy, C.; Vanhoucke, V.; Ioffe, S.; Shlens, J.; Wojna, Z. Rethinking the Inception Architecture for Computer Vision. In Proceedings of the IEEE Computer Society Conference on Computer Vision and Pattern Recognition, Las Vegas, NV, USA, 27–30 June 2016; pp. 2818–2826. [CrossRef]
58. Ruder, S. An Overview of Gradient Descent Optimization Algorithms. *arXiv* **2016**, arXiv:1609.04747.
59. Irmak, E. Multi-Classification of Brain Tumor MRI Images Using Deep Convolutional Neural Network with Fully Optimized Framework. *Iran. J. Sci. Technol. Trans. Electr. Eng.* **2021**, *45*, 1015–1036. [CrossRef]
60. Senan, E.M.; Jadhav, M.E.; Rassem, T.H.; Aljaloud, A.S.; Mohammed, B.A.; Al-Mekhlafi, Z.G. Early Diagnosis of Brain Tumour MRI Images Using Hybrid Techniques between Deep and Machine Learning. *Comput. Math. Methods Med.* **2022**, *2022*, 1–17. [CrossRef] [PubMed]
61. Alanazi, M.F.; Ali, M.U.; Hussain, S.J.; Zafar, A.; Mohatram, M.; Irfan, M.; AlRuwaili, R.; Alruwaili, M.; Ali, N.H.; Albarrak, A.M. Brain Tumor/Mass Classification Framework Using Magnetic-Resonance-Imaging-Based Isolated and Developed Transfer Deep-Learning Model. *Sensors* **2022**, *22*, 372. [CrossRef] [PubMed]
62. Kang, J.; Ullah, Z.; Gwak, J. Mri-based brain tumor classification using ensemble of deep features and machine learning classifiers. *Sensors* **2021**, *21*, 2222. [CrossRef]
63. Van Der Maaten, L.; Hinton, G. Visualizing Data using t-SNE. *J. Mach. Learn. Res.* **2008**, *9*, 2579–2605.

biomedicines

MDPI

*Article*

# Transfer Learning Approaches for Brain Metastases Screenings

**Minh Sao Khue Luu [1,*], Bair N. Tuchinov [1], Victor Suvorov [1], Roman M. Kenzhin [1], Evgeniya V. Amelina [2,4] and Andrey Yu. Letyagin [1,3]**

1 The Artificial Intelligence Research Center of Novosibirsk State University, Novosibirsk State University, 630090 Novosibirsk, Russia
2 FSBI Federal Neurosurgical Center, 630087 Novosibirsk, Russia
3 Research Institute of Clinical and Experimental Lymphology, Branch of the Institute of Cytology and Genetics, Siberian Branch of Russian Academy of Sciences, 630060 Novosibirsk, Russia
4 Center for New Functional Materials, Novosibirsk State University, 630090 Novosibirsk, Russia
* Correspondence: khue.luu@g.nsu.ru

**Abstract:** Background: In this study, we examined the effectiveness of transfer learning in improving automatic segmentation of brain metastases on magnetic resonance imaging scans, with potential applications in preventive exams and remote diagnostics. Methods: We trained three deep learning models on a public dataset from the ASNR-MICCAI Brain Metastasis Challenge 2024, fine-tuned them on a small private dataset, and compared their performance to models trained from scratch. Results: Results showed that models using transfer learning performed better than scratch-trained models, though the improvement was not statistically substantial. The custom Tversky and Binary Cross-Entropy loss function helped manage class imbalance and reduce false negatives, limiting missed tumor regions. Medical experts noted that, while fine-tuned models worked well with larger, well-defined tumors, they struggled with tiny, scattered tumors in complex cases. Conclusions: This study highlights the potential of transfer learning and tailored loss functions in medical imaging, while also pointing out the models' limitations in detecting very small tumors in challenging cases.

**Keywords:** transfer learning; brain metastases; segmentation

## 1. Introduction

Brain metastasis (BM) happens when cancer cells spread to the brain, creating secondary tumors that interfere with normal brain function [1,2]. These tumors usually indicate an advanced stage and high malignancy of cancer, making treatment more difficult and lowering the chances of long-term survival. BM is serious because even small brain tumors can be deadly by pressing inside the skull or causing serious problems if they are in critical areas [3]. Survival rates are generally low, especially without early detection and intervention. Even with treatment, the prognosis can vary depending on tumor characteristics (number and type of tumors), patient factors (age), and treatment options (surgery, radiation, and chemotherapy) [4].

Diagnosis and treatment of BMs often require advanced imaging techniques like magnetic resonance imaging (MRI) to identify the size, location, and number of tumors [5]. MRI is especially useful because it provides clear and detailed images with excellent soft tissue contrast, allowing doctors to detect small tumors, assess swelling around them, identify perinodular bleeding within the brain, and understand how they affect important parts of the brain. Moreover, MRI does not use ionizing radiation, making it safer for patients who need multiple scans [6]. Common MRI sequences, such as T1-weighted (T1W), post-contrast T1-weighted (T1C), and T2-weighted (T2W) scans, reveal the brain's structure, while Fluid-Attenuated Inversion Recovery (FLAIR) highlights areas of swelling and lesions near cerebrospinal fluid spaces. Advanced MRI techniques like diffusion-weighted imaging and perfusion imaging offer additional insights into tumor characteristics, such as cellularity and blood flow [6]. By using these sequences, MRI helps distinguish between healthy and abnormal tissues, guiding early personalized diagnosis and effective treatment planning for brain metastases. Using AI to automatically measure and track BM volumes following SRS treatment, this study showed a strong correlation between AI-driven measurements and the current clinically used method: manual axial diameter measurements [7].

Precise segmentation of BMs helps doctors accurately assess tumor size, location, and spread, which is essential in improving the results of surgery and radiation therapy [8]. By clearly defining tumor boundaries, treatments can be targeted more effectively while reducing harm to healthy brain tissue. Segmentation also allows doctors to track tumor changes over time, helping to evaluate how well treatments are working. However, manual segmentation is time-consuming, prone to errors, and can vary between doctors, leading to inconsistent results.

Deep learning (DL) [9] has made significant strides in medical image analysis by enabling automated segmentation of brain MRI [10–14]. A recent review demonstrated the role of machine learning and deep learning methods in lesion detection, diagnosis, and anatomical segmentation of various types of brain tumors, including metastases [15]. Building on these advancements, several novel approaches have been proposed to further improve the accuracy and reliability of brain metastasis segmentation. Grøvik et al. used a fully convolutional neural network based on a 2.5D GoogLeNet architecture to automatically detect and segment brain metastases from multisequence MRI, achieving an overall AUC of 0.98 and a Dice score of 0.79, with an average false-positive rate of 8.3 per patient [16]. Huang et al. proposed the DeepMedic+ network with a custom volume-level sensitivity-specificity loss, which significantly improved brain metastasis detection by increasing sensitivity from 0.853 to 0.975 and precision from 0.691 to 0.987, while reducing false positives by 0.444 [17]. Highlighting the importance of strategic modality selection and multi-stage processing, Sadegheih et al. [18] proposed a two-stage detection and segmentation model using T1C, T1W, and FLAIR modalities, which significantly improved brain metastasis segmentation accuracy compared to single-pass models. Yang et al. trained the 3D-TransUNet model for brain metastases segmentation, exploring both Encoder-only and Decoder-only configurations with Transformer self-attention, and found that the Encoder-only version with Masked-Autoencoder pre-training achieved a lesion-wise Dice score of 59.8% [19]. In addition to these model-based improvements, new tools have been developed to bring advanced segmentation methods into clinical practice. An open-source software, Raidionics (v1.2.3), was also developed to perform preoperative segmentation of major brain tumor types, allowing standardized clinical reports to be generated in about ten minutes. It achieved an average Dice of 0.85 with 0.95 recall and precision for preoperative segmentation, while postoperative performance was lower, with an average Dice of 0.41 [20].

Despite these advances, DL faces challenges, particularly the limited availability of high-quality annotated medical images. Transfer learning (TL) [21] can overcome these limitations by allowing models trained on large datasets to be adapted for specific tasks with less training data, based on the idea that many basic features are shared across different image types. This includes various techniques like instance-based, network-based, and adversarial-based transfer [22]. TL has been applied to a wide range of medical imaging tasks—such as segmentation, object identification, disease classification, and severity grading [23].

Several recent studies have introduced advanced deep learning techniques for brain tumor segmentation, using pre-trained models and improved architectures to boost performance. Wacker et al. enhanced the AlbuNet architecture for brain tumor segmentation using a 3D U-Net-based model with a ResNet34 encoder, pre-trained on ImageNet, and extended convolutional layers to process volumetric MRI data. The 3D model with pre-training achieved higher Dice scores and more stable training than the 2D model, but its performance was less consistent on a new clinical dataset due to data differences [24]. Tataei Sarshar et al. introduced a deep learning pipeline for brain tumor segmentation using a pre-trained ResNet50 model with a multi-modality approach, combined with cascade feature extraction, inception, and new mReLU blocks to enhance learning. The proposed method achieved mean Dice scores of 0.9211 for the tumor core, 0.8993 for the whole tumor, and 0.9223 for the enhancing region [25].

Further extending these innovations, a recent study by Messaoudi et al. [26] proposed embedding pre-trained 2D networks into higher-dimensional U-Nets for effective segmentation of 2D and 3D medical images, utilizing weight and dimensional transfer techniques, and demonstrated superior performance in benchmarks. Huang et al. presented a federated learning framework using Learning Without Forgetting to train deep learning models for brain metastasis segmentation across multiple medical centers without sharing raw data, achieving strong performance on diverse MRI datasets [27]. Pani and Chawla introduced a hybrid approach that integrates transfer learning and self-supervised learning within a 3D UNET architecture to improve brain tumor segmentation in MRI scans, effectively reducing the need for extensive annotated data while achieving high accuracy with a Dice score of 90.15 [28].

While deep learning models have shown promise in automatically segmenting brain metastases, their application in clinical settings is often limited by a lack of alignment with radiologist expertise. Most existing studies evaluate model performance based solely on quantitative metrics but do not consider feedback from medical experts, who are key to judging how useful and accurate the results are in practice. Our research addresses this by including radiologists' and physicians' feedback in evaluating fine-tuned models, ensuring that the results are both accurate and practical for clinical use.

## 2. Materials and Methods

### 2.1. Overall Process

We started by training three deep learning models on a public dataset of brain metastases MRI, using 5-fold cross-validation. The private dataset with a smaller number of images was divided into a training set (26 images) and a testing set (10 images). We fine-tuned the selected models on the private training set using pre-trained weights on the public dataset and also trained them from scratch on the private data to compare performance. Both the fine-tuned and scratch-trained models were tested on 10 unseen samples. We employed explainable techniques to visualize the segmentations and gathered detailed feedback from medical experts on the results, focusing on their clinical relevance and accuracy. This helped us assess how well the models performed in segmenting brain metastases after fine-tuning compared to training from scratch. Figure 1 depicts our overall process.

### 2.2. Data Sources and Preparation

In this study, we employed two distinct datasets. The first dataset, ASNR-MICCAI Brain Metastasis Challenge 2024 [29], included MRI scans of untreated brain metastases collected from multiple institutions under standard clinical conditions. This dataset provided four MRI sequences: T1W, T1C, T2W, and FLAIR. Initial segmentations were generated by automated algorithms, then manually refined by neuroradiologists, and finally approved by board-certified specialists. The dataset used a three-label segmentation system, classifying regions as non-enhancing tumor core (NETC), surrounding FLAIR hyperintensity (SNFH), and enhancing tumor (ET). The challenge provided a training dataset with annotations and a validation dataset without annotations. In this study, we selected 652 metastasis cases from the training dataset, which we referred to as the BraTS dataset.

The second dataset, the Siberian Brain Tumors (SBT), was sourced from the Federal Neurosurgical Center in Novosibirsk, Russia. This dataset included clinical data and MRI scans (T1W, T1C, T2W, FLAIR) from 496 patients, covering a range of diagnoses such as glioblastomas, astrocytomas, neurinomas, meningiomas, and metastases. For this study, only the metastasis cases (36 in total) were utilized. MRI scans were predominantly acquired using a 1.5T Siemens MAGNETOM Avanto MRI machine (manufactured by Siemens Healthineers, Erlangen, Germany), with some cases imaged at 3T Philips Ingenia

MRI system (manufactured by Philips Healthcare, Amsterdam, Netherlands). Annotations for this dataset included peritumoral edema, non-enhancing tumor regions, GD-enhancing tumor regions, and necrotic tumor cores. These annotations were manually conducted by two board-certified neuroradiologists following a strict protocol, with an independent expert reviewing and correcting any errors. Ground truth labels were cross verified with patient clinical histories, as well as histological and immunohistochemical data. All cases were formatted in NIFTI, with 192 slices of 512 × 432 pixels, co-registered to a standard template, and resampled to a $1 \times 0.5 \times 0.5\ mm^3$ resolution. The SBT data were divided into 26 training cases and 10 test cases.

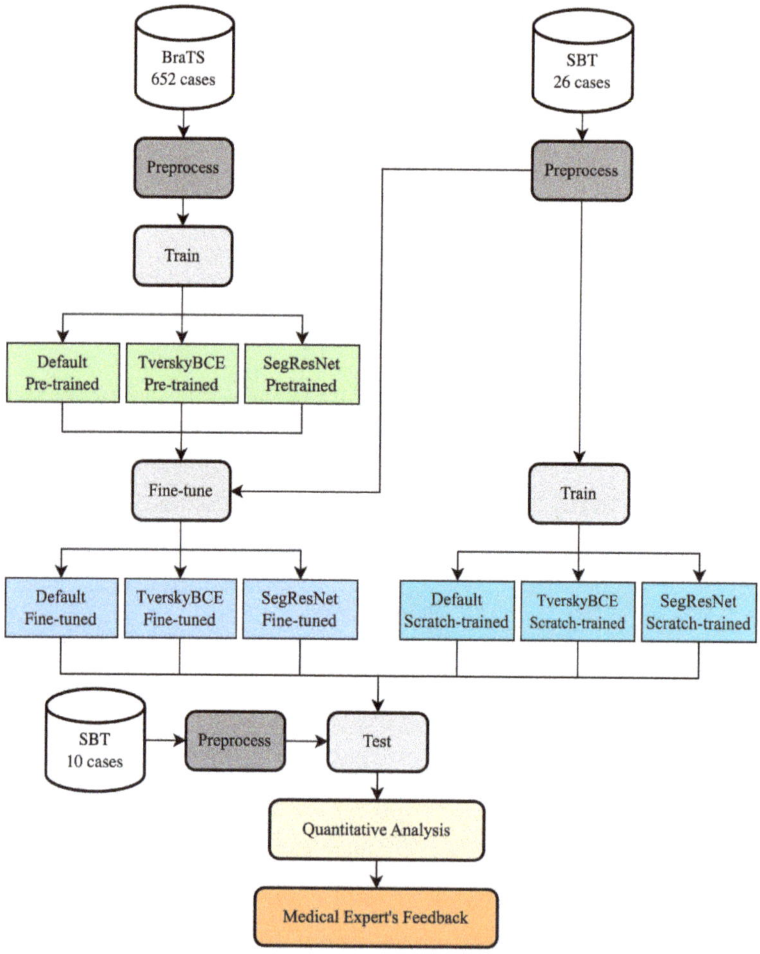


**Figure 1.** Overall process of training and fine-tuning models for brain metastasis segmentation.

The key feature is the availability of information about the source of the oncological process and medical imaging of metastases in the brain. During the anonymized preprocessing stage, unique IDs were assigned to each test sample, and these IDs were maintained across all experiments to facilitate verification and enable restoration of the original data when necessary. The test set IDs are 726, 727, 728, 729, 730, 731, 732, 733, 734, and 739.

In this study, the terms "case" and "sample" were used interchangeably, though with slight nuance: "case" referred to the overall patient, specifically a single patient diagnosed with brain metastasis, while "sample" denoted an individual instance within the dataset associated with that patient. Overall, when referring to a "case" or "sample", it did not indicate a single MRI scan but rather the complete set of four MRI sequences that collectively represented one patient with brain metastasis.

All models used the default nnU-Net preprocessing to ensure the input data were prepared consistently and effectively. This preprocessing included adaptive resampling, which adjusts depending on the dataset's characteristics, and dynamic intensity normalization, which changes the normalization method based on data variability. Foreground cropping was improved to better focus on the important areas, reducing extra computation. Padding and patch size were made flexible, adjusting to the GPU memory for maximum efficiency. Adaptive patch spacing was used to optimize the resolution during training, balancing efficiency and detail. Lastly, dynamic data sampling helped manage class imbalance, especially useful for datasets with different tumor sizes.

### 2.3. Model Architecture and Configuration

We employed the nnU-Net framework [30] for all model implementations, leveraging its self-configuring U-Net architecture. nnU-Net was chosen because of its ability to adapt automatically to different datasets by optimizing key architectural parameters such as depth, number of layers, and patch sizes, which are crucial in medical image segmentation tasks where data variability is high. The encoder-decoder structure, along with skip connections, ensures that spatial information is preserved, which is critical for precise tumor boundary detection.

For this study, we employed three models: (1) the default nnU-Net configuration (Default), (2) nnU-Net with a custom combined loss function (TverskyBCE), and (3) SegResNet [31]. Three graphs depicting the architectures used, including the dimensions of each layer, activation functions, normalization techniques, and other architectural details, were provided in the Supplementary Materials to offer a comprehensive understanding of the model configurations.

The Default model was used without modifications, dynamically adapting to the dataset by optimizing patch size, input modalities, and network depth. Its self-adjusting capability allowed it to achieve optimal performance across different datasets, justifying its use as a baseline. The initial learning rate was set to $1 \times 10^{-2}$, and the optimizer employed was stochastic gradient descent (SGD) with Nesterov momentum (momentum = 0.99). A polynomial learning rate scheduler was applied to gradually reduce the learning rate as training progressed, ensuring smooth convergence and preventing overshooting.

The TverskyBCE configuration modified the nnU-Net architecture by employing a combined loss function that merges Tversky loss [32] with Binary Cross-Entropy (BCE) loss. Tversky loss, designed to handle class imbalance, prioritized false negative reduction—critical for detecting small lesions. The Tversky loss parameters were set at $\alpha = 0.3$ and $\beta = 0.7$, emphasizing recall over precision. To further enhance sensitivity to small tumors, BCE loss was applied with a positive class weight (pos_w = 10), focusing more on correctly identifying tumor regions. The two losses were combined at a 1:1 ratio to optimize both small tumor detection and overall segmentation quality. The optimizer and learning rate scheduler used for the TverskyBCE model were identical to those of the Default model.

SegResNet was implemented with a U-Net-like architecture, enhanced by residual connections to improve gradient flow and prevent vanishing gradients, ensuring stability during the training of deeper networks. The encoder consisted of 4 stages, with residual blocks (1, 2, 2, 4) capturing progressively complex features, while the decoder used 1 residual block per stage for reconstruction. The model started with 32 filters, which increased in deeper layers to capture finer details. Input and output channels were configured based on the number of input modalities and segmentation classes. The Adam optimizer with a learning rate of $1 \times 10^{-4}$ and weight decay of $1 \times 10^{-5}$ was used to ensure efficient parameter updates. A polynomial learning rate scheduler was employed to gradually reduce the learning rate, preventing overfitting and ensuring smooth convergence.

### 2.4. Training and Fine-Tuning

All models were trained using GPU acceleration with the 3d_fullres configuration from the nnU-Net framework. The training spanned 200 epochs for pre-training and 100 epochs for fine-tuning. To enhance generalization and robustness, data augmentation included spatial transformations (such as rotations and scaling), intensity adjustments (including brightness, contrast, and gamma modifications), noise injections (e.g., Gaussian noise, Gaussian blur), and techniques like low-resolution simulation and mirroring.

For pre-training, models were trained on the BraTS dataset using training plans aligned with those of the SBT dataset, ensuring consistency and facilitating smoother transfer learning. During fine-tuning, models initialized with pre-trained weights from BraTS were further trained on the SBT dataset. Additionally, we trained models on the SBT dataset from scratch for 100 epochs to evaluate the benefits of fine-tuning compared to training from scratch.

All training was conducted on three workstations with GPUs (Quadro RTX 8000, NVIDIA GeForce GTX TITAN X, and NVIDIA RTX A4500), using Python 3.10.

### 2.5. Evaluation and Validation

To evaluate the performance of our segmentation models, we used several common metrics in medical image analysis, including Dice Similarity Coefficient (DSC), Hausdorff Distance (HD), Sensitivity (SEN), and Specificity (SPE) [33]. DSC measures overlap between predicted and ground-truth segmentations (0 to 1, higher is better). HD assesses boundary quality (lower is better). SEN and SPE measure the ability to correctly identify true positives and true negatives (both range from 0 to 1, higher is better). We also calculated true and predicted volumes to evaluate how well the model captures lesion size. For testing, we ensembled the predictions from all fold models and used a separate hold-out set of 10 images for an unbiased evaluation.

For the quantitative evaluation, we conducted a comparative analysis of the models using descriptive statistics, including the mean, standard deviation, and 95% confidence intervals. These statistical measures provided insights into the average performance, consistency, and reliability of the fine-tuned models relative to the scratch-trained models. To determine whether the data met the assumptions for parametric testing, we first applied the Shapiro–Wilk test [34] to assess whether the distribution of metrics for all models followed a normal distribution. Based on the results of the normality test, we observed that some of the metrics did not follow a normal distribution. As a result, we employed a one-sided Mann–Whitney U test [35], a non-parametric statistical test, to evaluate whether the performance scores of the fine-tuned models were stochastically greater than those of the scratch-trained models. The alternative hypothesis ($H_1$) posited that the fine-tuned models would outperform the scratch-trained models in terms of evaluation metrics. We reject the $H_1$ if the $p$-value from the Mann–Whitney U test is greater than or equal to the significance level of 0.05, indicating that there is insufficient evidence to support that the fine-tuned models outperform the scratch-trained models.

For the qualitative evaluation, we selected the two best models from our quantitative analysis and sent their predictions on 10 test cases to our medical experts (radiologists and physicians). The segmentation results were evaluated by medical experts using 3D Slicer v5.6.2 [36] software to visually assess the quality and accuracy of the predicted tumor boundaries. After their independent assessment, a follow-up meeting was held to discuss the models' strengths and weaknesses, particularly in handling complex tumor characteristics. This discussion helped identify areas for improvement to enhance clinical use.

### 2.6. Data and Code Availability

The BraTS dataset presented in the study is openly available at https://www.synapse.org/Synapse:syn59059764 (accessed on 15 July 2024).

The SBT dataset presented in this article is not readily available because it contains private health data collected in our Federal Neurosurgical Center and sharing it publicly would violate patient privacy and confidentiality agreements in accordance with ethical and regulatory standards.

The code to reproduce this study is available at https://github.com/luumsk/BrainMetaSeg.git (accessed on 5 November 2024).

### *2.7. Manuscript Preparation*

An AI-assisted tool, ChatGPT (powered by OpenAI's GPT-4o model), was used to assist in refining the text of the manuscript. The tool was employed exclusively to improve language clarity and grammar, without altering the scientific content or interpretations provided by the authors.

## 3. Results

### *3.1. Quantitative Analysis of Model Performance*

#### 3.1.1. Descriptive Statistics

Our analysis showed that fine-tuned models generally performed better than scratch-trained models in key areas such as segmentation accuracy, sensitivity, and boundary localization. Fine-tuned models achieved higher DSC, lower HD for enhancing the tumor and tumor core, and better SEN. They also demonstrated more consistent and reliable predictions, indicated by narrower confidence intervals and lower standard deviations. Although scratch-trained models had a slight advantage in specificity and boundary delineation for the whole tumor, fine-tuning proved to be more effective in the most important aspects of segmentation.

Specifically, fine-tuned models had higher DSC across all tumor regions: 0.905 compared to 0.902 for enhancing tumor, 0.936 compared to 0.934 for tumor core, and 0.914 compared to 0.907 for whole tumor. Figure 2 supports this trend, showing that fine-tuned models consistently achieved higher mean DSC across different configurations, with similar or narrower error bars, suggesting improved accuracy and stability. Although the differences were modest, the results suggest that fine-tuning offers clear benefits for segmentation accuracy and consistency, particularly for enhancing tumor and tumor core regions.

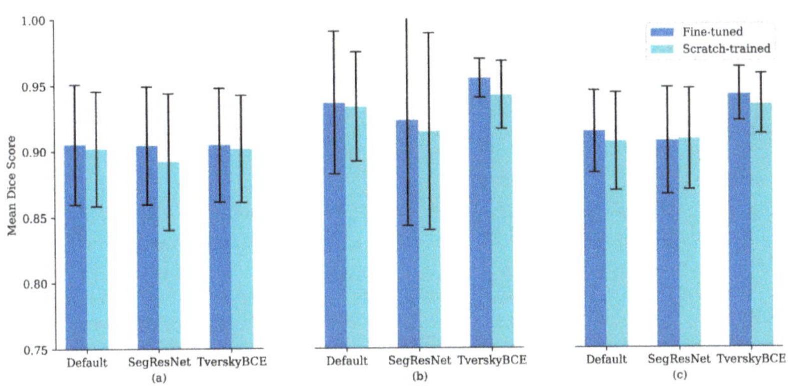


**Figure 2.** Comparison of mean Dice scores between fine-tuned and scratch-trained models on three brain tumor sub-regions: (**a**) enhancing tumor, (**b**) tumor core, and (**c**) whole tumor. The error bars indicate the 95% confidence interval.

Figure 3 further shows that fine-tuned models generally provided better boundary localization than scratch-trained models, especially for enhancing tumor and tumor core. The fine-tuned model had a mean HD of 2.76 mm for enhancing tumor and 1.87 mm for tumor core, compared to 2.78 mm and 5.50 mm, respectively, for the scratch-trained model.

For the whole tumor, scratch-trained models performed slightly better, with a mean HD of 3.23 mm versus 3.36 mm for fine-tuned models. Despite this, fine-tuned models showed lower variability for enhancing tumor and tumor core, indicating more consistent results.

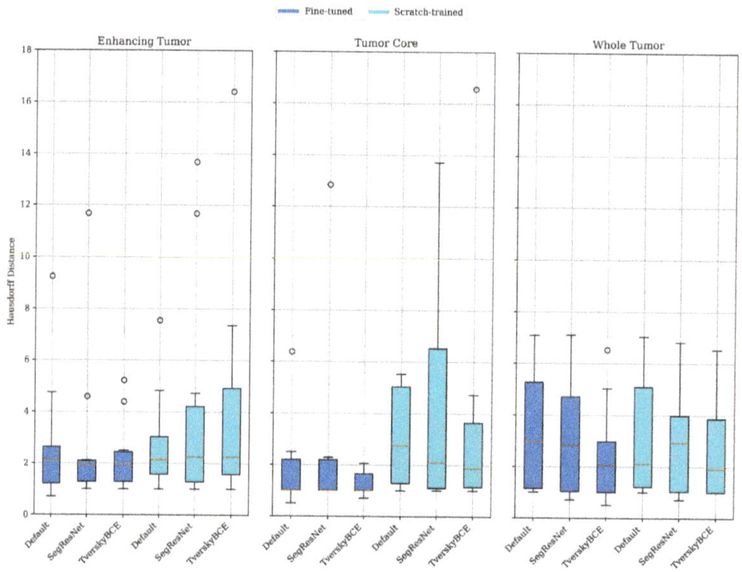


**Figure 3.** Comparison of mean Hausdorff Distances between fine-tuned and scratch-trained models on three brain tumor sub-regions. The circles indicate outliers (unusually high or low values), while the orange line within each box represents the median value, showing the central tendency of the Hausdorff Distances for each model.

Among the transfer learning approach with fine-tuned model variants, TverskyBCE demonstrated the best overall performance for brain metastasis segmentation. It achieved higher DSC and superior boundary localization compared to the Default model, showcasing a clear advantage in segmentation quality. While SegResNet also performed well, it lacked consistency, particularly in capturing fine boundary details. TverskyBCE offered the optimal balance of accuracy and stability, making it the preferred choice for enhancing segmentation quality in this context.

#### 3.1.2. Statistical Testing for Significance

The statistical test results indicated no significant difference between the fine-tuned and scratch-trained models across all metrics. However, it is important to note that the sample size in this study is relatively small, consisting of only 10 cases approved by the histological and immunohistochemical evaluations. This small sample size reduces the statistical power of the test, increasing the likelihood of failing to detect real performance differences that may exist between the models. Despite the lack of statistical significance, the fine-tuned models consistently showed slightly higher performance across all metrics, with improvements in DSC, lower HD, and better SEN and SPE. Also, all experiments were focused on evaluating the clinical significance and application of some explainable techniques. These slight improvements suggest that fine-tuning does contribute positively to segmentation performance, particularly in complex tasks such as brain tumor segmentation. In practical applications, where even small improvements in accuracy and boundary precision can be clinically meaningful, these results are still relevant and important meaning for the screening process.

To further compare model variants across sub-regions, we run the Mann–Whitney U test on pairs of models, with the alternative hypothesis that the scores of the first model were significantly greater than those of the second. The results indicated that the TverskyBCE model outperformed both SegResNet ($p = 0.04$) and Default ($p = 0.03$) in whole tumor segmentation. Additionally, TverskyBCE exhibited statistically significant higher SEN across all sub-regions compared to both SegResNet ($p = 0.02$) and Default ($p = 0.01$). These findings suggested that TverskyBCE offered superior performance in whole tumor segmentation and SEN metric.

### 3.2. *Individual Case Analysis and Visual Comparison*

For each model and metric, we identified the cases with the lowest DSC, SEN, SPE, and the highest HD, labeling them as challenging cases. To find the test samples that most models struggled with, we counted how often each sample was marked as a challenging case. Finally, we selected the top three most challenging cases (ID 731, 733, 739 from the test set SBT data) across all models for further visual analysis.

Test sample 733 was identified as the most challenging case, having been marked 28 times across models. The tumor is small, spans only a few slices, and is centrally located in the brain. Most models, including the TverskyBCE scratch-trained model, failed to correctly segment the tumor core. However, the TverskyBCE Fine-tuned model performed significantly better, as shown in Figure 4, achieving the highest DSC and the lowest HD for this case. The Tversky loss function helped the model focus on reducing false negatives, which is essential for capturing small tumor regions. However, the full potential of this loss function became evident only after fine-tuning. Fine-tuning allowed the model to further refine its weights, adapting more closely to the specific dataset and learning critical patterns necessary for detecting small, subtle lesions. The combination of the Tversky loss and fine-tuning enabled TverskyBCE Fine-tuned to successfully segment the tumor core, which was missed by the scratch-trained version and other models. This highlights the complementary roles of the Tversky loss function and fine-tuning in enhancing SEN to challenging cases.

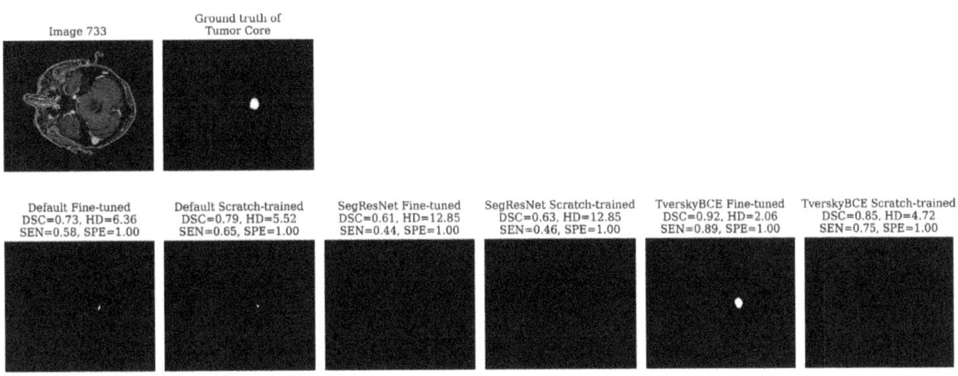


**Figure 4.** Segmentation results for test sample 733. TverskyBCE Fine-tuned achieved the highest DSC and lowest HD, successfully capturing the small tumor core missed by other models.

Test sample 731 was identified as the second most challenging case, marked 11 times across models. The tumor is large, located on one side of the brain with significant surrounding non-enhancing FLAIR hyperintensity. While all models detected the tumor boundary well, as indicated by the low HD for both the tumor core and whole tumor, most models struggled with the enhancing tumor, missing the middle region, which resulted in low SEN. The TverskyBCE Fine-tuned model performed best, with the smallest amount of missing tumor tissue. Although the DSC and SEN scores did not fully reflect the visual difference, the TverskyBCE Fine-tuned model clearly showed fewer missing regions in the images, as shown in Figure 5. This demonstrates that the combination of Tversky loss and fine-tuning was particularly effective at reducing false negatives and capturing more of the enhancing tumor.

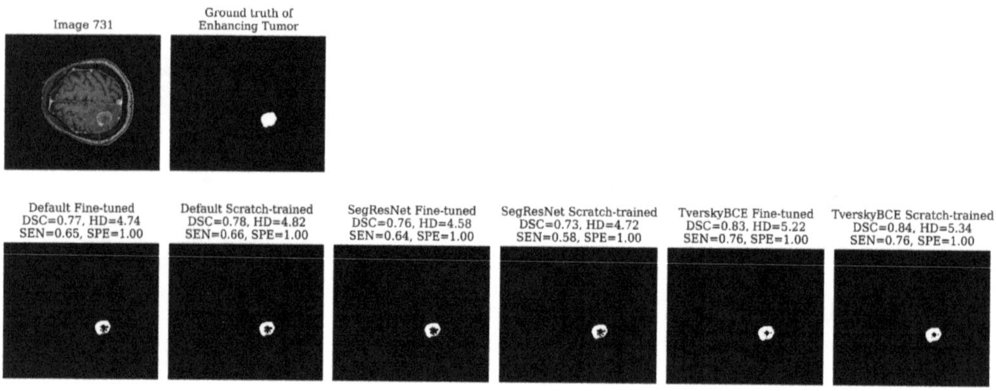


**Figure 5.** Segmentation results for Test sample 731. TverskyBCE Fine-tuned had the fewest missed regions for the enhancing tumor, with the highest SEN.

Test sample 739 was marked 11 times as a challenging case. The tumor is very large, covering a significant portion of the brain and spanning multiple slices. While most models segmented the boundary reasonably well for both the enhancing tumor and the tumor core, they struggled with the segmentation of the whole tumor, resulting in a high HD. The whole tumor's irregular, heterogeneous shape, with jagged edges and asymmetrical extensions, as shown in Figure 6, added to the difficulty. The TverskyBCE Fine-tuned model performed best in this case, achieving the highest DSC and SEN, along with the lowest HD for the whole tumor.

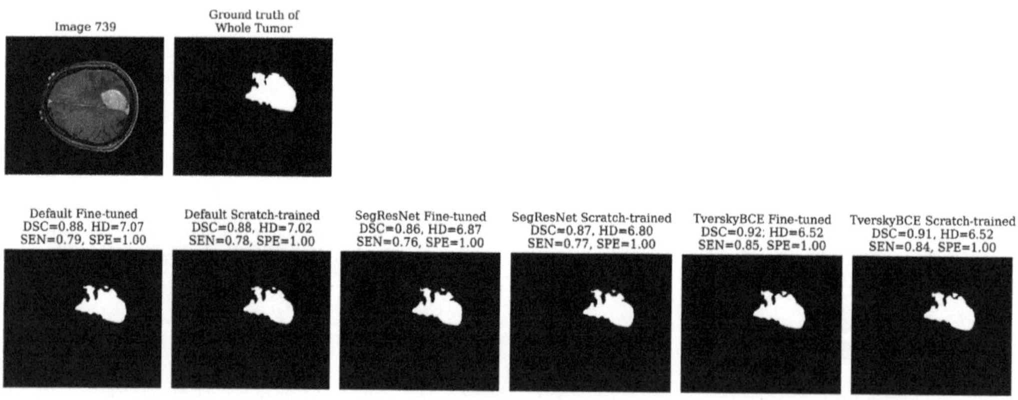


**Figure 6.** Segmentation results for Test sample 739. TverskyBCE Fine-tuned captured the large, irregular tumor more effectively, with the highest DSC and SEN.

The analysis showed that the TverskyBCE Fine-tuned model outperformed other models, particularly in challenging cases with small or irregular tumors. The combination of Tversky loss and fine-tuning was crucial in reducing false negatives and improving segmentation accuracy. Fine-tuning enabled the model to better adapt to the dataset, allowing it to capture complex tumor boundaries more effectively.

### 3.3. Expert Feedback and Clinical Relevance

After evaluating the predictions of the TverskyBCE Fine-tuned and SegResNet Fine-tuned models on 10 test cases, medical experts provided feedback based on medical visualization (evaluation) and comparisons with ground truth annotations. Overall, both models performed well. In cases with large, straightforward lesions, such as neuroendocrine metastases (test sample 730) and undifferentiated cancer metastases (test samples 734 and 739), there were minimal differences between the models' results. However, both models struggled with recognizing additional small tumors, often missing tiny lesions in more complex cases. Metastasis is grown from cancerous cells which spread to the brain from the affected area by blood. The imbalanced objects and regions issue affects the data-driven learning algorithm as the extracted features may be highly influenced by large tumors and additional small parts. For example, to the regions—the necrotic/non-enhancing tumor core region is much smaller than other regions; to the objects—a characteristic feature of metastases is their multicentrality, they can appear in several areas of the brain with different consequences (size of lesions).

In test sample 728 (breast metastases), displayed in Figure 7, both models achieved high quantitative metrics, including a high DSC (0.90–0.92) and low HD (0.8–2.0). However, upon review by medical experts, both models failed to detect tiny tumors, which are critical for accurate diagnosis and treatment. Missing these small tumors is especially concerning in metastatic breast cancer, as even tiny lesions can indicate early-stage metastasis or tumor progression, which directly impacts treatment decisions and prognosis. This discrepancy highlights the limitations of relying solely on quantitative metrics and indicates that comparative metrics may not fully capture the models' performance in detecting small, clinically important lesions.

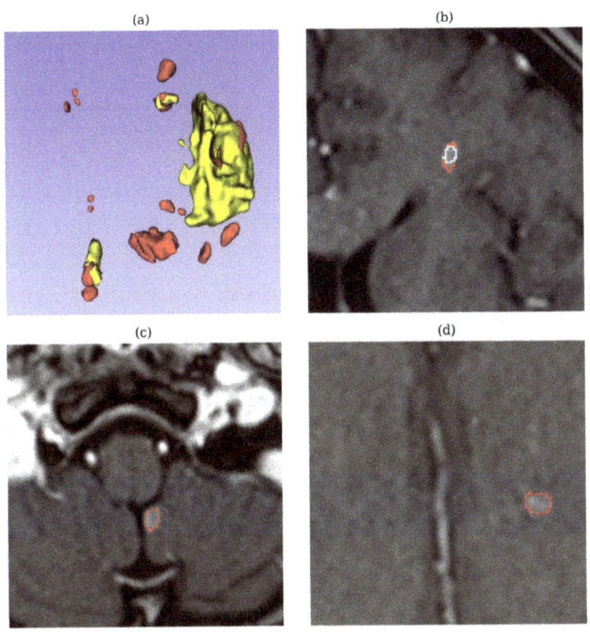


**Figure 7.** Segmentation results for case 728 (breast metastasis). (**a**) Ground truth 3D image with annotations marking multiple tiny, scattered tumors (yellow for peritumoral edema, red for the GD-enhancing tumor); (**b**) a 2D slice showing TverskyBCE segmentation, with red outlines for expert annotations and cyan for TverskyBCE's detection of a small metastasis missed by SegResNet; (**c**,**d**) 2D slices showing tiny tumor regions missed by both models, with red outlines added by medical experts to mark these undetected areas.

In test sample 729 (adenocarcinoma metastases), TverskyBCE outperformed SegResNet in segmenting a cerebellar tumor. While SegResNet showed less edema in its segmentation, TverskyBCE's result was closer to the ground truth, capturing the tumor's characteristics more accurately. Similarly, in test sample 733 (low-grade cancer metastases), TverskyBCE successfully detected the necrotic tumor core, a critical sub-region that SegResNet entirely missed. Necrosis within a tumor is crucial for diagnosis and prognostic evaluation. This comparison is visualized in Figure 8, where the differences between the models are shown.

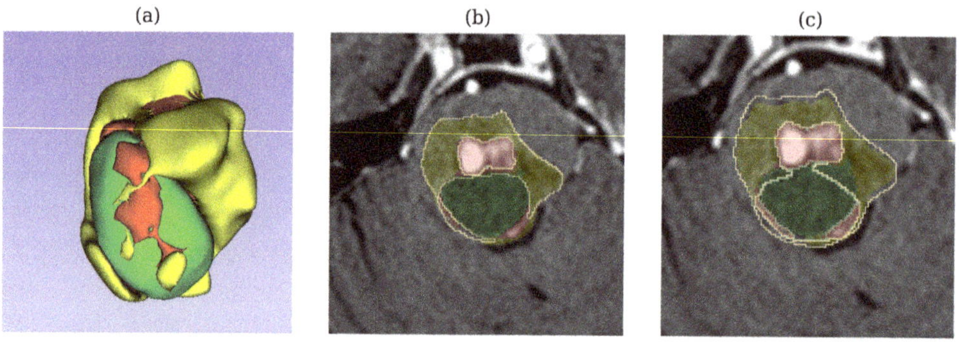


**Figure 8.** Visual comparison of segmentation results for test sample 729 (adenocarcinoma metastases) showing (**a**) the ground truth annotated by medical experts, (**b**) SegResNet segmentation where the necrotic tumor core (the green area) was missed, and (**c**) TverskyBCE segmentation successfully detecting the necrotic core. Solid-colored regions represent the ground truth annotations, while the outlined regions depict the model predictions. The annotations use yellow for peritumoral edema, red for the GD-enhancing tumor, and green for the necrotic tumor core.

In summary, expert feedback indicated that while both models performed reliably in simpler cases, TverskyBCE demonstrated a clear advantage in handling more complex scenarios, particularly in recognizing the necrotic core and accurately managing intricate tumor features. These findings highlight the strengths of each model and point to areas for improvement, especially in detecting small tumors and complex tissue structures.

## 4. Discussion

In this study, we assessed the performance of transfer learning models for brain metastasis segmentation. We pre-trained nnU-Net (with default settings and with Tversky and BCE loss) and SegResNet on the BraTS Metastases 2024 dataset, and then fine-tuned these models on our private SBT dataset. We compared the fine-tuned models with those trained from scratch, using both quantitative and qualitative analyses, to obtain feedback from medical experts on the segmentations produced. This expert input was crucial for evaluating clinical accuracy and ensuring the models were fit for practical use.

Our main objective was to develop methodologies for diagnosing and screening metastases. Brain metastases occur when cancer cells spread to the brain from the affected area. Any cancer can spread to the brain, but lung cancer, breast cancer, colon cancer, kidney cancer, and melanoma are the most likely to cause brain metastases. Tumors that experts call high-grade gliomas are tumors of the central nervous system (CNS). They are solid tumors and appear due to a mutation of brain or spinal cord cells. Since these tumors grow in the central nervous system, they are also called primary CNS tumors. That is, they are not metastases from other malignant tumors that have grown in other organs, and their cancer cells have penetrated the central nervous system. An additional feature is that the lesions can be small—difficult to detect with the human eye. The diagnosis is highly dependent on the quality of the image, the thickness of the slices, the scanning parameters, and the expertise of the doctor.

Throughout this study, several key challenges were identified in the context of brain metastasis segmentation:

- Location and morphological uncertainty: Metastases are grown from cancer cells which spread to the brain from the affected area by blood. Due to the wide spatial distribution of cancer cells, lung cancer, breast cancer, colon cancer, kidney cancer, and melanoma may appear at any location inside the brain. The shape and size of different brain tumors vary with large morphology uncertainty. Each sub-region of a meta may also vary in shape and size.
- Low contrast: High resolution and high contrast images are expected to contain diverse image information. Due to the image projection and tomography process, MRI images may be of low quality and low contrast. The boundary between biological tissues tends to be blurred and hard to detect.
- Annotation bias: Manual annotation highly depends on individual experience, which can introduce an annotation bias during data labeling. The annotation biases have a huge impact on the AI algorithm during the learning process.
- Imbalanced region and object issues: The imbalanced issue affects the data-driven learning algorithm as the extracted features may be highly influenced by large tumor regions. For example, to the regions—the necrotic/non-enhancing tumor core (NCR/ECT) region is much smaller than another region; to the objects—a characteristic feature of metastases is their multicentrality, they can appear in several areas of the brain with different consequences (size of lesions).

Our results showed that fine-tuning the SBT dataset led to a slight improvement in metrics compared to training from scratch. This suggests that transfer learning is helpful when labeled data are limited, as pre-trained models use features learned from larger datasets. However, the performance differences were not statistically significant, likely due to the small test size (10 samples). Despite this, qualitative analysis showed that the fine-tuned models, especially TverskyBCE, performed better in challenging cases. TverskyBCE detected small or irregularly shaped tumors and did not miss the tumor core, unlike other models. This suggests that a customized loss function like Tversky with Binary Cross-Entropy helps handle class imbalance and improves sensitivity for smaller lesions. Medical experts also confirmed that TverskyBCE detected critical tumors in some cases where SegResNet did not.

However, all models, including TverskyBCE and SegResNet, struggled with detecting tiny tumors. Experts emphasized the need to improve the detection of smaller lesions and boundary accuracy, despite the models achieving good overall segmentation quality. Even with high quantitative metrics, both models failed to detect small tumors, revealing that metrics alone are insufficient to capture clinically important issues. The involvement of medical experts is crucial in identifying these gaps, as missing tiny tumors can significantly impact treatment decisions and prognosis. Our findings not only demonstrate the potential of transfer learning for brain metastasis segmentation but also highlight the importance of medical expert input in developing AI algorithms. Their insights help bridge the gap between quantitative performance and real-world clinical needs, ensuring AI models are refined to detect smaller tumors and improve boundary segmentation for more accurate treatment planning.

Looking ahead, future research could address these limitations by applying advanced post-processing techniques to improve boundary delineation and using synthetic data generation or data augmentation to enhance tiny tumor detection. Incorporating multi-modal imaging data, such as combining MRI with CT or PET, could also provide more comprehensive information and boost model performance. Additionally, exploring new model architectures or hybrids could further improve segmentation, particularly for small or complex lesions. Another important direction would be to consider including edema in segmentation, as brain edema often surrounds metastases and could provide a critical context for more accurate tumor delineation. Finally, ongoing collaboration with medical experts will be key to refining models and ensuring they meet clinical standards, with expert feedback incorporated into the model development process to continuously improve performance in real-world applications.

**Supplementary Materials:** All images, including network architecture graphs, and detailed test scores are available in the GitHub repository at https://github.com/luumsk/BrainMetaSeg (accessed on 5 November 2024).

**Author Contributions:** Conceptualization, B.N.T. and A.Y.L.; methodology, A.Y.L.; software, V.S. and M.S.K.L.; validation, E.V.A., A.Y.L. and B.N.T.; formal analysis, M.S.K.L.; investigation, M.S.K.L.; resources, B.N.T.; data curation, E.V.A.; writing—original draft preparation, M.S.K.L. and R.M.K.; writing—review and editing, B.N.T. and A.Y.L.; visualization, R.M.K.; supervision, A.Y.L.; project administration, B.N.T.; funding acquisition, B.N.T. All authors have read and agreed to the published version of the manuscript.

**Funding:** This work was supported by a grant for research centers, provided by the Analytical Center for the Government of the Russian Federation in accordance with the subsidy agreement (agreement identifier 000000D730324P540002) and the agreement with the Novosibirsk State University dated 27 December 2023 No. 70-2023-001318.

**Institutional Review Board Statement:** Not applicable.

**Informed Consent Statement:** Our partner FSBI Federal Neurosurgical Center specializing in neurosurgical operations (including brain cancers) collected all documents, including the Consent Statement from the patients.

**Data Availability Statement:** The BraTS Metastases 2024 dataset presented in this study is openly available at https://www.synapse.org/Synapse:syn59059764 (accessed on 15 July 2024). The Siberian Brain Tumor dataset presented in this article is not readily available due to privacy and ethical restrictions, as it contains private health data collected from our Federal Neurosurgical Center. Sharing it publicly would violate patient privacy and confidentiality agreements in accordance with ethical and regulatory standards. Requests for access to this dataset may be considered on a case-by-case basis, contingent upon appropriate ethical review and agreements.

**Acknowledgments:** We acknowledge the use of ChatGPT (OpenAI) for editorial support in language enhancement during the preparation of this manuscript.

**Conflicts of Interest:** The authors declare no conflicts of interest.

## References

1. Achrol, A.S.; Rennert, R.C.; Anders, C.; Soffietti, R.; Ahluwalia, M.S.; Nayak, L.; Peters, S.; Arvold, N.D.; Harsh, G.R.; Steeg, P.S.; et al. Brain Metastases. *Nat. Rev. Dis. Prim.* **2019**, *5*, 5. [CrossRef] [PubMed]
2. Gavrilovic, I.T.; Posner, J.B. Brain Metastases: Epidemiology and Pathophysiology. *J. Neurooncol.* **2005**, *75*, 5–14. [CrossRef] [PubMed]
3. Lassman, A.B.; DeAngelis, L.M. Brain Metastases. *Neurol. Clin.* **2003**, *21*, 1–23. [CrossRef] [PubMed]
4. Hall, W.; Djalilian, H.; Nussbaum, E.; Cho, K. Long-Term Survival with Metastatic Cancer to the Brain. *Med. Oncol.* **2000**, *17*, 279–286. [CrossRef]
5. Lin, X.; DeAngelis, L.M. Treatment of Brain Metastases. *J. Clin. Oncol.* **2015**, *33*, 3475–3484. [CrossRef]
6. Derks, S.H.A.E.; Van Der Veldt, A.A.M.; Smits, M. Brain Metastases: The Role of Clinical Imaging. *Br. J. Radiol.* **2022**, *95*, 20210944. [CrossRef]
7. Prezelski, K.; Hsu, D.G.; Del Balzo, L.; Heller, E.; Ma, J.; Pike, L.R.G.; Ballangrud, Å.; Aristophanous, M. Artificial-Intelligence-Driven Measurements of Brain Metastases' Response to SRS Compare Favorably with Current Manual Standards of Assessment. *Neuro-Oncol. Adv.* **2024**, *6*, vdae015. [CrossRef]
8. Nabors, L.B.; Portnow, J.; Ahluwalia, M.; Baehring, J.; Brem, H.; Brem, S.; Butowski, N.; Campian, J.L.; Clark, S.W.; Fabiano, A.J.; et al. Central Nervous System Cancers, Version 3.2020, NCCN Clinical Practice Guidelines in Oncology. *J. Natl. Compr. Cancer Netw.* **2020**, *18*, 1537–1570. [CrossRef]
9. LeCun, Y.; Bengio, Y.; Hinton, G. Deep Learning. *Nature* **2015**, *521*, 436–444. [CrossRef]
10. Akkus, Z.; Galimzianova, A.; Hoogi, A.; Rubin, D.L.; Erickson, B.J. Deep Learning for Brain MRI Segmentation: State of the Art and Future Directions. *J. Digit. Imaging* **2017**, *30*, 449–459. [CrossRef]
11. Zhao, F.; Wu, Z.; Li, G. Deep Learning in Cortical Surface-Based Neuroimage Analysis: A Systematic Review. *Intell. Med.* **2023**, *3*, 46–58. [CrossRef]
12. Jyothi, P.; Singh, A.R. Deep Learning Models and Traditional Automated Techniques for Brain Tumor Segmentation in MRI: A Review. *Artif. Intell. Rev.* **2023**, *56*, 2923–2969. [CrossRef]

13. Singh, S.P.; Wang, L.; Gupta, S.; Goli, H.; Padmanabhan, P.; Gulyás, B. 3D Deep Learning on Medical Images: A Review. *Sensors* **2020**, *20*, 5097. [CrossRef] [PubMed]
14. Litjens, G.; Kooi, T.; Bejnordi, B.E.; Setio, A.A.A.; Ciompi, F.; Ghafoorian, M.; Van Der Laak, J.A.W.M.; Van Ginneken, B.; Sánchez, C.I. A Survey on Deep Learning in Medical Image Analysis. *Med. Image Anal.* **2017**, *42*, 60–88. [CrossRef]
15. Chukwujindu, E.; Faiz, H.; Al-Douri, S.; Faiz, K.; De Sequeira, A. Role of Artificial Intelligence in Brain Tumour Imaging. *Eur. J. Radiol.* **2024**, *176*, 111509. [CrossRef]
16. Grøvik, E.; Yi, D.; Iv, M.; Tong, E.; Rubin, D.; Zaharchuk, G. Deep Learning Enables Automatic Detection and Segmentation of Brain Metastases on Multisequence MRI. *J. Magn. Reson. Imaging* **2020**, *51*, 175–182. [CrossRef]
17. Huang, Y.; Bert, C.; Sommer, P.; Frey, B.; Gaipl, U.; Distel, L.V.; Weissmann, T.; Uder, M.; Schmidt, M.A.; Dörfler, A.; et al. Deep Learning for Brain Metastasis Detection and Segmentation in Longitudinal MRI Data. *Med. Phys.* **2022**, *49*, 5773–5786. [CrossRef]
18. Sadegheih, Y.; Merhof, D. Segmentation of Brain Metastases in MRI: A Two-Stage Deep Learning Approach with Modality Impact Study. In *Predictive Intelligence in Medicine*; Springer: Cham, Switzerland, 2024.
19. Yang, S.; Li, X.; Mei, J.; Chen, J.; Xie, C.; Zhou, Y. 3D-TransUNet for Brain Metastases Segmentation in the BraTS2023 Challenge. *arXiv* **2024**, arXiv:2403.15735. [CrossRef]
20. Bouget, D.; Alsinan, D.; Gaitan, V.; Helland, R.H.; Pedersen, A.; Solheim, O.; Reinertsen, I. Raidionics: An Open Software for Pre- and Postoperative Central Nervous System Tumor Segmentation and Standardized Reporting. *Sci. Rep.* **2023**, *13*, 15570. [CrossRef]
21. Bozinovski, S. Reminder of the First Paper on Transfer Learning in Neural Networks, 1976. *Informatica* **2020**, *44*, 291–302. [CrossRef]
22. Tan, C.; Sun, F.; Kong, T.; Zhang, W.; Yang, C.; Liu, C. A Survey on Deep Transfer Learning. In *Artificial Neural Networks and Machine Learning—ICANN 2018*; Kůrková, V., Manolopoulos, Y., Hammer, B., Iliadis, L., Maglogiannis, I., Eds.; Lecture Notes in Computer Science; Springer International Publishing: Cham, Switzerland, 2018; Volume 11141, pp. 270–279, ISBN 978-3-030-01423-0.
23. Kora, P.; Ooi, C.P.; Faust, O.; Raghavendra, U.; Gudigar, A.; Chan, W.Y.; Meenakshi, K.; Swaraja, K.; Plawiak, P.; Rajendra Acharya, U. Transfer Learning Techniques for Medical Image Analysis: A Review. *Biocybern. Biomed. Eng.* **2022**, *42*, 79–107. [CrossRef]
24. Wacker, J.; Ladeira, M.; Nascimento, J.E.V. Transfer Learning for Brain Tumor Segmentation. In *Brainlesion: Glioma, Multiple Sclerosis, Stroke and Traumatic Brain Injuries*; Crimi, A., Bakas, S., Eds.; Lecture Notes in Computer Science; Springer International Publishing: Cham, Switzerland, 2021; Volume 12658, pp. 241–251, ISBN 978-3-030-72083-4.
25. Tataei Sarshar, N.; Ranjbarzadeh, R.; Jafarzadeh Ghoushchi, S.; De Oliveira, G.G.; Anari, S.; Parhizkar, M.; Bendechache, M. Glioma Brain Tumor Segmentation in Four MRI Modalities Using a Convolutional Neural Network and Based on a Transfer Learning Method. In *Proceedings of the 7th Brazilian Technology Symposium (BTSym'21), Campinas, Brazil, 19–21 October 2021*; Lano, Y., Saotome, O., Kemper Vásquez, G.L., Cotrim Pezzuto, C., Arthur, R., Gomes De Oliveira, G., Eds.; Smart Innovation, Systems and Technologies; Springer International Publishing: Cham, Switzerland, 2023; Volume 207, pp. 386–402, ISBN 978-3-031-04434-2.
26. Messaoudi, H.; Belaid, A.; Ben Salem, D.; Conze, P.-H. Cross-Dimensional Transfer Learning in Medical Image Segmentation with Deep Learning. *Med. Image Anal.* **2023**, *88*, 102868. [CrossRef] [PubMed]
27. Huang, Y.; Khodabakhshi, Z.; Gomaa, A.; Schmidt, M.; Fietkau, R.; Guckenberger, M.; Andratschke, N.; Bert, C.; Tanadini-Lang, S.; Putz, F. Multicenter Privacy-Preserving Model Training for Deep Learning Brain Metastases Autosegmentation. *arXiv* **2024**, arXiv:2405.10870. [CrossRef] [PubMed]
28. Pani, K.; Chawla, I. A Hybrid Approach for Multi Modal Brain Tumor Segmentation Using Two Phase Transfer Learning, SSL and a Hybrid 3DUNET. *Comput. Electr. Eng.* **2024**, *118*, 109418. [CrossRef]
29. De Verdier, M.C.; Saluja, R.; Gagnon, L.; LaBella, D.; Baid, U.; Tahon, N.H.; Foltyn-Dumitru, M.; Zhang, J.; Alafif, M.; Baig, S.; et al. The 2024 Brain Tumor Segmentation (BraTS) Challenge: Glioma Segmentation on Post-Treatment MRI. *arXiv* **2024**, arXiv:2405.18368. [CrossRef]
30. Isensee, F.; Jaeger, P.F.; Kohl, S.A.A.; Petersen, J.; Maier-Hein, K.H. nnU-Net: A Self-Configuring Method for Deep Learning-Based Biomedical Image Segmentation. *Nat. Methods* **2021**, *18*, 203–211. [CrossRef]
31. Myronenko, A. 3D MRI Brain Tumor Segmentation Using Autoencoder Regularization. *arXiv* **2018**, arXiv:1810.11654. [CrossRef]
32. Salehi, S.S.M.; Erdogmus, D.; Gholipour, A. Tversky Loss Function for Image Segmentation Using 3D Fully Convolutional Deep Networks. *arXiv* **2017**, arXiv:1706.05721. [CrossRef]
33. Taha, A.A.; Hanbury, A. Metrics for Evaluating 3D Medical Image Segmentation: Analysis, Selection, and Tool. *BMC Med. Imaging* **2015**, *15*, 29. [CrossRef]
34. Shapiro, S.S.; Wilk, M.B. An Analysis of Variance Test for Normality (Complete Samples). *Biometrika* **1965**, *52*, 591–611. [CrossRef]
35. Mann, H.B.; Whitney, D.R. On a Test of Whether One of Two Random Variables Is Stochastically Larger than the Other. *Ann. Math. Stat.* **1947**, *18*, 50–60. [CrossRef]
36. Fedorov, A.; Beichel, R.; Kalpathy-Cramer, J.; Finet, J.; Fillion-Robin, J.-C.; Pujol, S.; Bauer, C.; Jennings, D.; Fennessy, F.; Sonka, M.; et al. 3D Slicer as an Image Computing Platform for the Quantitative Imaging Network. *Magn. Reson. Imaging* **2012**, *30*, 1323–1341. [CrossRef]

*biomedicines*

MDPI

*Review*

# Advancements in Glioma Care: Focus on Emerging Neurosurgical Techniques

**Matteo De Simone [1,*], Valeria Conti [1,2], Giuseppina Palermo [1], Lucio De Maria [3,4] and Giorgio Iaconetta [1,5]**

[1] Department of Medicine, Surgery and Dentistry "Scuola Medica Salernitana", University of Salerno, Via S. Allende, 84081 Baronissi, Italy; vconti@unisa.it (V.C.); g.palermo10@studenti.unisa.it (G.P.); giaconetta@unisa.it (G.I.)

[2] Clinical Pharmacology and Pharmacogenetics Unit, University Hospital "San Giovanni di Dio e Ruggi, D'Aragona", 84131 Salerno, Italy

[3] Unit of Neurosurgery, Department of Surgical Specialties, Radiological Sciences, and Public Health, University of Brescia, 25123 Brescia, Italy; lucio.demaria@hcuge.ch

[4] Unit of Neurosurgery, Department of Clinical Neuroscience, Geneva University Hospitals (HUG), 1205 Geneva, Switzerland

[5] Neurosurgery Unit, University Hospital "San Giovanni di Dio e Ruggi, D'Aragona", 84131 Salerno, Italy

* Correspondence: m.desimone82@studenti.unisa.it

**Abstract:** Background: Despite significant advances in understanding the molecular pathways of glioma, translating this knowledge into effective long-term solutions remains a challenge. Indeed, gliomas pose a significant challenge to neurosurgical oncology because of their diverse histopathological features, genetic heterogeneity, and clinical manifestations. Relevant sections: This study focuses on glioma complexity by reviewing recent advances in their management, also considering new classification systems and emerging neurosurgical techniques. To bridge the gap between new neurosurgical approaches and standards of care, the importance of molecular diagnosis and the use of techniques such as laser interstitial thermal therapy (LITT) and focused ultrasound (FUS) are emphasized, exploring how the integration of molecular knowledge with emerging neurosurgical approaches can personalize and improve the treatment of gliomas. Conclusions: The choice between LITT and FUS should be tailored to each case, considering factors such as tumor characteristics and patient health. LITT is favored for larger, complex tumors, while FUS is standard for smaller, deep-seated ones. Both techniques are equally effective for small and superficial tumors. Our study provides clear guidance for treating pediatric low-grade gliomas and highlights the crucial roles of LITT and FUS in managing high-grade gliomas in adults. This research sets the stage for improved patient care and future developments in the field of neurosurgery.

**Keywords:** gliomas; LITT; brain tumors; FUS; classification; functional neurosurgery

## 1. Introduction

Gliomas, primary brain tumors that originate from glial cells, represent a significant challenge in the field of neuro-oncology. These intra-axial tumors exhibit high variability in their histopathological features, genetic profiles, and clinical manifestations, aspects that make it crucial to take an integrated, multidisciplinary approach to effectively combat their annihilating effect on patients [1].

Over the years, extensive research has shed light on the molecular basis of gliomas, offering promising possibilities for targeted therapies. In addition, advances in neurosurgical techniques have opened new horizons in the management of these tumors, providing more precise and personalized therapeutic options [1,2].

Despite these significant advances in understanding the complexity of glioma molecular pathways, the current standard of care, which includes maximal safe resection followed by radiotherapy and chemotherapy, often fails to provide patients with long-term survival and optimal quality of life. New therapeutic strategies are still needed to address the complexity of glioma biology and improve patient outcomes [2].

This study aimed to comprehensively review the most recent advances in the management of gliomas, focusing on the new classification system and emerging functional neurosurgical techniques. By integrating cutting-edge molecular knowledge with emerging and innovative neurosurgical approaches, it will be possible to lay the foundation for a more holistic and tailored glioma treatment paradigm [2].

## 2. Epidemiology and Classification of Gliomas

Gliomas account for about 80% of all malignant brain tumors, with an incidence that increases with age and peaks in individuals older than 65 years. In the United States, the age-adjusted glioma incidence rate is 6.16 per 100,000 person-years in subjects aged 65 and older, compared with 0.50 per 100,000 person-years in subjects aged 20–44 years [3,4]. Another relevant factor is the origin of the patients. In Europe, the highest incidence rates have been reported in Denmark and Finland, with age-standardized rates of 6.8 and 5.5 per 100,000 person-years, respectively [5]. In the United States, the incidence of gliomas is higher among whites compared with other racial/ethnic groups [6]. Genetic predispositions, such as specific markers associated with glioma susceptibility, contribute to these disparities [7]. Environmental factors, including exposure to ionizing radiation, are implicated in the etiology of GBM [8].

Differences in health infrastructure and diagnostic accessibility also influence reported incidence rates. Developed countries with advanced diagnostic capabilities can detect and report cases more accurately, potentially contributing to higher incidence rates. Conversely, underdiagnosis in some regions may lead to biased reporting [9].

Lifestyle and diet-related factors may further contribute to the complex epidemiological landscape of GBM. Emerging research suggests potential links between certain dietary components and glioblastoma risk. [10]

In essence, the global distribution of glioblastoma involves a complex interplay of genetic, environmental, and health factors. Unraveling these complexities is essential to advancing our knowledge of glioma epidemiology and ultimately improving prevention and treatment strategies on a global scale.

Gliomas can be classified into grades based on their histological features and molecular characteristics. Grade I gliomas are considered benign tumors, while grades II, III, and IV are malignant. The most common malignant gliomas, approximately 50% of all, are grade IV glioblastomas (GBMs), which are most frequently diagnosed in individuals older than 65 years. In the United States, the age-adjusted incidence rate of GBM is 3.21 per 100,000 person-years in subjects aged 65 and older, compared with 0.23 per 100,000 person-years in individuals aged 20–44 years [3].

Recent advances in molecular profiling have led to a better understanding of the underlying genetic alterations that drive the development of gliomas and can be used to classify them into molecular subtypes, which are characterized by different clinical outcomes and responses to treatment.

Among the most important genetic variants, mutations in the isocitrate dehydrogenase (IDH) gene identify IDH-mutant gliomas that demonstrate a better prognosis than wildtype IDH gliomas [11,12]. Other molecular alterations include mutations in the tumor-suppressor gene tumor protein 53 (TP53), the alpha-thalassemia/mental retardation x-linked (ATRX) syndrome, and the epidermal growth factor receptor (EGFR) pathway [13]. However, even taking into account this valuable genetic profiling, establishing glioma prognosis is more complicated because it is associated with a complex network of factors that contribute to determining the likelihood of patient survival [14].

### 2.1. Classification of Gliomas

In 2021, the World Health Organization (WHO) provided a new classification of tumors of the central nervous system, which is based on a more comprehensive understanding of the molecular and genetic characteristics of tumors and suggests the use of a combination of histology and molecular markers to predict patient outcomes and guide the choice of appropriate treatments (Table 1).

**Table 1.** Comparison of overall survival and progression-free survival of various types of gliomas.

| Glioma Classification | Molecular Markers | Histology | Malignancy | OS | PFS | References |
|---|---|---|---|---|---|---|
| Diffuse astrocytic and oligodendroglial tumors | IDH-mutant | Diffuse astrocytoma | Low grade | >5 years | Variable | [15] |
| | | Anaplastic astrocytoma | Intermediate grade | 2–5 years | 6–12 months | [15] |
| | | Oligodendroglioma with 1p/19q co-deletion | Low grade | >5 years | Variable | [15,16] |
| Anaplastic astrocytic and oligodendroglial tumors | IDH-mutant | Anaplastic astrocytoma | Intermediate grade | 2–5 years | 6–12 months | [15,16] |
| | | Anaplastic oligodendroglioma with 1p/19q co-deletion | Intermediate grade | 2–5 years | Variable | [15,16] |
| Glioblastoma | IDH-wildtype | Glioblastoma | High grade | 15 months | 7–9 months | [15] |
| | IDH-mutant | Glioblastoma | High grade | 31–46 months | 11–20 months | [15] |

Abbreviations: OS, overall survival; PFS, progression-free survival.

In particular, the document identifies four categories of gliomas: (1) diffuse astrocytic and oligodendroglial tumors, which include diffuse astrocytoma IDH-mutant, anaplastic astrocytoma IDH-mutant, oligodendroglioma IDH-mutant, and 1p/19q co-deleted tumors; (2) anaplastic astrocytic and oligodendroglial tumors, including anaplastic astrocytoma IDH-mutant, anaplastic oligodendroglioma IDH-mutant, and 1p/19q co-deleted tumors; (3) GBMs, including GBM IDH-wildtype and GBM IDH-mutant; (4) and other gliomas, such as ependymoma, choroid plexus tumors, and embryonal tumors.

The WHO classification does not automatically associate histological type with the grade of malignancy. For example, in the past, anaplastic astrocytoma was automatically considered a grade III glioma, as was anaplastic meningioma.

Therefore, it was expected that these two tumors, although biologically different, would have similar survival times. However, this is a simplistic approach, and today, the method is to stratify the various histological types as much as possible and to study their characteristics in more detail [15].

### 2.2. Molecular Signature of Gliomas

Accurately classifying the different types of gliomas by combining specific genetic and molecular features is critical to mitigating the difficulties of treating such a heterogeneous group of malignancies [16,17] and maximizing the chances of success [18].

Indeed, the molecular signature is strongly associated with the pathogenesis and prognosis of several tumors. Glioma pathogenesis is no exception, as it is closely dependent on genetic and epigenetic alterations, cellular signaling pathways, and the tumor microenvironment. The PI3K-Akt-mTOR pathway, which modulates cell growth, proliferation, survival, and metabolism, is one of the most important signaling pathways in glioma pathogenesis. As suggested by Wang et al., this pathway leads to increased activity in downstream effectors that promote glioma growth and progression through the amplification of growth factor receptors or the loss of negative regulators, promoting the processes of invasion and metastasis and resistance to chemotherapy and radiotherapy [19]. Of note, according to Cancer Genome Atlas (TCGA) data, approximately 88% of diffuse gliomas, which include GBM and lower-grade gliomas (LGGs), have genetic alterations in at least one component of the PI3K-Akt-mTOR pathway [15]. Specifically, mutations in the gene encoding the catalytic subunit of PI3K (PIK3CA) and the regulatory subunit of PI3K (PIK3R1) were found with a frequency of 17% and 18%, respectively. Less frequent are mutations in the downstream effector Akt (AKT1) and the gene encoding the component of the mTOR complex (MTOR), identified in 2% and 3% of diffuse gliomas, respectively. In addition, amplification of the gene encoding the epithelial growth factor receptor (EGFR), a potent activator of the PI3K-Akt-mTOR pathway, observed in approximately 50% of GBMs, further underscores the importance of this pathway in glioma pathogenesis [20–22]. As a result, targeting this signaling pathway is now considered a promising therapeutic strategy for treating gliomas, with several drugs currently in clinical trials [23,24]. Table 2 summarizes the most studied mutations in gliomas with their relative frequency and etiopathogenetic roles. The frequency of each genetic variant is represented by a wide range because of the different grades and histological types of gliomas considered.

**Table 2.** Genes affecting GBM tumor growth with their related mutation frequencies.

| Gene | Role in GBM Tumor Growth | Mutation Frequency | References |
|---|---|---|---|
| EGFR | Amplification or mutation of EGFR leads to increased proliferation and invasion of tumor cells. | 40–60% | [22–24] |
| PTEN | Loss of PTEN function promotes tumor cell survival and proliferation. | 20–40% | [15,25,26] |
| TP53 | Mutation of TP53 is associated with a more aggressive phenotype. | 25–30% | [27–29] |
| IDH1 | Mutation of IDH1 is associated with a better prognosis. | 5–10% | [25,30]. |
| MGMT | Methylation of the MGMT promoter is associated with increased sensitivity to chemotherapy and better patient outcomes. | 30–40% | [31–33] |
| VEGF | Overexpression of VEGF promotes angiogenesis and tumor growth in GBM. | 50–80% | [34–36] |
| PDGF | Overexpression of PDGF and its receptor promotes tumor cell proliferation and migration. | 15–20% | [37,38] |
| CDK4 | Amplification of CDK4 promotes cell cycle progression and tumor growth. | 5–10% | [39,40] |

Abbreviations: OS, overall survival; PFS, progression-free survival; EGFR, epidermal growth factor receptor; PTEN, phosphatase and tensin homolog; TP53, tumor protein 53; IDH1, isocitrate dehydrogenase 1; MGMT, O6-methylguanine-DNA methyltransferase; VEGF, vascular endothelial growth factor; PDGF, platelet-derived growth factor; CDK4, cyclin-dependent kinase 4.

Lei et al. developed a model using transcriptome data from two cohorts of patients with GBM. They identified 341 metabolic genes that showed significant differences between normal brain and GBM tissues, among which, 56 genes were found to be correlated with the patients' overall survival (OS). In the end, the model was constructed using a Lasso regression model with 18 genes and showed high accuracy in predicting the OS of the patients. In particular, the high-risk group of patients included in this model had a significantly shorter OS than the low-risk group in the training cohort ($p < 0.0001$) and in the independent external validation ($p < 0.001$). The study by Lei et al. showed that the prognosis of GBM is closely related to metabolic pathways and that, by using a model, it is possible to predict the prognosis of patients with GBM [41]. As is evident in contemporary neuro-oncology, searching for better diagnostic and therapeutic strategies for glioblastoma has led to a deep exploration of prognostic and predictive biomarkers. Increasingly, the evidence allows us to make a targeted choice regarding the type of treatment in relation to the molecular profile of the tumor.

In fact, these biomarkers may now influence the choice of therapeutic approach. For example, the presence of the mutated IDH1 gene is associated with a better prognosis, and patients carrying this mutation may benefit from more aggressive surgical resection aimed at maximal tumor excision [42].

MGMT promoter methylation status is another crucial biomarker. Glioblastoma patients with MGMT promoter methylation tend to respond better to chemotherapy [43], and knowledge of this status may influence the decision on the extent of surgical resection and the subsequent use of adjuvant therapies. The expression of certain molecular markers, such as EGFR, may also impact the choice of treatment. Elevated EGFR expression may suggest a more aggressive tumor phenotype, influencing the decision for a more extensive surgical approach or the inclusion of targeted therapies [44]. Montano et al. observed significantly longer OS in GBM patients with high levels of EGFR treated with total resection and standard radiochemotherapy involving temozolomide [45].

An innovative approach to glioblastoma can be achieved not only through research on new biomarkers but also through new treatment technologies.

Among the innovative avenues gaining traction is the study of PROteolysis TAgeting Chimeras (PROTACs) as potential game-changers in glioblastoma therapy [46].

PROTACs represent a paradigm shift in drug development, harnessing the cellular machinery to induce targeted protein degradation. In the context of glioblastoma, this approach holds the promise of selectively eliminating specific oncogenic proteins, thereby disrupting key pathways implicated in tumor progression. Yang et al. used a therapeutic nanosystem created by combining the BRD4-degrading PROTAC ARV-825 with a complex micelle. This micelle was able to penetrate the blood–brain barrier and target brain tumors. The drug released by this system shows antitumor effects by reducing cell proliferation, inducing apoptosis, and suppressing M2 macrophage polarization. These effects are achieved through the inhibition of IRF4 promoter transcription and the phosphorylation of STAT6, STAT3, and AKT [47].

To contextualize these advancements, a comprehensive understanding of the molecular landscape of glioblastoma subtypes is imperative. This molecular intricacy necessitates a tailored approach, recognizing that one-size-fits-all interventions may fall short in addressing the diverse manifestations of the disease.

Neurosurgical approaches and treatment modalities are intrinsically linked to the molecular profile of the glioblastoma, and understanding these connections is pivotal in optimizing patient outcomes. The choice between maximal safe resection, adjuvant therapies, and targeted interventions hinges on a nuanced appreciation of the specific molecular signatures at play [48]. Given the complex interplay between biomarkers, neurosurgical strategies, and other therapeutic options, we approach a future in which personalized and precise interventions will redefine the trajectory of glioblastoma management.

## 3. Overview of Treatments for Gliomas

Surgery has been an important component of the management of gliomas since the late 19th century. The first surgical resection was performed in 1884 by Victor Horsley, who removed a frontal lobe glioma from a patient who had been experiencing seizures. On 25 November 1884, Mr. Rickman J. Godlee performed the first recognized resection of a primary brain tumor. The surgery was performed at the Hospital for Epilepsy and Paralysis in London, UK. The patient died postoperatively from apparent meningitis, but postmortem examination revealed no residuals of the excised glioma [49].

In the following decades, surgical techniques have continued to evolve, with the development of new tools and methods to access and remove brain tumors.

In the early 20th century, Harvey Cushing became one of the most prominent neurosurgeons contributing significantly to glioma surgery. He is credited with developing the transsphenoidal approach, a minimally invasive method of accessing tumors located at the base of the skull. He also introduced the use of the operating microscope, which enabled more precise and controlled surgical resections [50].

In the mid-20th century, the use of neuroimaging techniques such as computed tomography (CT) and magnetic resonance imaging (MRI) revolutionized the diagnosis and management of gliomas. These techniques enabled more accurate preoperative planning and intraoperative visualization of the tumor and surrounding brain tissue [51].

Many advances in glioma surgery have been made in recent decades, including the development of intraoperative neurophysiological monitoring, which enables the real-time monitoring of neurological function during surgery. This technique has helped to minimize the risk of neurological deficits associated with the surgical resection of gliomas [52].

In addition, image-guided navigation systems and endoscopic techniques have expanded the range of accessible tumors and increased the precision of surgical resection [53,54].

Despite these advances, surgical resection of gliomas remains a complex and challenging procedure, particularly for high-grade tumors located in critical or eloquent areas of the brain. Furthermore, for these cases, the development of new surgical tools, such as ultrasonic aspirators and lasers, has enabled more effective and efficient tumor removal. In recent years, interest has grown in the use of less invasive approaches, such as stereotactic radiosurgery and laser interstitial thermal therapy (LITT). These approaches may offer a less invasive alternative to surgical resection for certain types of tumors [55,56].

The primary goal of surgery is to safely remove as much of the glioma as possible while preserving neurological function. The extent of surgical resection is a major determinant of patient survival and functional outcomes [57].

Several factors should be considered when deciding on the optimal surgical approach to gliomas, including the location, size, and grade of the tumor, as well as the patient's age and general health status [58]. In general, surgical resection is recommended for patients with low-grade gliomas and for those with high-grade gliomas who can tolerate the procedure [59].

The goal of surgery for high-grade gliomas is typically to achieve a safe maximal resection (SMR), which involves removing as much of the visible tumor as possible while preserving neurological function [60]. The extent of resection is typically assessed using MRI and is classified as gross total resection (GTR), subtotal resection (STR), or partial resection (PR) [61,62].

Several studies have demonstrated that achieving a GTR is associated with improved OS and progression-free survival (PFS) in patients with high-grade gliomas. A meta-analysis of 37 studies found that patients who underwent GTR probably increased the likelihood of 1-year survival compared with STR by about 61% and increased the likelihood of 2-year survival by about 19% (GTR compared with STR at 1 year: RR, 0.62; 95% CI, 0.56–0.69; $p < 0.001$) [63].

However, achieving a GTR is not always possible or safe. Indeed, in the case of tumors located in critical or eloquent areas of the brain case, the goal of surgery may be to obtain a biopsy or to remove the tumor to alleviate symptoms. Increasing evidence has shown that the use of fluorescence-guided surgery, which involves the administration of a fluorescent contrast agent that accumulates in tumor tissue, can improve the extent of glioma resection and increase PFS in patients with high-grade gliomas [64,65]

These studies have demonstrated that patients undergoing fluorescence-guided surgery have a significantly higher rate of complete tumor resection than those undergoing conventional white-light surgery.

Another recent study investigated the impact of intraoperative MRI (iMRI) on the extent of resection and patient outcomes in glioma surgery [60]. The study found that the use of iMRI can improve the extent of tumor resection and increase PFS in patients with high-grade gliomas. In addition, the use of iMRI is associated with a lower risk of postoperative neurological deficits.

Advances in imaging technology have also led to the development of new tools for preoperative planning and intraoperative navigation. For example, diffusion tensor imaging (DTI) has been shown to provide valuable information on the location and orientation of white matter tracts in the brain, which can help surgeons avoid damaging these critical structures during tumor resection [66].

There has also been growing interest in the use of minimally invasive approaches for the treatment of gliomas, such as LITT and stereotactic radiosurgery (SRS). A recent study compared the effectiveness of LITT and SRS for the treatment of recurrent high-grade gliomas and found that both approaches were associated with similar rates of tumor control and survival [67]. The study suggested that LITT may be a less invasive alternative to SRS for certain types of recurrent gliomas.

In summary, recent studies have focused on refining surgical techniques for glioma resection and identifying the most effective surgical strategies to improve patient outcomes. Advances in imaging technology and the development of new tools for operative navigation have also provided valuable insights into improving surgical precision and minimizing damage to surrounding brain tissue.

*The Standard of Care for Gliomas*

The standard of care for gliomas depends on various factors, including the grade of the tumor, the location and size of the tumor, and the patient's health status. However, the current standard of care typically involves a combination of surgery, radiation therapy, and chemotherapy.

Surgery is usually the first-line treatment and aims to remove as much of the tumor as possible while preserving critical brain functions. The extent of tumor resection is an important predictor of patient outcomes, and efforts to maximize resection while minimizing damage to the surrounding brain tissue have led to the development of various surgical techniques and technologies, such as intraoperative imaging and fluorescence-guided surgery [68].

Radiation therapy is typically used following surgery to kill any remaining tumor cells and prevent tumor regrowth. Various types of radiation therapy can be used for gliomas, including external beam radiation therapy, brachytherapy, and stereotactic radiosurgery [69]. The use of concurrent chemotherapy and radiation therapy has also been shown to improve outcomes in certain cases, such as for patients with high-grade gliomas [70].

Chemotherapy is usually reserved for cases where surgery and radiotherapy alone are not sufficient or in cases where the tumor recurs after initial treatment. Several chemotherapy agents can be used for gliomas, such as the DNA alkylating agents temozolomide (TMZ) and carmustine (BCNU) [71] TMZ is the most used drug in the treatment of high-risk gliomas after surgical resection. However, chemoresistance occurs in many patients, representing a substantial obstacle to successful therapy. A crucial role in chemoresistance is played by genes encoding for DNA repair proteins, such as DNA mismatch repair (MMR)-related proteins, O-6-methylguanine-DNA methyltransferase (MGMT), base excision repair (BER)-related proteins, AlkB homolog 2, alpha-ketoglutarate-dependent dioxygenase (ALKBH2), and proteins involved in homologous recombinational repair/non-homologous end joining (HRR/NHEJ), all of which correlate with TMZ efficacy [72].

Fortunately, today, it is possible to use a pharmacogenetically guided approach to predict the degree of sensitivity or resistance to a specific drug [73]. Although there are still barriers to overcome, pharmacogenetics, which studies responses to drug therapy based on a patient's genetic background, is very useful in personalizing drug therapy in all medical areas [73,74]. Regarding TMZ, pharmacogenetic testing based on the determination of MGMT methylation status is a useful tool for predicting responses to TMZ [71].

In 2016, Buckner et al. analyzed the clinical outcomes of 254 patients younger than 40 years randomized to receive radiotherapy (n. 128) or radiotherapy with chemotherapy (n. 126). The patients had astrocytoma, oligodendroglioma, or oligoastrocytoma and, after STR, were randomly separated into two treatment groups.

The results showed that, regardless of histologic type, patients who received radiotherapy with chemotherapy had a longer median OS than those who received chemotherapy alone (13.3 years vs. 7.8 years; $p = 0.003$). The two OS curves did not separate immediately but rather after one year in astrocytoma and about three years in oligoastrocytoma and oligodendroglioma.

The two curves, in all cases, remained well separated and distinct until the end of observation, about 12 years. The study showed that disease-free survival was also higher in patients who received chemotherapy and radiotherapy than those who received only radiotherapy, but the data are less reliable because these patients were promptly treated in various ways to improve prognosis [75].

The standard of care for gliomas is continuously evolving as new treatment options and strategies are developed. Recent studies have investigated the potential benefits of immunotherapy, targeted therapy, and other therapeutic approaches [76], and clinical trials are ongoing to determine the safety and efficacy of these new treatment options [77].

We report in Table 3 a brief description of each of the treatments for gliomas.

**Table 3.** Treatments for gliomas.

| Treatment | Description | Reference |
|---|---|---|
| Surgery | First-line treatment to remove as much of the tumor as possible while preserving critical brain functions. | [68] |
| LITT | A minimally invasive procedure that uses laser energy to heat and destroy tumor cells. It can be used as an alternative to traditional surgery or as a salvage treatment option for recurrent tumors. | [75] |
| FUS | A non-invasive procedure that uses focused ultrasound waves to heat and destroy tumor cells. It can be used as an alternative to traditional surgery or as a salvage treatment option for recurrent tumors. | [76] |
| Radiation therapy | Used following surgery to kill any remaining tumor cells and prevent tumor regrowth. Various types of radiation therapy can be used, including external beam radiation therapy, brachytherapy, and stereotactic radiosurgery. | [69] |
| Chemotherapy | Reserved for cases where surgery and radiation therapy alone are insufficient or for cases where the tumor recurs following initial treatment. Various chemotherapy agents can be used, such as temozolomide and carmustine (BCNU). | [71] |
| Concurrent chemotherapy and radiation therapy | Used in certain cases, such as for patients with high-grade gliomas, to improve outcomes. | [70] |
| Immunotherapy | Investigated as a potential treatment option for gliomas. | [77] |
| Targeted therapy | Investigated as a potential treatment option for gliomas. | [56] |

Abbreviations: LITT, laser interstitial thermal therapy; FUS, magnetic resonance-guided focused ultrasound.

## 4. Functional Neurosurgery

Functional neurosurgery is a rapidly advancing subspecialty that aims to manage neurological disorders through precise surgical interventions targeting specific brain circuits and regions. Unlike conventional neurosurgery, which often focuses on removing lesions or repairing structural abnormalities, functional neurosurgery primarily aims to modulate neural activity or disrupt malfunctioning circuits to restore normal brain function or mitigate symptoms. Two examples of innovative treatments applicable to brain tumors are focused ultrasound surgery (FUS) and LITT [78].

FUS is a non-invasive medical procedure that uses ultrasound waves to heat and destroy targeted tissues within the body without the need for traditional surgery. Focused ultrasound has been researched and developed as a potential treatment option for various medical conditions, including brain tumors. During the procedure, multiple intersecting beams of ultrasound energy are focused precisely on the tumor, generating enough heat to destroy the abnormal tissue while leaving surrounding healthy tissue relatively unharmed [79].

LITT has emerged as a promising treatment option for gliomas, which are difficult to treat with traditional surgery, radiation, and chemotherapy because they often invade surrounding brain tissue. LITT for gliomas is a relatively new approach with promising results. It is often used as an adjunct to traditional treatments, such as surgery and radiotherapy, or as a primary treatment option for smaller tumors. However, not all patients are candidates for LITT, and careful evaluation is necessary to determine the most appropriate treatment plan [80].

### 4.1. Laser Interstitial Thermal Therapy (LITT)

LITT is a minimally invasive surgical technique that uses a laser to create a controlled lesion within the brain and to destroy cancerous or abnormal tissue. It is a type of thermal ablation therapy that delivers controlled heat to the tumor, causing the cancer cells to die.

LITT is performed under general anesthesia to ensure patient comfort and safety during the procedure. The head is placed in a rigid frame, and an MRI is performed. An appropriate trajectory is established, and then, a small incision and straw-sized hole are made in the skull. A thin, flexible laser probe is inserted into the lesion. The laser is then used to heat and destroy the tumor while sparing healthy brain tissue. During the procedure, real-time MRI is used to guide laser placement and monitor treatment progress. After surgery, the patient is closely monitored for neurological deficits or complications. Depending on the case, the patient may require further treatment, such as radiotherapy or chemotherapy [81,82].

The use of lasers for medical applications dates back to the 1960s, but LITT as a therapeutic technique for brain tumors was first proposed in 1983 [83]. However, its use is limited because of an inability to monitor tissue temperature during laser treatment and, thus, control the extent of ablation [84].The first clinical application of LITT for brain tumors was reported in 1992, and since then, the technique has been increasingly used in neurosurgical practice [80]. However, it should be mentioned that LITT is used on many other organs.

The development of LITT has been driven by advancements in laser technology, imaging technology, and computational modeling. The use of MRI or CT to guide laser placement and monitor thermal ablation has enabled the accurate and precise treatment of brain tumors with minimal damage to surrounding tissue. The development of computer simulations of laser–tissue interactions has also improved the safety and efficacy of LITT [81]. In each laser application, absorbed light is converted into heat, causing changes in tissues. Wavelengths between 620 and 1200nm provide low absorption and deep penetration, which are optimal for LITT.

LITT is effective for the treatment of a variety of brain tumors, including GBMs, metastatic tumors, and low-grade gliomas. Studies have demonstrated that LITT is associated with a low complication rate and shorter hospital stays compared with traditional brain surgery. In addition, LITT has been used in the treatment of epilepsy, Parkinson's disease, and other neurological conditions.

Although LITT has shown promise as a minimally invasive alternative to traditional brain surgery, further research is needed to fully establish its safety and efficacy. Ongoing clinical trials are exploring the use of LITT for the treatment of other neurological conditions and recurrent brain tumors [85].

### 4.2. Indications for LITT

LITT is a relatively new procedure, and guidelines, which vary depending on the institution and the specific case, are still evolving [86–94]. Table 4 summarizes some guidelines for LITT.

First, a detailed preoperative evaluation must be performed. On the one hand, it is mandatory to characterize the tumor. LITT is usually considered for patients with small or deep brain tumors, which are not good candidates for traditional surgery or radiotherapy. On the other hand, it is also important to define the patient's clinical features, paying special attention to their neurological status. This evaluation usually includes a thorough history, physical examination, and imaging studies, such as MRI or CT scans [85].

**Table 4.** Indications for LITT.

| Patient Age and Glioma Grade | Indications for LITT | References |
|---|---|---|
| Pediatric Low-Grade Gliomas | Recurrent or residual tumors that are not amenable to resection or conventional radiation therapy. | [86,87] |
| | Newly diagnosed tumor that is not surgically accessible. | [88] |
| Adult Low-Grade Gliomas | Small or deep-seated tumors that are not amenable to resection or as an adjunct to surgical resection to improve the extent of resection. | [89–91] |
| High-Grade Gliomas (Grade III/IV) | As a salvage therapy for recurrent tumors or as a palliative therapy for patients with poor performance status. | [92–94] |

#### 4.2.1. LITT in Low-Grade Gliomas

Treatments of low-grade gliomas (LGGs) must consider how best to manage symptoms and remove or shrink the tumor. The optimal treatment of LGGs, particularly the timing, is controversial, and decision making must balance the benefits of therapy with potential treatment-related complications.

Although diffuse LGGs account for only 15% of gliomas, they have received increasing attention in the past decade [95]. The standard of care is surgical resection, but it can be difficult to perform radical surgery because of the anatomical location and surrounding functional and vascular structures. Such anatomical complexity often makes nonnegligible surgical sequelae highly likely, so less invasive techniques may confer a more optimal balance between cytoreduction and surgical complications [89–91].

In such cases, alternative treatments, including LITT, are preferred. Indeed, even in recurrences, it has already been evident for several years that LITT, given its minimal invasiveness, is the therapy of choice and may be preferred over reoperation [96].

In 2013, Patel et al. set out to analyze changes in lesion volume during the period following LITT using polygonal fusion tracing. Sixteen patients with intracranial neoplasms participated in the study. Using the OsiriX DICOM Viewer, three evaluators calculated lesion volumes as follows: pre-ablation (PreA), immediate post-ablation (IPA), 24 h post-ablation (24PA), and first post-ablation follow-up (FPA), which ranged from 4 to 11 weeks after ablation. The observed sizes showed an acute increase in volume at IPA with a decrease in the size at 24PA. Therefore, it is recommended that conclusions about the size of post-LITT intracranial lesions be drawn at least 24 h after LITT rather than immediately thereafter [97].

In a 2012 study, Jethwa et al. selected 20 patients who had previously failed conventional therapies; were unable to tolerate an open cranial procedure; or, finally, had a tumor considered otherwise inoperable. The patients then underwent LITT. The result was a particularly short hospitalization period of 2.27 days, with most patients going home on the first day postoperative. Complications, on the other hand, occurred in only four patients, generally those who were in poor health before surgery. Thus, the study argues that laser-induced heat therapy is a procedure for difficult intracranial neoplasms. Like any other procedure, it can provide excellent results if patient and lesion selection is performed carefully [98].

This evidence suggests that, even in LGGs, LITT plays a nonnegligible role, so it could not only be an alternative for inoperable cases but also a first choice in the neurosurgical armamentarium, especially in recurrences.

#### 4.2.2. LITT in GBMs

GBM is the most common and aggressive malignant primary brain tumor. Several treatment options are available today, but patients have a median survival of 12–15 months [13]. Surgery is generally considered the mainstay of treatment for patients with surgically accessible tumors; however, GBM has a strong tendency to recur [99].

GBM recurrence is a major problem because subsequent surgical resection is not always feasible and, even if possible, does not always benefit the patient. In other circumstances, patients are generally managed with chemo-radiotherapy, with a worse prognosis. The advent of LITT has provided a potentially viable salvage therapy for patients with recurrent or newly diagnosed lesions inaccessible to conventional surgical approaches. In a study by Traylor et al., 69 patients with a median age of 56 years with GBMs who underwent LITT were retrospectively analyzed. OS from the time of LITT was the primary endpoint measured.

The median tumor volume was 10.4 $cm^3$ (range, 1.0–64.0 $cm^3$). The Kaplan–Meier estimate from the time of LITT was 12 months (95% confidence interval of 8–16 months). The median volume uncovered by the ablation beam was 1.31 $cm^3$ (range, 0–41.2 $cm^3$). The median hospital length of stay was two days (range, 0–47 days), and the median PFS from the time of LITT was four months (95% confidence interval, 3–7 months). Adjuvant chemotherapy significantly prolonged PFS and OS (both, $p < 0.01$). Total gross ablation was not significantly associated with PFS ($p = 0.09$).

The results of this study suggest that LITT may confer a survival benefit over the nonoperative management of newly diagnosed GBM. However, the treatment of traditional inoperable GBMs with surgical therapy represents a challenge for physicians, and larger studies are needed to establish the impact of this surgical technique on GBM management [94].

## 5. Focus Ultrasound and LITT: Similarities and Differences

The effectiveness of LITT has already been described, so now, we will report the evidence related to FUS; the possible limitations of the two techniques; and finally, the choice between FUS and LITT with their respective indications.

Although FUS and LITT share some similarities in mechanisms of action, there are also obvious and understandable differences between the two techniques. FUS uses focused ultrasound waves to generate high levels of thermal energy at a precise location in the brain, creating a small lesion that can destroy tumor tissue. The procedure is non-invasive and is performed while the patient is awake, with real-time MRI or ultrasounds to guide the procedure [100]. The clinical indications for FUS are given in Table 5.

Several studies have examined the effectiveness of these surgical approaches in terms of PFS and OS in patients with gliomas. The available literature suggests that LITT and FUS are both safe and effective treatments for such tumors [101].

The literature offers a lot of evidence. Among the most recent, we have selected two highly indicative studies. A study by Coluccia et al. published in 2020 found that patients with GBMs who had not received previous chemotherapy had a longer PFS and OS after treatment with FUS [102].

Furthermore, another study performed in the same years by Deng et al. also reported that patients who had not received previous bevacizumab therapy had a longer PFS and OS after FUS treatment [103].

Since both techniques are effective, the therapeutic choice should be based on the possible difficulties of adopting the techniques. It has emerged from various studies that the effectiveness of LITT is strongly influenced by the size and location of the tumor, making it less effective for large and deep tumors and more effective for small and deep tumors. The location of the tumor means not only a greater or lesser depth within the brain but also proximity to particularly critical anatomical structures or eloquent areas of the brain.

Regarding tumor dimensions, a study by Chang et al. (2021) found that patients with tumors larger than 30 cc had significantly worse PFS and OS after FUS treatment compared with those with smaller tumors [104].

Overall, LITT may be preferable for larger or more diffuse tumors [104], especially those near critical brain structures [105] or in areas responsible for essential functions [98]. FUS is typically chosen for smaller or deep-seated tumors in anatomically less complex regions that cannot be surgically removed [100,106–109]. FUS and LITT show similar effectiveness for tumors close to the brain's surface, as well as small ones [100,105].

**Table 5.** Indications for FUS.

| Patient Age and Glioma Grade | Indications for FUS | References |
|---|---|---|
| Pediatric Low-Grade Gliomas | Recurrent or residual tumors that are not amenable to resection or conventional radiation therapy | [106,107] |
| Adult Low-Grade Gliomas | Small or deep-seated tumors that are not amenable to resection or as an adjunct to surgical resection to improve the extent of resection | [108] |
| High-Grade Gliomas (Grade III/IV) | As a salvage therapy for recurrent tumors or as a palliative therapy for patients with poor performance status | [109] |

In conclusion, both FUS and LITT are generally well tolerated, with relatively low rates of adverse effects such as infections, hemorrhages, or neurological deficits. However, as with any medical procedure, there are potential risks associated with each technique.

## 6. Conclusions and Future Directions

Through advances in molecular diagnosis, clinicians can better understand the underlying genetic and molecular characteristics of gliomas, leading to more personalized treatment strategies. Targeted therapies and precision medicine offer promising avenues for improving the effectiveness of glioma treatment while minimizing side effects.

The emergence of functional neurosurgery techniques, particularly FUS and LITT, represents a significant step forward in the management of gliomas. These minimally invasive procedures offer the potential for precise and controlled tumor ablation, reducing damage to surrounding healthy brain tissue and improving patient recovery.

However, while both have shown promise in achieving local tumor control and improving outcomes for patients with gliomas, further research is needed to determine their long-term effects on OS and to fully understand their risks and benefits. Ultimately, the choice of technique to be used should be made on a case-by-case basis, taking into account several factors such as tumor location, size, and grade, as well as the general health of the patient and treatment goal. Overall, the literature suggests that LITT is preferred for larger, diffuse tumors near critical brain structures or essential functions, while FUS is typically used for smaller, deep-seated tumors in less complex areas. FUS and LITT are equally effective for superficial and small tumors. As a general guideline, both techniques are used in the treatment of pediatric low-grade gliomas for recurrent or inoperable cases.

In adults, these techniques can play a crucial role in the management of high-grade (grade III/IV) gliomas, serving as salvage therapies for recurrent tumors and providing palliative care for patients with reduced performance status.

**Author Contributions:** Conceptualization, M.D.S.; methodology, M.D.S. and V.C.; writing—original draft preparation, M.D.S., V.C. and G.P.; writing—review and editing, V.C. and L.D.M.; supervision, L.D.M. and G.I. All authors have read and agreed to the published version of the manuscript.

**Funding:** This research received no external funding.

**Institutional Review Board Statement:** Not applicable.

**Informed Consent Statement:** Not applicable.

**Data Availability Statement:** Not applicable.

**Acknowledgments:** We deeply thank Marco Fontanella for his support, vision, and generosity in building bridges for knowledge and progress in the field of neurosurgery.

**Conflicts of Interest:** The authors declare no conflict of interest.

## References


1. Weller, M.; Wick, W.; Aldape, K.; Brada, M.; Berger, M.; Pfister, S.M.; Nishikawa, R.; Rosenthal, M.; Wen, P.Y.; Stupp, R.; et al. Glioma. *Nat. Rev. Dis. Prim.* **2015**, *1*, 15017. [CrossRef] [PubMed]
2. Lee, J.H.; Wee, C.W. Treatment of Adult Gliomas: A Current Update. *Brain Neurorehabilit* **2022**, *15*, e24. [CrossRef]
3. Kim, M.; Ladomersky, E.; Mozny, A.; Kocherginsky, M.; O'shea, K.; Reinstein, Z.Z.; Zhai, L.; Bell, A.; Lauing, K.L.; Bollu, L.; et al. Glioblastoma as an age-related neurological disorder in adults. *Neuro-Oncol. Adv.* **2021**, *3*, vdab125. [CrossRef] [PubMed]
4. Kim, H.J.; Park, J.W.; Lee, J.H. Genetic Architectures and Cell-of-Origin in Glioblastoma. *Front. Oncol.* **2021**, *10*, 615400. [CrossRef] [PubMed]
5. Crocetti, E.; Trama, A.; Stiller, C.; Caldarella, A.; Soffietti, R.; Jaal, J.; Weber, D.C.; Ricardi, U.; Slowinski, J.; Brandes, A. Epidemiology of glial and non-glial brain tumours in Europe. *Eur. J. Cancer* **2012**, *48*, 1532–1542. [CrossRef]
6. Ostrom, Q.T.; Cote, D.J.; Ascha, M.; Kruchko, C.; Barnholtz-Sloan, J.S. Adult Glioma Incidence and Survival by Race or Ethnicity in the United States From 2000 to 2014. *JAMA Oncol.* **2018**, *4*, 1254–1262. [CrossRef]
7. Mo, Z.; Xin, J.; Chai, R.; Woo, P.Y.; Chan, D.T.; Wang, J. Epidemiological characteristics and genetic alterations in adult diffuse glioma in East Asian populations. *Cancer Biol. Med.* **2022**, *19*, 1440–1459. [CrossRef]
8. Braganza, M.Z.; Kitahara, C.M.; de González, A.B.; Inskip, P.D.; Johnson, K.J.; Rajaraman, P. Ionizing radiation and the risk of brain and central nervous system tumors: A systematic review. *Neuro-Oncology* **2012**, *14*, 1316–1324. [CrossRef]
9. Pramesh, C.S.; Badwe, R.A.; Bhoo-Pathy, N.; Booth, C.M.; Chinnaswamy, G.; Dare, A.J.; de Andrade, V.P.; Hunter, D.J.; Gopal, S.; Gospodarowicz, M.; et al. Priorities for cancer research in low- and middle-income countries: A global perspective. *Nat. Med.* **2022**, *28*, 649–657. [CrossRef]
10. Shu, L.; Yu, D.; Jin, F. Healthy dietary patterns, foods, and risk of glioma: A systematic review and meta-analysis of observational studies. *Front. Nutr.* **2023**, *9*, 1077452. [CrossRef]
11. Villani, V.; Casini, B.; Tanzilli, A.; Lecce, M.; Rasile, F.; Telera, S.; Pace, A.; Piludu, F.; Terrenato, I.; Rollo, F.; et al. The Glioma-IRE project—Molecular profiling in patients with glioma: Steps toward an individualized diagnostic and therapeutic approach. *J. Transl. Med.* **2023**, *21*, 215. [CrossRef]
12. Cairncross, J.G.; Wang, M.; Jenkins, R.B.; Shaw, E.G.; Giannini, C.; Brachman, D.G.; Buckner, J.C.; Fink, K.L.; Souhami, L.; Laperriere, N.J.; et al. Benefit From Procarbazine, Lomustine, and Vincristine in Oligodendroglial Tumors Is Associated With Mutation of *IDH. J. Clin. Oncol.* **2014**, *32*, 783–790. [CrossRef]
13. Manjunath, M.; Yan, J.; Youn, Y.; Drucker, K.L.; Kollmeyer, T.M.; McKinney, A.M.; Zazubovich, V.; Zhang, Y.; Costello, J.F.; Eckel-Passow, J.; et al. Functional analysis of low-grade glioma genetic variants predicts key target genes and transcription factors. *Neuro-Oncology* **2021**, *23*, 638–649. [CrossRef] [PubMed]
14. Lee, K.; Kim, S.-I.; Kim, E.E.; Shim, Y.-M.; Won, J.-K.; Park, C.-K.; Choi, S.H.; Yun, H.; Lee, H.; Park, S.-H. Genomic profiles of IDH-mutant gliomas: MYCN-amplified IDH-mutant astrocytoma had the worst prognosis. *Sci. Rep.* **2023**, *13*, 6761. [CrossRef] [PubMed]
15. Louis, D.N.; Perry, A.; Wesseling, P.; Brat, D.J.; Cree, I.A.; Figarella-Branger, D.; Hawkins, C.; Ng, H.K.; Pfister, S.M.; Reifenberger, G.; et al. The 2021 WHO Classification of Tumors of the Central Nervous System: A summary. *Neuro-Oncology* **2021**, *23*, 1231–1251. [CrossRef] [PubMed]
16. Liu, D.; Chen, J.; Hu, X.; Yang, K.; Liu, Y.; Hu, G.; Ge, H.; Zhang, W.; Liu, H. Imaging-Genomics in Glioblastoma: Combining Molecular and Imaging Signatures. *Front. Oncol.* **2021**, *11*, 699265. [CrossRef] [PubMed]
17. McMahon, D.J.; Gleeson, J.P.; O'reilly, S.; Bambury, R.M. Management of newly diagnosed glioblastoma multiforme: Current state of the art and emerging therapeutic approaches. *Med. Oncol.* **2022**, *39*, 129. [CrossRef]
18. Barthel, L.; Hadamitzky, M.; Dammann, P.; Schedlowski, M.; Sure, U.; Thakur, B.K.; Hetze, S. Glioma: Molecular signature and crossroads with tumor microenvironment. *Cancer Metastasis Rev.* **2021**, *41*, 53–75. [CrossRef]
19. Mohamed, E.; Kumar, A.; Zhang, Y.; Wang, A.S.; Chen, K.; Lim, Y.; Shai, A.; Taylor, J.W.; Clarke, J.; Hilz, S.; et al. PI3K/AKT/mTOR signaling pathway activity in IDH-mutant diffuse glioma and clinical implications. *Neuro-Oncology* **2022**, *24*, 1471–1481. [CrossRef]
20. Hersh, A.M.; Gaitsch, H.; Alomari, S.; Lubelski, D.; Tyler, B.M. Molecular Pathways and Genomic Landscape of Glioblastoma Stem Cells: Opportunities for Targeted Therapy. *Cancers* **2022**, *14*, 3743. [CrossRef]
21. Rudà, R.; Capper, D.; Waldman, A.D.; Pallud, J.; Minniti, G.; Kaley, T.J.; Bouffet, E.; Tabatabai, G.; Aronica, E.; Jakola, A.S.; et al. EANO—EURACAN—SNO Guidelines on circumscribed astrocytic gliomas, glioneuronal, and neuronal tumors. *Neuro-Oncology* **2022**, *24*, 2015–2034. [CrossRef] [PubMed]
22. Gilard, V.; Tebani, A.; Dabaj, I.; Laquerrière, A.; Fontanilles, M.; Derrey, S.; Marret, S.; Bekri, S. Diagnosis and Management of Glioblastoma: A Comprehensive Perspective. *J. Pers. Med.* **2021**, *11*, 258. [CrossRef]
23. Li, X.; Wu, C.; Chen, N.; Gu, H.; Yen, A.; Cao, L.; Wang, E.; Wang, L. PI3K/Akt/mTOR signaling pathway and targeted therapy for glioblastoma. *Oncotarget* **2016**, *7*, 33440–33450. [CrossRef] [PubMed]
24. Colardo, M.; Segatto, M.; Di Bartolomeo, S. Targeting RTK-PI3K-mTOR Axis in Gliomas: An Update. *Int. J. Mol. Sci.* **2021**, *22*, 4899. [CrossRef]
25. Parsons, D.W.; Jones, S.; Zhang, X.; Lin, J.C.-H.; Leary, R.J.; Angenendt, P.; Mankoo, P.; Carter, H.; Siu, I.-M.; Gallia, G.L.; et al. An Integrated Genomic Analysis of Human Glioblastoma Multiforme. *Science* **2008**, *321*, 1807–1812. [CrossRef]

26. Verhaak, R.G.W.; Hoadley, K.A.; Purdom, E.; Wang, V.; Wilkerson, M.D.; Miller, C.R.; Ding, L.; Golub, T.; Jill, P.; Alexe, G.; et al. Integrated Genomic Analysis Identifies Clinically Relevant Subtypes of Glioblastoma Characterized by Abnormalities in PDGFRA, IDH1, EGFR, and NF1. *Cancer Cell* **2010**, *17*, 98–110. [CrossRef]
27. Ohgaki, H.; Kleihues, P. Genetic alterations and signaling pathways in the evolution of gliomas. *Cancer Sci.* **2009**, *100*, 2235–2241. [CrossRef]
28. Mellinghoff, I.K.; Wang, M.Y.; Vivanco, I.; Haas-Kogan, D.A.; Zhu, S.; Dia, E.Q.; Lu, K.V.; Yoshimoto, K.; Huang, J.H.Y.; Chute, D.J.; et al. Molecular Determinants of the Response of Glioblastomas to EGFR Kinase Inhibitors. *N. Engl. J. Med.* **2005**, *353*, 2012–2024. [CrossRef]
29. Garcia-Fabiani, M.B.; Haase, S.; Comba, A.; Carney, S.; McClellan, B.; Banerjee, K.; Alghamri, M.S.; Syed, F.; Kadiyala, P.; Nunez, F.J.; et al. Genetic Alterations in Gliomas Remodel the Tumor Immune Microenvironment and Impact Immune-Mediated Therapies. *Front. Oncol.* **2021**, *11*, 631037. [CrossRef]
30. Cohen, A.L.; Holmen, S.L.; Colman, H. IDH1 and IDH2 Mutations in Gliomas. *Curr. Neurol. Neurosci. Rep.* **2013**, *13*, 345. [CrossRef]
31. Hegi, M.E.; Diserens, A.-C.; Gorlia, T.; Hamou, M.-F.; De Tribolet, N.; Weller, M.; Kros, J.M.; Hainfellner, J.A.; Mason, W.; Mariani, L.; et al. MGMT Gene Silencing and Benefit from Temozolomide in Glioblastoma. *N. Engl. J. Med.* **2005**, *352*, 997–1003. [CrossRef] [PubMed]
32. Wick, W.; Weller, M.; Van Den Bent, M.; Sanson, M.; Weiler, M.; Von Deimling, A.; Plass, C.; Hegi, M.; Platten, M.; Reifenberger, G. MGMT testing—The challenges for biomarker-based glioma treatment. *Nat. Rev. Neurol.* **2014**, *10*, 372–385. [CrossRef] [PubMed]
33. Weller, M.; Stupp, R.; Reifenberger, G.; Brandes, A.A.; Bent, M.J.V.D.; Wick, W.; Hegi, M.E. MGMT promoter methylation in malignant gliomas: Ready for personalized medicine? *Nat. Rev. Neurol.* **2009**, *6*, 39–51. [CrossRef] [PubMed]
34. Carmeliet, P.; Jain, R.K. Angiogenesis in cancer and other diseases. *Nature* **2000**, *407*, 249–257. [CrossRef]
35. Plate, K.H.; Risau, W. Angiogenesis in malignant gliomas. *Glia* **1995**, *15*, 339–347. [CrossRef]
36. Folkman, J.; Klagsbrun, M. Angiogenic factors. *Science* **1987**, *235*, 442–447. [CrossRef]
37. Lindberg, N.; Holland, E.C. PDGF in gliomas: More than just a growth factor? *Upsala J. Med. Sci.* **2012**, *117*, 92–98. [CrossRef] [PubMed]
38. Cantanhede, I.G.; de Oliveira, J.R.M. PDGF Family Expression in Glioblastoma Multiforme: Data Compilation from Ivy Glioblastoma Atlas Project Database. *Sci. Rep.* **2017**, *7*, 15271. [CrossRef]
39. Rollbrocker, B.; Waha, A.; Louis, D.N.; Wiestler, O.D.; von Deimling, A. Amplification of the cyclin-dependent kinase 4 ( CDK4 ) gene is associated with high cdk4 protein levels in glioblastoma multiforme. *Acta Neuropathol.* **1996**, *92*, 70–74. [CrossRef]
40. Higa, N.; Akahane, T.; Yokoyama, S.; Yonezawa, H.; Uchida, H.; Takajo, T.; Otsuji, R.; Hamada, T.; Matsuo, K.; Kirishima, M.; et al. Prognostic impact of *PDGFRA* gain/amplification and *MGMT* promoter methylation status in patients with *IDH* wild-type glioblastoma. *Neuro-Oncol. Adv.* **2022**, *4*, vdac097. [CrossRef]
41. Lei, C.; Chen, W.; Wang, Y.; Zhao, B.; Liu, P.; Kong, Z.; Liu, D.; Dai, C.; Wang, Y.; Wang, Y.; et al. Prognostic Prediction Model for Glioblastoma: A Metabolic Gene Signature and Independent External Validation. *J. Cancer* **2021**, *12*, 3796–3808. [CrossRef] [PubMed]
42. Beiko, J.; Suki, D.; Hess, K.R.; Fox, B.D.; Cheung, V.; Cabral, M.; Shonka, N.; Gilbert, M.R.; Sawaya, R.; Prabhu, S.S.; et al. IDH1 mutant malignant astrocytomas are more amenable to surgical resection and have a survival benefit associated with maximal surgical resection. *Neuro-Oncology* **2013**, *16*, 81–91. [CrossRef] [PubMed]
43. Richard, S.; Tachon, G.; Milin, S.; Wager, M.; Karayan-Tapon, L. Dual *MGMT* inactivation by promoter hypermethylation and loss of the long arm of chromosome 10 in glioblastoma. *Cancer Med.* **2020**, *9*, 6344–6353. [CrossRef]
44. Shinojima, N.; Tada, K.; Shiraishi, S.; Kamiryo, T.; Kochi, M.; Nakamura, H.; Makino, K.; Saya, H.; Hirano, H.; Kuratsu, J.-I.; et al. Prognostic value of epidermal growth factor receptor in patients with glioblastoma multiforme. *Cancer Res.* **2003**, *63*, 6962–6970.
45. Montano, N.; Cenci, T.; Martini, M.; D'alessandris, Q.G.; Pelacchi, F.; Ricci-Vitiani, L.; Maira, G.; De Maria, R.; Larocca, L.M.; Pallini, R. Expression of EGFRvIII in Glioblastoma: Prognostic Significance Revisited. *Neoplasia* **2011**, *13*, 1113–1121. [CrossRef] [PubMed]
46. Severini, L.L.; Bufalieri, F.; Infante, P.; Di Marcotullio, L. Proteolysis-Targeting Chimera (PROTAC): Is the Technology Looking at the Treatment of Brain Tumors? *Front. Cell Dev. Biol.* **2022**, *10*, 854352. [CrossRef]
47. Yang, T.; Hu, Y.; Miao, J.; Chen, J.; Liu, J.; Cheng, Y.; Gao, X. A BRD4 PROTAC nanodrug for glioma therapy via the intervention of tumor cells proliferation, apoptosis and M2 macrophages polarization. *Acta Pharm. Sin. B* **2022**, *12*, 2658–2671. [CrossRef]
48. Thon, N.; Tonn, J.-C.; Kreth, F.-W. The surgical perspective in precision treatment of diffuse gliomas. *OncoTargets Ther.* **2019**, *12*, 1497–1508. [CrossRef]
49. Kirkpatrick, D.B. The first primary brain-tumor operation. *J. Neurosurg.* **1984**, *61*, 809–813. [CrossRef]
50. Ellis, H. Harvey Cushing: A founding father of neurosurgery. *Br. J. Hosp. Med.* **2009**, *70*, 600. [CrossRef]
51. Ramaglia, A.; Tortora, D.; Mankad, K.; Lequin, M.; Severino, M.; D'arco, F.; Löbel, U.; Benenati, M.; de Leng, W.W.J.; De Marco, P.; et al. Role of diffusion weighted imaging for differentiating cerebral pilocytic astrocytoma and ganglioglioma BRAF V600E-mutant from wild type. *Neuroradiology* **2019**, *62*, 71–80. [CrossRef] [PubMed]
52. Ghatol, D.; Widrich, J. Intraoperative Neurophysiological Monitoring. In *StatPearls [Internet]*; StatPearls Publishing: Treasure Island, FL, USA, 2022.

53. Maybody, M.; Stevenson, C.; Solomon, S.B. Overview of Navigation Systems in Image-Guided Interventions. *Technol. Vasc. Interv. Radiol.* **2013**, *16*, 136–143. [CrossRef] [PubMed]
54. Harris, L.W. Endoscopic techniques in neurosurgery. *Microsurgery* **1994**, *15*, 541–546. [CrossRef]
55. Ostrom, Q.T.; Patil, N.; Cioffi, G.; Waite, K.; Kruchko, C.; Barnholtz-Sloan, J.S. CBTRUS Statistical Report: Primary Brain and Other Central Nervous System Tumors Diagnosed in the United States in 2013–2017. *Neuro-Oncology* **2020**, *22* (Suppl. 2), iv1–iv96. [CrossRef] [PubMed]
56. Wen, P.Y.; Kesari, S. Malignant Gliomas in Adults. *N. Engl. J. Med.* **2008**, *359*, 492–507. [CrossRef]
57. McGirt, M.J.; Chaichana, K.L.; Attenello, F.J.; Weingart, J.D.; Than, K.; Burger, P.C.; Olivi, A.; Brem, H.; Quinoñes-Hinojosa, A. Extent of surgical resection is independently associated with survival in patients with hemispheric infiltrating low-grade gliomas. *Neurosurgery* **2008**, *63*, 700–708. [CrossRef] [PubMed]
58. Kuhnt, D.; Becker, A.; Ganslandt, O.; Bauer, M.; Buchfelder, M.; Nimsky, C. Correlation of the extent of tumor volume resection and patient survival in surgery of glioblastoma multiforme with high-field intraoperative MRI guidance. *Neuro-Oncology* **2011**, *13*, 1339–1348, Erratum in: *Neuro-Oncology* **2014**, *16*, 1429. [CrossRef]
59. Hervey-Jumper, S.L.; Berger, M.S. Role of Surgical Resection in Low- and High-Grade Gliomas. *Curr. Treat. Opt. Neurol.* **2014**, *16*, 284. [CrossRef]
60. Bonosi, L.; Marrone, S.; Benigno, U.E.; Buscemi, F.; Musso, S.; Porzio, M.; Silven, M.P.; Torregrossa, F.; Grasso, G. Maximal Safe Resection in Glioblastoma Surgery: A Systematic Review of Advanced Intraoperative Image-Guided Techniques. *Brain Sci.* **2023**, *13*, 216. [CrossRef]
61. Lacroix, M.; Abi-Said, D.; Fourney, D.R.; Gokaslan, Z.L.; Shi, W.; DeMonte, F.; Lang, F.F.; McCutcheon, I.E.; Hassenbusch, S.J.; Holland, E.; et al. A multivariate analysis of 416 patients with glioblastoma multiforme: Prognosis, extent of resection, and survival. *J. Neurosurg.* **2001**, *95*, 190–198. [CrossRef]
62. Han, Q.; Liang, H.; Cheng, P.; Yang, H.; Zhao, P. Gross Total vs. Subtotal Resection on Survival Outcomes in Elderly Patients With High-Grade Glioma: A Systematic Review and Meta-Analysis. *Front. Oncol.* **2020**, *10*, 151. [CrossRef] [PubMed]
63. Brown, T.J.; Brennan, M.C.; Li, M.; Church, E.W.; Brandmeir, N.J.; Rakszawski, K.L.; Patel, A.S.; Rizk, E.B.; Suki, D.; Sawaya, R.; et al. Association of the Extent of Resection With Survival in Glioblastoma: A systematic review and meta-analysis. *JAMA Oncol.* **2016**, *2*, 1460–1469. [CrossRef] [PubMed]
64. Sanai, N.; Eschbacher, J.; Hattendorf, G.; Coons, S.W.; Preul, M.C.; Smith, K.A.; Nakaji, P.; Spetzler, R.F. Intraoperative Confocal Microscopy for Brain Tumors: A Feasibility Analysis in Humans. *Neurosurg.* **2011**, *68*, ons282–ons290. [CrossRef] [PubMed]
65. Falco, J.; Cavallo, C.; Vetrano, I.G.; de Laurentis, C.; Siozos, L.; Schiariti, M.; Broggi, M.; Ferroli, P.; Acerbi, F. Fluorescein Application in Cranial and Spinal Tumors Enhancing at Preoperative MRI and Operated With a Dedicated Filter on the Surgical Microscope: Preliminary Results in 279 Patients Enrolled in the FLUOCERTUM Prospective Study. *Front. Surg.* **2019**, *6*, 49. [CrossRef] [PubMed]
66. Shalan, M.E.; Soliman, A.Y.; Nassar, I.A.; Alarabawy, R.A. Surgical planning in patients with brain glioma using diffusion tensor MR imaging and tractography. *Egypt. J. Radiol. Nucl. Med.* **2021**, *52*, 110. [CrossRef]
67. Lee, I.; Kalkanis, S.; Hadjipanayis, C.G. Stereotactic Laser Interstitial Thermal Therapy for Recurrent High-Grade Gliomas. *Neurosurgery* **2016**, *79*, S24–S34. [CrossRef] [PubMed]
68. Sanai, N.; Berger, M.S. Glioma extent of resection and its impact on patient outcome. *Neurosurgery* **2008**, *62*, 753–766. [CrossRef]
69. Weller, M.; van den Bent, M.; Hopkins, K.; Tonn, J.C.; Stupp, R.; Falini, A.; Cohen-Jonathan-Moyal, E.; Frappaz, D.; Henriksson, R.; Balana, C.; et al. EANO guideline for the diagnosis and treatment of anaplastic gliomas and glioblastoma. *Lancet Oncol.* **2014**, *15*, e395–e403. [CrossRef]
70. Stupp, R.; Mason, W.P.; van den Bent, M.J.; Weller, M.; Fisher, B.; Taphoorn, M.J.B.; Belanger, K.; Brandes, A.A.; Marosi, C.; Bogdahn, U.; et al. Radiotherapy plus Concomitant and Adjuvant Temozolomide for Glioblastoma. *N. Engl. J. Med.* **2005**, *352*, 987–996. [CrossRef]
71. Bent, M.v.D.; Keime-Guibert, F.; Brandes, A.; Taphoorn, M.; Kros, J.; Eskens, F.; Carpentier, A. Temozolomide chemotherapy in recurrent oligodendroglioma. *Neurology* **2001**, *57*, 340–342. [CrossRef]
72. Nagel, Z.D.; Kitange, G.J.; Gupta, S.K.; Joughin, B.A.; Chaim, I.A.; Mazzucato, P.; Lauffenburger, D.A.; Sarkaria, J.N.; Samson, L.D. DNA Repair Capacity in Multiple Pathways Predicts Chemoresistance in Glioblastoma Multiforme. *Cancer Res.* **2017**, *77*, 198–206. [CrossRef] [PubMed]
73. Roden, D.M.; McLeod, H.L.; Relling, M.V.; Williams, M.S.; Mensah, G.A.; Peterson, J.F.; Van Driest, S.L. Pharmacogenomics. *Lancet* **2019**, *394*, 521–532. [CrossRef] [PubMed]
74. Conti, V.; Corbi, G.; Manzo, V.; Sellitto, C.; Iannello, F.; Esposito, S.; De Bellis, E.; Iannaccone, T.; Filippelli, A. The Role of Pharmacogenetics in Antithrombotic Therapy Management: New Achievements and Barriers Yet to Overcome. *Curr. Med. Chem.* **2021**, *28*, 6675–6703. [CrossRef]
75. Melnick, K.; Shin, D.; Dastmalchi, F.; Kabeer, Z.; Rahman, M.; Tran, D.; Ghiaseddin, A. Role of Laser Interstitial Thermal Therapy in the Management of Primary and Metastatic Brain Tumors. *Curr. Treat. Opt. Oncol.* **2021**, *22*, 108. [CrossRef] [PubMed]
76. Wei, H.-J.; Upadhyayula, P.S.; Pouliopoulos, A.N.; Englander, Z.K.; Zhang, X.; Jan, C.-I.; Guo, J.; Mela, A.; Zhang, Z.; Wang, T.J.; et al. Focused Ultrasound-Mediated Blood-Brain Barrier Opening Increases Delivery and Efficacy of Etoposide for Glioblastoma Treatment. *Endocrine* **2020**, *110*, 539–550. [CrossRef] [PubMed]

77. Lim, M.; Xia, Y.; Bettegowda, C.; Weller, M. Current state of immunotherapy for glioblastoma. *Nat. Rev. Clin. Oncol.* **2018**, *15*, 422–442. [CrossRef] [PubMed]
78. Lev-Tov, L.; Barbosa, D.A.N.; Ghanouni, P.; Halpern, C.H.; Buch, V.P. Focused ultrasound for functional neurosurgery. *J. Neuro-Oncology* **2021**, *156*, 17–22. [CrossRef]
79. Bretsztajn, L.; Gedroyc, W. Brain-focussed ultrasound: What's the "FUS" all about? A review of current and emerging neurological applications. *Br. J. Radiol.* **2018**, *91*, 20170481. [CrossRef]
80. de Groot, J.F.; Kim, A.H.; Prabhu, S.; Rao, G.; Laxton, A.W.; Fecci, P.E.; O'brien, B.J.; Sloan, A.; Chiang, V.; Tatter, S.B.; et al. Efficacy of laser interstitial thermal therapy (LITT) for newly diagnosed and recurrent *IDH* wild-type glioblastoma. *Neuro-Oncol. Adv.* **2022**, *4*, vdac040. [CrossRef]
81. Torres-Reveron, J.; Tomasiewicz, H.C.; Shetty, A.; Amankulor, N.M.; Chiang, V.L. Stereotactic laser induced thermotherapy (LITT): A novel treatment for brain lesions regrowing after radiosurgery. *J. Neuro-Oncol.* **2013**, *113*, 495–503. [CrossRef]
82. Bown, S.G. Phototherapy of tumors. *Mol. Med.* **1983**, *7*, 700–709. [CrossRef] [PubMed]
83. Rosomoff, H.L.; Carroll, F. Reaction of Neoplasm and Brain to Laser. *Arch. Neurol.* **1966**, *14*, 143–148. [CrossRef] [PubMed]
84. Hawasli, A.H.; Kim, A.H.; Dunn, G.P.; Tran, D.D.; Leuthardt, E.C. Stereotactic laser ablation of high-grade gliomas. *Neurosurg. Focus* **2014**, *37*, E1. [CrossRef] [PubMed]
85. Muir, M.; Traylor, J.I.; Gadot, R.; Patel, R.; Prabhu, S.S. Repeat laser interstitial thermal therapy for recurrent primary and metastatic intracranial tumors. *Surg. Neurol. Int.* **2022**, *13*, 311. [CrossRef] [PubMed]
86. Sloan, A.E.; Ahluwalia, M.S. Laser interstitial thermal therapy for the management of brain metastases. *Neuro-Oncol. Pract.* **2019**, *6*, 169–177.
87. Kuo, C.-H.; Feroze, A.H.; Poliachik, S.L.; Hauptman, J.S.; Novotny, E.J.; Ojemann, J.G. Laser Ablation Therapy for Pediatric Patients with Intracranial Lesions in Eloquent Areas. *World Neurosurg.* **2019**, *121*, e191–e199. [CrossRef]
88. Tovar-Spinoza, Z.; Choi, H. MRI-guided laser interstitial thermal therapy for the treatment of low-grade gliomas in children: A case-series review, description of the current technologies and perspectives. *Child's Nerv. Syst.* **2016**, *32*, 1947–1956. [CrossRef]
89. Kang, J.Y.; Wu, C.; Tracy, J.; Lorenzo, M.; Evans, J.; Nei, M.; Skidmore, C.; Mintzer, S.; Sharan, A.D.; Sperling, M.R. Laser interstitial thermal therapy for medically intractable mesial temporal lobe epilepsy. *Epilepsia* **2016**, *57*, 325–334. [CrossRef]
90. El Majdoub, F.; Simon, T.; Hoevels, M.; Berthold, F.; Sturm, V.; Maarouf, M. Interstitial Brachytherapy using Stereotactic Implanted 125Iodine Seeds for Recurrent Medulloblastoma. *Clin. Oncol.* **2011**, *23*, 532–537. [CrossRef]
91. Hafez, D.M.; Liekweg, C.; Leuthardt, E.C. Staged Laser Interstitial Thermal Therapy (LITT) Treatments to Left Insular Low-Grade Glioma. *Neurosurgery* **2019**, *86*, E337–E342. [CrossRef]
92. Montemurro, N.; Anania, Y.; Cagnazzo, F.; Perrini, P. Survival outcomes in patients with recurrent glioblastoma treated with Laser Interstitial Thermal Therapy (LITT): A systematic review. *Clin. Neurol. Neurosurg.* **2020**, *195*, 105942. [CrossRef]
93. Arsiwala, T.; Sprowls, S.; Blethen, K.; Adkins, C.; Saralkar, P.; Fladeland, R.; Pentz, W.; Gabriele, A.; Kielkowski, B.; Mehta, R.; et al. Ultrasound-mediated disruption of the blood tumor barrier for improved therapeutic delivery. *Neoplasia* **2021**, *23*, 676–691. [CrossRef]
94. Traylor, J.I.; Patel, R.; Muir, M.; Bastos, D.C.d.A.; Ravikumar, V.; Kamiya-Matsuoka, C.; Rao, G.; Thomas, J.G.; Kew, Y.; Prabhu, S.S. Laser Interstitial Thermal Therapy for Glioblastoma: A Single-Center Experience. *World Neurosurg.* **2021**, *149*, e244–e252. [CrossRef] [PubMed]
95. Jooma, R.; Waqas, M.; Khan, I. Diffuse low-grade glioma—Changing concepts in diagnosis and management: A review. *Asian J. Neurosurg.* **2019**, *14*, 356–363. [CrossRef]
96. Leonardi, M.A.; Lumenta, C.B. Stereotactic Guided Laser-Induced Interstitial Thermotherapy (SLITT) in Gliomas with Intraoperative Morphologic Monitoring in an Open MR: Clinical Expierence. *Minim. Invasive Neurosurg.* **2002**, *45*, 201–207. [CrossRef] [PubMed]
97. Patel, N.V.; Jethwa, P.R.; Barrese, J.C.; Hargreaves, E.L.; Danish, S.F. Volumetric trends associated with MRI-guided laser-induced thermal therapy (LITT) for intracranial tumors. *Lasers Surg. Med.* **2013**, *45*, 362–369. [CrossRef] [PubMed]
98. Jethwa, P.R.; Barrese, J.C.; Gowda, A.; Shetty, A.; Danish, S.F. Magnetic Resonance Thermometry-Guided Laser-Induced Thermal Therapy for Intracranial Neoplasms. *Neurosurg.* **2012**, *71*, ons133–ons145. [CrossRef]
99. Manrique-Guzmán, S.; Herrada-Pineda, T.; Revilla-Pacheco, F. Surgical Management of Glioblastoma. In *Glioblastoma*; De Vleeschouwer, S., Ed.; Codon Publications: Brisbane, Australia, 2017; Chapter 12.
100. Hersh, A.M.; Bhimreddy, M.; Weber-Levine, C.; Jiang, K.; Alomari, S.; Theodore, N.; Manbachi, A.; Tyler, B.M. Applications of Focused Ultrasound for the Treatment of Glioblastoma: A New Frontier. *Cancers* **2022**, *14*, 4920. [CrossRef]
101. Coluccia, D.; Figueiredo, C.A.; Wu, M.Y.; Riemenschneider, A.N.; Diaz, R.; Luck, A.; Smith, C.; Das, S.; Ackerley, C.; O'reilly, M.; et al. Enhancing glioblastoma treatment using cisplatin-gold-nanoparticle conjugates and targeted delivery with magnetic resonance-guided focused ultrasound. *Nanomedicine* **2018**, *14*, 1137–1148. [CrossRef]
102. Lee-Chang, C.; Miska, J.; Hou, D.; Rashidi, A.; Zhang, P.; Burga, R.A.; Jusué-Torres, I.; Xiao, T.; Arrieta, V.A.; Zhang, D.Y.; et al. Activation of 4-1BBL+ B cells with CD40 agonism and IFN$\gamma$ elicits potent immunity against glioblastoma. *J. Exp. Med.* **2021**, *218*, e20200913. [CrossRef]
103. Deng, Z.; Sheng, Z.; Yan, F. Ultrasound-Induced Blood-Brain-Barrier Opening Enhances Anticancer Efficacy in the Treatment of Glioblastoma: Current Status and Future Prospects. *J. Oncol.* **2019**, *2019*, 2345203. [CrossRef]

104. Fabiano, A.J.; Alberico, R.A. Laser-Interstitial Thermal Therapy for Refractory Cerebral Edema from Post-Radiosurgery Metastasis. *World Neurosurg.* **2014**, *81*, 652.e1–652.e4. [CrossRef]
105. Carpentier, A.; Canney, M.; Vignot, A.; Reina, V.; Beccaria, K.; Horodyckid, C.; Karachi, C.; Leclercq, D.; Lafon, C.; Chapelon, J.-Y.; et al. Clinical trial of blood-brain barrier disruption by pulsed ultrasound. *Sci. Transl. Med.* **2016**, *8*, 343re2. [CrossRef]
106. Beccaria, K.; Canney, M.; Bouchoux, G.; Desseaux, C.; Grill, J.; Heimberger, A.B.; Carpentier, A. Ultrasound-induced blood-brain barrier disruption for the treatment of gliomas and other primary CNS tumors. *Cancer Lett.* **2020**, *479*, 13–22. [CrossRef] [PubMed]
107. Jung, N.Y.; Chang, J.W. Magnetic Resonance-Guided Focused Ultrasound in Neurosurgery: Taking Lessons from the Past to Inform the Future. *J. Korean Med. Sci.* **2018**, *33*, e279. [CrossRef] [PubMed]
108. Rubino, F.; Eichberg, D.G.; Cordeiro, J.G.; Di, L.; Eliahu, K.; Shah, A.H.; Luther, E.M.; Lu, V.M.; Komotar, R.J.; Ivan, M.E. Robotic guidance platform for laser interstitial thermal ablation and stereotactic needle biopsies: A single center experience. *J. Robot. Surg.* **2022**, *16*, 549–557. [CrossRef]
109. McDannold, N.; Clement, G.T.; Black, P.; Jolesz, F.; Hynynen, K. Transcranial Magnetic Resonance Imaging– Guided Focused Ultrasound Surgery of Brain Tumors. *Neurosurgery* **2010**, *66*, 323–332. [CrossRef]

biomedicines

MDPI

*Review*

# Advances on Liquid Biopsy Analysis for Glioma Diagnosis

**Panagiotis Skouras [1,2], Mariam Markouli [3], Theodosis Kalamatianos [2], George Stranjalis [2], Penelope Korkolopoulou [4] and Christina Piperi [1,*]**

1 Department of Biological Chemistry, Medical School, National and Kapodistrian University of Athens, 11527 Athens, Greece; panosskouras@hotmail.gr
2 1st Department of Neurosurgery, Evangelismos Hospital, National and Kapodistrian University of Athens, 11527 Athens, Greece; tkalamatian@med.uoa.gr (T.K.); gstranjalis@med.uoa.gr (G.S.)
3 Department of Medicine, Boston Medical Center, Boston University School of Medicine, Boston, MA 02118, USA; myriam.markouli@bmc.org
4 Department of Pathology, Medical School, National and Kapodistrian University of Athens, 75 M. Asias Street, 11527 Athens, Greece; pkorkol@med.uoa.gr
* Correspondence: cpiperi@med.uoa.gr; Tel.: +30-210-7462610

**Abstract:** Gliomas comprise the most frequent primary central nervous system (CNS) tumors, characterized by remarkable genetic and epigenetic heterogeneity, difficulty in monitoring, and increased relapse and mortality rates. Tissue biopsy is an established method of tumor cell collection and analysis that enables diagnosis, classification of different tumor types, and prediction of prognosis upon confirmation of tumor's location for surgical removal. However, it is an invasive and often challenging procedure that cannot be used for frequent patient screening, detection of mutations, disease monitoring, or resistance to therapy. To this end, the minimally invasive procedure of liquid biopsy has emerged, allowing effortless tumor sampling and enabling continuous monitoring. It is considered a novel preferable way to obtain faster data on potential tumor risk, personalized diagnosis, prognosis, and recurrence evaluation. The purpose of this review is to describe the advances on liquid biopsy for glioma diagnosis and management, indicating several biomarkers that can be utilized to analyze tumor characteristics, such as cell-free DNA (cfDNA), cell-free RNA (cfRNA), circulating proteins, circulating tumor cells (CTCs), and exosomes. It further addresses the benefit of combining liquid biopsy with radiogenomics to facilitate early and accurate diagnoses, enable precise prognostic assessments, and facilitate real-time disease monitoring, aiming towards more optimal treatment decisions.

**Keywords:** glioma; glioblastoma; tissue biopsy; liquid biopsy; cfDNA; circulating tumor cells; exosomes; LINE-1; epigenetics; radiogenomics

## 1. Introduction

Gliomas comprise the most frequent primary tumor of the central nervous system (CNS), characterized by increased heterogeneity and aggressiveness, as well as high relapse and mortality rates [1]. More specifically, gliomas are responsible for 24.7% and 74.6% of all primary and malignant brain tumors, respectively [2]. Glioblastoma (GB), the most malignant glioma subtype is a highly aggressive, infiltrating, and poorly managed brain tumor. Currently available therapeutical approaches include surgical resection, followed by radiotherapy, as well as temozolomide (TMZ) chemotherapy. Thus far, limited drug penetration to the CNS, along with the rapid development of chemotherapy resistance, present ongoing challenging issues that contribute to poor prognosis [3,4].

Until now, tissue sample acquisitions from biopsies have been used for histological and molecular tumor analysis to confirm diagnosis, classify tumor types, detect tumor-specific mutations, and guide therapeutic protocols [5]. Tissue biopsy, although of major importance, is, however, an invasive and oftentimes challenging procedure. A surgical brain tissue biopsy increases the risk of bleeding and the likelihood of damaging an important brain area (e.g., glioma infiltrating a sensitive area of the frontal lobe), which can lead to neurological deficits, especially in the cases of the developing pediatric brain [6].

Moreover, invasive procedures are characterized by limitations in acquiring tumor samples in both efficient quantity and quality of tumor material. Sequential collection of tissue biopsies during treatment to monitor tumor response and relapse also poses a major challenge in tumor profiling [7]. It often provides limited material to perform all necessary molecular tests that are required to achieve a reliable diagnosis. In this context, the detection of mutations, genetic and epigenetic defects, or polymorphisms is not always possible.

In addition, tissue biopsy may not enable satisfactory detection of early-stage tumor or residual lesions and is not suitable for screening [8]. Furthermore, temporal and spatial tumor heterogeneity may decrease the practical utility of tissue biopsies to be used as tools for monitoring tumor progression and evaluating response to therapy [5]. In multifocal cases in particular, multiple biopsies may be necessary in order to obtain an accurate representation of the tumor, since invasive tumors constantly evolve both spatially and temporally over time or in response to treatment [7].

To this end, the minimal invasive procedure of liquid biopsy has surfaced as a solution to these issues, constantly progressing to enable first tumor diagnosis and collecting information on risk, prognosis, and recurrence potential.

In this review, we describe recent advances on liquid biopsy analysis for glioma patients, pointing towards the most promising biomarkers that can be used to explore tumor characteristics. Moreover, the combination of liquid biopsy with radiogenomics is suggested not only as a potential approach to achieve early and accurate diagnoses and prognostic assessments but also to enable real-time disease monitoring and optimal treatment decisions.

## 2. Basic Principles and Applications of Liquid Biopsy

Tumor cells may release multiple components into the circulation that can travel to distant organs throughout the body. These may include intact cancer cells and also tumor cell DNA and RNA, proteins, or exosomes that can be exploited as blood-circulating cancer biomarkers and may provide us with similar information to that of tissue biopsy, pinpointing the primary origin site and playing a key role in routine monitoring of tumor progression or treatment efficacy [5].

Given the increasing sensitivity of techniques that are used to study nucleic acids, it is nowadays possible to analyze tumor-released nucleic acids and blood-circulating cells. Genomic and epigenomic analysis of the circulating tumor DNA, for example, is an important novel tool regarding treatment. By using liquid biopsy (LB), clinicians can obtain data related to risk and disease prognosis and can predict chances of recurrence [9]. In more detail, LB is a minimally invasive procedure (e.g., venipuncture) that enables detection of several different analytes [6] in a variety of biological fluids to monitor disease progression [10,11]. Since most tumors encounter blood circulation, LB usually involves blood sampling. However, other body fluids, such as mucosa, fluid from pleural effusions, cerebrospinal fluid (CSF), and urine, may also be analyzed [12,13].

### *2.1. Analytes Detected in Biofluids by Liquid Biopsy*

In 1996, the National Comprehensive Cancer Network (NCCN) introduced several biomarkers into clinical practice, based on clinical and/or technical factors for diagnostic and prognostic purposes [14]. More recently, in 2011, the NCCN updated and established specific biomarkers as standard of care in some cancer types, including gliomas [15]. Analytes with a biomarker potential that can be detected through LB include cell-free DNA (cfDNA), cell-free RNA (cfRNA) and circulating proteins, circulating tumor cells (CTCs), and extracellular vesicles (EVs) such as exosomes, which can encapsule tumor DNA and other nanomolecules, such as mRNAs/miRNAs (Figure 1) [16,17].

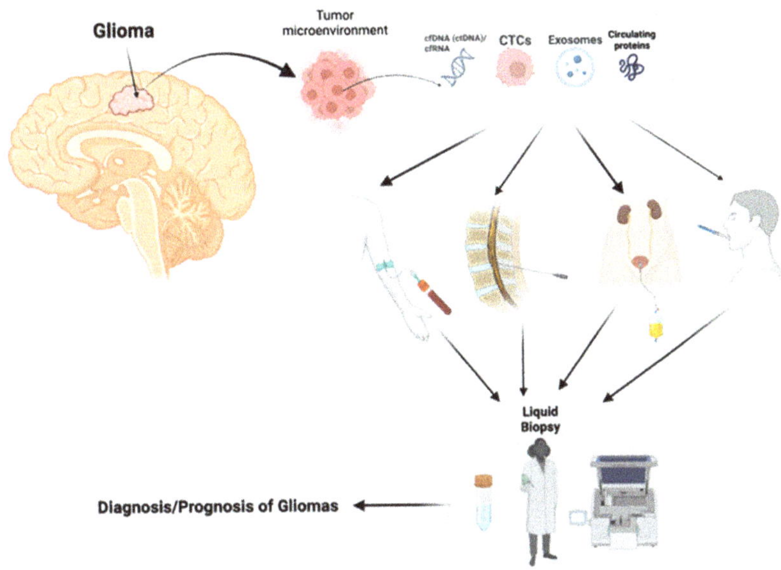


**Figure 1.** Detection of various tumor analytes in a variety of biofluids (CSF, urine, blood) with liquid biopsy. Several types of biofluids can be collected by using liquid biopsy and tumor-derived analytes (cfDNA, cfRNA, exosomes) measured using genomic processes to aid in tumor diagnosis, prognosis, and monitoring response to therapy (created with BioRender access on 24 June 2023).

#### 2.1.1. Cell-Free DNA (cfDNA), Cell-Free RNA (cfRNA), and Circulating Proteins

Circulating cfDNA refers to DNA fragments of the tumor cell DNA that can be detected in the circulation. Tumor cells that are being necrotized and/or undergoing apoptosis release cfDNA into the bloodstream, which is further digested by DNases or other enzymes [18]. Based on this evidence, patient's blood may contain cfDNA that entails fragments of the circulating tumor DNA derived from tumor cells. This may be further analyzed on a molecular level to fully reveal the molecular profile of the cancer [19]. With regards to cfDNA characteristics, it has a half-life of almost two hours and it can be detected during that period after the isolation of the sample [20]. Based on the aforementioned data, cfDNA may be utilized as an indicator of tumor molecular dynamics in real time [18]. In patients with glioma, most of the cell-free DNA pool comprises non-tumor-derived cfDNA [21], most of which is believed to originate from various cellular events, including apoptosis, necrosis, and other cellular secretory processes. The specific mechanism of cfDNA release is dependent on a range of biological and environmental factors, including age, gender, body mass index, organ health, and the presence of infections or systemic inflammatory conditions [22,23].

Circulating tumor DNA (ctDNA) analysis involves the detection and analysis of DNA fragments released by tumor cells into the bloodstream. By isolating and sequencing ctDNA, specific genetic alterations associated with brain tumor can be identified and analyzed. cfDNA technologies can identify specific mutations in ctDNA that are indicative of brain neoplasm. These mutations may include single nucleotide variants (SNVs), insertions/deletions (indels), or larger structural variations, such as gene fusions or chromosomal rearrangements [24]. Copy number variation (CNV) analysis of cfDNA allows the detection of alterations in the DNA copy number, such as amplifications or deletions, which can provide insights into the genomic landscape of brain tumors [25,26]. These advanced technologies enable the real-time monitoring of tumor dynamics and treatment response by analyzing ctDNA at multiple time points during the course of treatment. This allows the detection of emerging genetic alterations and resistance mechanisms, guiding treatment decisions [27].

CSF has been proposed as the most suitable biological sample for cfDNA assessment in brain tumors; its collection, however, requires an invasive procedure that cannot be used routinely to monitor responses to therapy [28]. Therefore, there is a growing need for the utilization of alternative body fluids, such as blood, saliva, and urine, which can also be used in specific collection tubes and stabilization buffer/solutions, such as ethylenediaminetetraacetic acid (EDTA).

Saliva, on the other hand, serves as a vital body fluid that possesses advantages over blood, primarily because it does not coagulate. This characteristic makes it easier to collect, transport, and store. However, it is important to note that saliva primarily comprises water, mucins, enzymes, and proteins, all of which can have an impact on downstream analyses. Additionally, salivary proteins, cfDNAs, and cfRNAs encapsulated within exosomes may rapidly degrade when they are exposed to an environment outside their natural milieu [29]. Furthermore, it is worth noting that saliva has a viscous nature and can contain food residues or other particles, which can pose challenges when analyzing cfDNAs, cfRNAs, or exosomes. However, it is important to consider that exosomes, being larger than proteins, may be inadvertently lost during the filtration process [30].

In addition, while urine holds great promise as a body fluid, our understanding of urinary cell-free Nucleic acids (cfNAs) and their processing methods remains limited. Urinary cfDNA, unlike cfDNA derived from blood, exhibits a shorter half-life [31] due to the elevated activity of DNase I and II [32,33]. It is crucial to inhibit DNases to prevent degradation and preserve the integrity of urinary cfDNA. One effective approach is to treat urine samples with EDTA and store them at −70 °C or lower temperatures for long-term preservation. Alternatively, sterile containers can be used to store samples at 4 °C or on ice, with immediate processing being preferable [34]. However, even under these conditions, there is a risk of cfDNA loss [35].

Similar to cfDNAs, cfRNAs are RNA fragments that are distinguished into coding and noncoding RNAs (ncRNAs) [36]. ncRNAs are further categorized by transcript length to smaller molecules that are less than 200 nucleotides in size and longer molecules that exceed 200 nucleotides. The group of small ncRNAs comprises microRNAs (miRNAs), short interfering RNAs (siRNAs), piwi-interacting RNAs (piRNAs), and small nucleolar RNAs (snoRNAs) [37,38].

The cf-ncRNAs molecules are characterized by high stability since they exist in vesicles or are associated with proteins [39]. They participate in cellular communication and may have a regulating role or alter tumor microenvironment, influencing tumor progress and invasion [40,41]. cf-ncRNAs present excellent biomarker candidates for diagnosing several diseases [41–43] but are also suitable for tumor screening and monitoring and studying resistance to therapy [44].

Finally, circulating proteins, oftentimes carried in exosomes and extracellular vesicles, are usually membrane receptors, receptor ligands, growth factors, or cytokines [45]. While numerous studies have successfully identified a range of protein cargo present in extracellular vesicles (EVs) derived from glioblastoma (GB), a comprehensive and detailed understanding of EV protein profiles remains yet to be explored [46].

#### 2.1.2. Circulating Tumor Cells (CTCs)

CTCs are cells that stem from a primary or secondary (metastatic) tumor. CTCs may be implicated in the metastatic process; however, cellular migration is extremely complex and its pathogenetic mechanisms need to be fully elucidated. Also, it remains to be clarified whether CTCs are derived from central tumor subpopulations or if they represent the entire tumor [47]. These cells enter the stream of different biological fluids such as blood, CSF, and urine. It is, however, important to mention that CTCs, in cases of brain tumors, are more conveniently collected from the CSF rather than from the bloodstream. Moreover, cisternal puncture seems to give more accurate results compared with lumbar puncture, although lumbar puncture seems to be less invasive and less dangerous regarding central nervous system complications. It also needs to be noted that CSF is recycled and fully exchanged every three to five days [48]. Therefore, a universally accepted method for CTC identification and collection needs yet to be discovered [49].

In this context, it is interesting to mention that circulating benign epithelial cells can also be found in inflammatory bowel diseases, which points out the need for a molecular characterization of CTC testing [50]. In a wide array of studies, CTCs seem to be useful as biomarkers for the prognosis of different cancer types, such as lung cancer, melanoma, etc. [51]. When it comes to brain tumors, and especially GB, it has been shown that CTCs can be detected in patients' bloodstreams at a percentage of 20 to 40% [52,53].

#### 2.1.3. Extracellular Vesicles (EVs)/Exosomes and miRNAs

Exosomes were originally considered to be cellular waste products but are now known to serve as signaling vesicles between cells and coordinators of cellular communication through proteins and nucleotide transfer [54]. Exosome uptake depends on the acceptor cell type and primarily occurs through phago- or endocytosis [55,56].

Exosomes play a key role in disease pathophysiology, as well as cellular development, homeostasis, and immune response [57,58]. Cancer cells release a high number of exosomes into the extracellular space, the majority of which are made of functional biomolecules [59]. In this context, exosomes were initially researched as non-cellular therapeutic antigens for the development of vaccines against tumors or infectious diseases [60,61]. However, exosomes may also include miRNAs and other compounds that can mirror the progression of various brain diseases, which suggests that they can serve as representative tools for the molecular profile of the respective tumor [58,62] (Table 1).

**Table 1.** Main analytes detected by liquid biopsies.

| LB Sample | Sample Source | Biomarker Isolation Technique | Benefit | Reference |
|---|---|---|---|---|
| cfDNA (ctDNA) | Blood, CSF, Saliva | MS/digital PCR, sequencing, LOH ddPCR, MAF, TAS/WES, NGS | Molecular diagnosis, tumor growth, and therapy response monitoring | [27,28,30,63–67] |
| cfRNA | Blood, CSF | Microarray, qPCR | Tumor growth, therapy response monitoring, and molecular diagnosis | [41–44,62] |
| Circulating proteins | Blood, Urine, CSF, Saliva | qPCR, NGS, microarray | Diagnosis and prognosis | [16,45,46] |
| CTCs | Blood, CSF | EpCAM Immunomagnetic isolation, FISH, density gradient centrifugation/ telomerase activity | Molecular diagnosis, tumor Growth, and therapy response monitoring | [52,53,68–70] |
| EVs | Blood, Saliva | qPCR, NGS, microarray | Molecular diagnosis and disease prognosis | [30,58,71–73] |

MS PCR—methylation-specific PCR; NGS—next generation sequencing; TAS—targeted analysis sequencing; WES—whole exome sequencing; WGS—whole genome sequencing, ddPCR—droplet digital PCR; MAF—mutant allelic frequency; EpCAM—epithelial cell adhesion molecule; FISH—fluorescence in situ hybridization; qPCR—quantitative polymerase chain reaction.

## 3. Biomarkers Detected through Liquid Biopsy Analysis for the Diagnosis of Gliomas

The ability of a liquid biopsy to analyze tumor products that are found in body fluids has led to its increasing utilization in different types of tumors. This technique provides vital diagnostic and prognostic information, as well as real-time updates on tumor status. In the case of gliomas, the use of a liquid biopsy is extremely promising [74]. Below, we discuss additional cfDNA biomarkers detected in different studies of liquid biopsy analysis for the diagnosis of gliomas.

### 3.1. cfDNA, cfRNA, and Circulating Proteins in the Diagnosis of Gliomas

#### 3.1.1. cfDNA

LB has allowed the detection of cfDNA in cancer patients, as mentioned above, which has created the potential of its incorporation into clinical practice, with the goal of identifying both genetic and epigenetic alterations in tumors. cfDNA may be a suitable molecular marker, indicative of tumor status, which will enable disease monitoring and distinction between tumor-free individuals and brain tumor patients [75].

One study demonstrated that the serum of control subjects was consistently characterized by low levels of cfDNA, in contrast to GB patients who had higher cfDNA concentrations. This may suggest that cfDNA could serve as a useful biomarker for differentiating GB patients from healthy individuals and serve as an indicator for tumor progression [20].

cfDNA may also help in the quantitative measurement of genetic (e.g., *IDH* mutations) or epigenetic (*MGMT* methylation) alterations [75,76]. In the cfDNA derived from glioma patients, there are various genes and epigenetic alterations that can be detected and used as biomarkers to diagnose/monitor the disease. A prominent example is the *H3K27M* mutation characteristic of diffuse midline gliomas (DMG). In the study of Daphne Li et al., cfDNA was obtained from two types of specimens: H3.3K27M mutant and H3 wildtype (H3WT). The specimens included H3.3K27M tumor tissues (four samples), CSF (six samples), plasma (four samples), and human primary pediatric glioma cells. The researchers observed a sensitivity and specificity of 100% in detecting mutations in both matched DMG tissue and CSF samples [77]. Another study demonstrated that H3K27M was detected in the CSF and plasma of 88% of patients with DMG. Among the two, CSF exhibited the highest concentration of ctDNA [78].

Global DNA methylation (*LINE-1* or *L1*) may also play a pivotal role in the epigenetic status of glioma cells, enabling diagnosis through cfDNA analysis. The addition of a methyl group to cytosine, leading to the formation of 5-methylcytosine, is the most well-known mechanism of DNA modification through epigenetic processes, playing a significant role in the development and progression of cancer and, specifically, of gliomas [79,80]. The dysregulation of cells at a systemic level, often associated with tumor development, can result from either DNA hypomethylation or increased activity of *LINE-1* transposons [81]. The majority of transposition events occurring in the human genome are carried out by non-LTR (long-terminal repeat) retrotransposons, with long interspersed elements (LINEs) being the most prevalent. Of all LINEs, *LINE-1* accounts for approximately one-sixth of genomic transposition and exists as multiple copies in cell-free blood samples. Due to its abundance and unique characteristics, such as methylation status, it has the potential to serve as a valuable epigenetic biomarker for tumor detection in liquid biopsy samples [79].

Numerous studies have reported the detection of *LINE-1* retrotransposition events in various cell types, such as neural precursor cells (NPCs) derived from both human embryonic stem cells (ESCs) and fetal brain stem cells. Retrotransposition assays performed on these cell types confirmed the occurrence of *LINE-1* activity under both conditions [82].

In addition, quantitative polymerase chain reaction (qPCR) analysis has revealed a notably elevated number of *LINE-1* ORF2 copies in multiple regions of healthy adult human brain samples, indicating the presence of *LINE-1* activity within the brain. The utilization of retrotransposon capture sequencing (RC-Seq) also allowed the authors to identify somatic *LINE-1* insertions in protein-coding genes that were differentially expressed and actively functioning within the brain [83].

One study assessed *LINE-1* methylation levels through the quantitative bisulfite pyrosequencing of four CpG sites in healthy brain samples, as well as frozen tumor tissues, including grade 2/3 astrocytoma, primary/secondary GB, and grade 2/3 oligodendroglioma. A relatively higher degree of *LINE-1* methylation was observed in astrocytoma, oligodendroglioma, and oligoastrocytoma samples, while GB tumors exhibited the greatest variability in *LINE-1* methylation levels and the lowest *LINE-1* scores (indicating a global hypomethylation) compared with other subtypes. Additionally, the gliomas that were classified as having a class 1 pattern of gene methylation displayed significantly elevated *LINE-1* methylation levels [84].

Finally, normal brain tissues and LGGs exhibited higher *LINE-1* methylation levels compared with primary and secondary GBs and increased *LINE-1* methylation functions as a positive prognostic indicator in primary GBs. *LINE-1* methylation appeared to be present in all samples, including normal brain tissues, LGGs, and primary and secondary GBs [85]. This may suggest that *LINE-1* plays a key role in glioma diagnosis and/or prognosis.

Sabedot et al. utilized genome-wide DNA methylation profiling to detect unique methylation patterns in tumor tissue and cfDNA from glioma patients. They employed a score metric system named the "glioma-epigenetic liquid biopsy score", or GeLB, to assess cfDNA methylome and diagnose gliomas. This non-invasive technique revealed a methylation signature in cfDNA associated with glioma presence and immune features. Independent cohort testing validated GeLB as a highly effective tool, with 100% sensitivity and 97.78% specificity for distinguishing glioma-positive and glioma-negative patients. GeLB score changes during surveillance reflected the clinicopathological conditions and treatment effects, suggesting its potential for diagnosis and monitoring of glioma patients [86].

Furthermore, Nassiri et al. showcased that plasma-based DNA methylation profiles exhibited distinctive signatures capable of effectively detecting and accurately distinguishing between common primary intracranial tumors. These tumors often shared cell-of-origin lineages, making them challenging to differentiate using standard imaging methods. The researchers utilized a technique called cell-free methylated DNA immunoprecipitation and high-throughput sequencing (cfMeDIP-seq) to successfully detect ctDNA derived from brain tumors [87].

Another study, by Martínez-Ricarte et al., performed genomic analysis of patients with diffuse gliomas that involved the examination of mutations in seven genes: *IDH1*, *IDH2*, *TP53*, *TERT*, *ATRX*, *H3F3A*, and *HIST1H3B* [88]. By analyzing ctDNA derived from CSF, the study successfully diagnosed diffuse gliomas and provided valuable support for surgical and clinical management decisions [88]. An additional study highlighted the importance of liquid biopsy in diagnosing and monitoring the evolving molecular landscape of tumors during treatment. The study aimed to identify *TERT* promoter mutations (C228T and C250T) in the cfDNA of glioma patients, providing evidence of the practical aspects of detecting circulating cfDNA *TERT* promoter mutations in patients with glioma, exhibiting clinically relevant sensitivity and specificity [89].

Fujioka et al. further developed a novel, highly sensitive, and specific molecular diagnostic method utilizing a chip-based digital PCR system to target ctDNA derived from CSF obtained from 11 patients with high-grade gliomas. The extracted ctDNA was analyzed for diagnostic point mutations in IDH1 R132H *TERT* promoter (C228T and C250T) and *H3F3A* (K27M). Diagnostic mutations were detected in tumor DNA samples from 28 out of 34 patients. Among these cases, precise molecular diagnoses were achieved using intracranial CSF in 20 patients, accounting for 71% of the cases [90].

The study of Husain et al. examined pre-radiotherapy cfDNA levels using next-generation sequencing (NGS) and compared the results to immunohistochemical (IHC) analysis, showing a high degree of accordance. This approach may prove beneficial in cases where surgery is not possible or in cases of adult diffuse glioma recurrence [91].

Finally, in the study of Palande et al., gene mutations and gene–gene fusions were detected in the circulating cfDNA of patients with GB. This study utilized the ChiTaRS 5.0 gene–gene fusion database to identify gene–gene fusions in cfDNA and tumor DNA samples. Notably, fusions involving *PDGFRA* were found in 44% of GB samples, which can be effectively targeted using tyrosine kinase inhibitors. Other gene fusion events, such as *BCR-ABL1*, *COL1A1-PDGFB*, *NIN-PDGFRB*, and *FGFR1-BCR*, were also detected in cfDNA and could be targeted with imatinib analogs. Additionally, *ROS1* fusions were identified in 8% of cfDNA samples and could potentially be targeted using specific drugs [92].

#### 3.1.2. cfRNA

When it comes to cfRNAs and their use in gliomas, Dong et al. discovered that the serum of GB patients exhibited a significant decrease in 24 miRNAs and a substantial increase in 115 miRNAs, which was not observed in the serum of healthy controls [93]. Furthermore, Wang et al. identified three downregulated miRNAs, namely miR-128, miR-485-3p, and miR-342-3p, in patients compared with controls. These miRNAs demonstrated a correlation with GB grades and served as biomarkers for assessing tumor grading and monitoring treatment response [94].

MicroRNA-21 is the most extensively studied miRNA in cancer and its overexpression has been consistently observed both in tissue and plasma of GB patients. This elevated expression is closely linked to lower overall survival rates and higher tumor grading [95]. miR-21 functions as an anti-apoptotic factor in glioblastoma cell lines by exerting its effects through caspase inhibition. The suppression of miR-21 activity consequently impedes cell growth, enhances apoptosis, and diminishes the proliferation of GB cancer cells [95,96].

It is worth noting a study that identified TP73-AS1 as a clinically relevant long non-coding RNA (lncRNA) in GB. The researchers observed a significant upregulation of TP73-AS1 in primary GB samples. They provided evidence that TP73-AS1 serves as a robust prognostic biomarker, as this lncRNA promotes tumor aggressiveness and confers resistance to TMZ treatment in GB cancer stem cells [97].

In the study of Shen et al., increased levels of the long noncoding RNA HOX transcript antisense intergenic RNA (HOTAIR) and decreased levels of GAS5 were observed in the serum of GB patients. These expression patterns were associated with a reduced likelihood of 2-year survival [98]. Additionally, it was found that the transfer of long noncoding RNA HOTAIR via serum exosomes showed promising diagnostic value in GB [99].

#### 3.1.3. Circulating Proteins

Given the remarkable patient and disease heterogeneity, as well as the broad protein involvement in multiple processes, it is difficult to find diagnostic plasma/serum markers for glioblastoma with sufficient clinical value. However, extensive research has identified the levels of several circulating proteins to be associated with the presence of GB. One example is the plasma glycoprotein, haptoglobin, which is an acute phase protein involved in the protection of tissue damage and oxidative stress [100]. Haptoglobin's levels change during pathologies and, although its single estimation may not be specific and sensitive enough, several proteoforms of haptoglobin ($\alpha$2 and $\beta$-chain) have been detected upregulated in GB patients' plasma, indicating its potential use as a GB-specific blood biomarker [100].

In addition, a quantitative proteomic analysis showed that several plasma proteins, such as carnosinase 1 (CNDP1) that regulates carnosine levels, the inflammatory marker ferritin light chain (FTL), and the $Ca^{2+}$ signaling protein S100A9, were found to be altered in GB tissues, with CNDP1 been suggested as a potential drug target. This confirms that plasma-based tests for initial diagnosis or recurrence monitoring can be highly useful for GB [101,102]. Moreover, the study of Pérez-Larraya et al. performed a comprehensive analysis of insulin-like growth factor binding protein 2 (IGFBP-2), the intermediate filament (IF) III protein GFAP, and the glycoprotein YKL-40 plasma levels as a supplementary diagnostic and prognostic tool [103]. This approach holds great promise, particularly for patients with inoperable brain lesions [103]. Interestingly, the presence of the glycoprotein's fetuin-A auto-antibodies was also revealed as indicators of GB development [104]. Ectopic fetuin-A levels have been suggested to play a role in tumor progression, including GB. Therefore, evaluation of fetuin-A auto-antibodies in the serum of GB patients was proposed to potentially serve as a screening tool, presenting one of the earliest indicators for GB growth [104].

Naryzhny and colleagues aimed to identify common exosomal proteins in various cell lines and explore potential biomarkers for glioblastoma in exosomes. Through proteomic analysis of exosomes, they identified 133 proteins, such as pyruvate kinase PKM (KPYM), annexin A1 (ANXA1), transitional endoplasmic reticulum ATPase (TERA), alpha-enolase

(ENOA), nucleophosmin (NPM), and cofilin (COF1), which were consistently present in all samples. The study established a correlation between certain proteins overexpressed in glial cells and their detection in exosomes, confirming the presence of numerous potential protein biomarkers for GB in exosomes [105]. Another investigation involving SWATH mass spectrometry and quantitative targeted absolute proteomics, where GB patients and healthy controls were studied. The findings revealed notable positive associations between leucine-rich alpha-2-glycoprotein (LRG1), complement component C9, and C-reactive protein (CRP) levels and tumor size [106]. In a meta-analysis conducted by Qin et al., a robust correlation was discovered between elevated YKL-40 expression and unfavorable overall survival among GB patients. This finding positions YKL-40 as a promising prognostic biomarker with significant disease-specific potential [107].

### 3.2. CTCs in Glioma Diagnosis

Numerous studies have confirmed the significant role of CTCs in CNS tumors. CTCs provide a less invasive way of obtaining tumor samples for cancer detection; certain CTC genotypes may indicate primary tumor progression and changes in malignant genetic information during disease relapse. GB has been reported to have a high prevalence of CTCs, which is estimated to be around 75% [108].

CTC concentrations have been utilized for the detection of various conditions. According to Santos et al., CTCs exhibit promising results in the early detection of colorectal cancer as they can be identified in the blood of patients who have recently developed the disease. As a result, CTC testing could be utilized to diagnose colorectal cancer [109].

When it comes to gliomas, the blood brain barrier (BBB) may restrict the quantity of CTCs which are found in the bloodstream. Despite this, circulating cells have been identified in glioma patients' blood and several studies have reported the detection of CTCs in the bloodstream of glioma patients through the use of various isolation techniques. After using the technology of CTC-iChip to enrich CTCs and staining them with antibodies, such as SOX2, EGFR, tubulin b-3, c-MET, and A2B5, Sullivan et al. identified CTCs in GB patients' blood with a frequency of 39%. They also discovered that the CTCs exhibited a mesenchymal molecular signature [48]. Muller et al. conducted a study where they employed Ficoll–Paque gradients through differential centrifugation, followed by GFAP staining, to identify 20.6% of CTCs [108]. In addition, Gao et al. demonstrated that CTCs were present in 77% of blood samples from patients with brain cancer and in 82% of patients with GB [52].

An interesting study by MacArthur et al. suggested that CTCs could also serve as a prognostic factor. Using density gradient centrifugation, they isolated CTCs and analyzed the activity of the telomerase enzyme in brain cancer patients. The study revealed that before radiotherapy treatment, CTCs were isolated in 72% of patients versus only in 8% after treatment [70]. Moreover, the anti-glioma aptamers Gli-233 and Gli-55 have been used to detect CTCs in LBs in order to increase diagnostic specificity [110]. Aptamers are nucleic acid molecules that bind exclusively to their targets, such as cancer-related proteins, with high affinity and selectivity [111]. They have therefore already been used in biosensing, bioregulation, and bioimaging as reliable recognition ligands [112]. Kichkailo et al. also demonstrated that aptamers may specifically bind to glial tumor cells for CNS tumor detection [110]. Although more research is necessary to uncover the full potential of CTCs in diagnosing gliomas and GB, their use as prognostic factors for gliomas is very promising.

### 3.3. Exosomes in Glioma Diagnosis

Another important characteristic of GB cells is their ability to use invadopodia, through which they are capable of invading adjacent cells, specifically astrocytes. Invadopodia are GB cell membrane-derived extensions that adhere to neighboring tissues and promote the proteolytical degradation of the matrix, thus allowing invasion [113,114]. Several proteins excreted from GB-derived exosomes are linked to invadopodia biogenesis and subsequently GB-invasive potential, among which are annexin A1 (ANXA1), actin-related protein 3 (ACTR3), integrin β1 (ITGB1), calreticulin (CALR), and programmed cell death 6-interacting protein (PDCD6IP) [114]. Hallal et al. also observed the formation of podosomes in astrocytes and degradation of the matrix after their interaction with exosomes derived from GB cells, a process that appears to be favored by decreased p53 levels. Exosomes thereby exhibit carcinogenic potential and promote the neighboring astrocytes to become tumorigenic [115].

In this context, GB exosomes can be detected both in the blood and the CSF by using the minimally invasive technique of liquid biopsy. A robust number of tumor-derived exosomes may be found in the CSF. It is important to note that CSF is not contaminated with blood-related EVs, for example, platelet-derived exosomes. However, on the other hand, blood samples are more conveniently and less invasively collected compared with the CSF [113,116].

Furthermore, Shao et al. described how a nuclear magnetic resonance system can detect GB-expelled microvesicles in blood samples [117]. Studies have shown that exosome-carried molecules may promote tumor development and therapy resistance through the formation of a tumor-friendly microenvironment. Therefore, exosomes have been suggested as promising tools in GB diagnosis and prognosis, enabling a more specific characterization of the tumor [118,119].

Despite being carriers of many miRNA types, including miR-21/-29a/-221/-222, etc., as demonstrated in several studies (in vitro and microarray analyses), exosomes also play a key role in boosting proliferation and inhibiting GB cell apoptosis [58,120]. They could therefore be utilized as a promising delivery vehicle for tumor-suppressive miRNAs that target gliomas [121]. Approximately 1000 proteins have been identified in GB exosomes using mass spectrometry analysis; they mostly act as pro-angiogenic factors in normal endothelial cells of the brain, such as angiogenin, IL-6/-8, and tissue inhibitor of metalloproteinase-1 and 2 (TIMP-1, and TIMP-2), capable of promoting malignancy by causing hypoxia [122,123].

Exosomes may also transfer receptors with tumorigenic features, such as epidermal growth factor receptor vIII (EGFRvIII), human epidermal growth factor receptor 2 (HER2), and platelet-derived growth factor (PDGFR), which promote GB proliferation to healthy stromal cells [123]. Furthermore, exosomes deliver phosphatase and tensin homolog (PTEN), which is present in the nucleus or cytoplasm and whose absence has been associated with tumorigenesis [124] (Table 2). Ndfifip1 protein plays a major role in exosome internalization. Interestingly, the Ndfifip1 protein is repressed in GB; subsequently, the intranuclear concentration of PTEN is also suppressed, allowing tumor cells to proliferate and survive [124].

**Table 2.** Glioma-derived exosomes biomarkers.

| Sample | Cargo (Biomarker) | Model | Benefit | Reference |
|---|---|---|---|---|
| CSF/Plasma | miRNA-21 | Patients with glioma (grade I to IV) | Characterization of stable versus progressive disease | [120,125,126] |
| CSF | miR-21/miR-103/miR-24/miR-125 | Patients with GB | Disease diagnosis | [126,127] |
| CSF | miR-151a | Patients with GB who received TMZ treatment | Prediction of treatment response | [128] |
| Blood | EGFR/EGFRvIII | In vitro (μNMR) | Characterization of molecular status | [123,129] |
| Blood | PTRF | In vivo mice xenograft | Disease diagnosis | [130] |
| Blood | MGMT, APNG | Patients with GB (microfluidic chip) | Disease prognosis | [76,131] |
| Serum | miR-320/miR-574 3p/RNU6-1 | Patients with GB | Disease diagnosis | [132] |
| Serum | miR-301a | In vitro | Disease prognosis | [133] |

**Table 2.** *Cont.*

| Sample | Cargo (Biomarker) | Model | Benefit | Reference |
|---|---|---|---|---|
| Serum | miR-15b-5p, miR-16-5p, miR-19a-3p, miR-19b-3p, miR-20a-5p, miR-106a-5p, miR-130-3p, miR-181b-5p, miR-208a-3p | Astrocytoma (grade II to IV) (TaqMan low-density array) | Disease prognosis | [134,135] |
| Serum | miR-497, miR-125b | Patients with GB | Disease prognosis | [136] |
| Plasma | miR-221/222 | Patients with high-grade gliomas | Disease diagnosis, Prediction of poorer prognosis and response to treatment | [137] |
| Plasma | PDGFR, CAV1, IL-8 | In vivo glioma mice xenograft | Disease diagnosis and prognosis | [138] |
| Blood | miR-100 | Patients with GB | Disease diagnosis | [139] |

## 4. Combination of Radiogenomics and Liquid Biopsy for Glioma Diagnosis

Radiogenomics is a field that integrates large amounts of quantitative data obtained from medical images with individual genomic phenotypes and uses deep learning techniques to develop a predictive model. This model is useful for stratifying patients, guiding therapeutic strategies, and assessing clinical outcomes [140].

The combination of liquid biopsy and radiomics/radiogenomics can offer a promising avenue for non-invasive disease diagnosis and provide valuable insights for treatment planning (Figure 2). Both approaches have individually shown great potential, but their combination can lead to even more accurate and comprehensive diagnoses. The study of images using artificial intelligence (AI) and machine and deep learning techniques has progressed significantly in recent years. These methods involve an increasingly complex analysis of computational processes and imaging features, with the potential to accurately predict the molecular alterations necessary for correct disease diagnosis. As a result, they represent a promising approach for improving the accuracy and precision of glioma diagnosis [141,142]. In more detail, the field of radiogenomics offers a unique opportunity to predict genetic mutations, the status of molecular markers, and chromosomal aberrations through the analysis of imaging features. This approach utilizes imaging data as a substitute for the presence of genetic alterations, providing a non-invasive and efficient way of diagnosing and monitoring gliomas [143].

Radiogenomics has already been employed in other forms of cancer, such as breast [144] and lung cancer [145]. These studies have demonstrated the potential of radiogenomics to accurately predict the genetic characteristics of tumors through the analysis of imaging data. The extension of this approach to gliomas represents a significant opportunity to improve the accuracy and efficiency of non-invasive diagnosis and monitoring.

In the study of Li Y et al., for example, that used MRI radiomics (machine learning), the authors were able to predict ATRX mutations in low-grade gliomas [146]. In another study, the same mutations were identified in gliomas with deep learning, with 94% sensitivity and 92% specificity [147]. Diffuse midline gliomas with H3F3A histone mutations have greater enhancement in the T2 sequence, while tumors without the mutation have a poor identification of the non-contrast-enhancing tumor (NCET) margin. Furthermore, non-mutated tumors present with sound edema and cortical invasion [148]. A study of radiomics analysis (multiparameter MRI) managed to identify that tumors harboring TERT mutations have more necrotic areas, since TERT mutations suggest a high-grade profile [149].

Moon et al. also identified several characteristics of high-grade gliomas relative to *MGMT* promoter methylation [150]. When it comes to other brain tumors and, particularly, oligodendrogliomas, which are subcategorized to IDH-mutant and 1p/19q co-deleted [151], studies have been able to predict the 1p/19q codeletion with high specificity and sensitivity using, for example, a textural analysis of the T2 sequence [152–154]. It therefore becomes evident that the combination of liquid biopsy with the genomics component of radiogenomics profiling represents a promising approach to revolutionize cancer treatment. This combination may have the potential to facilitate early diagnosis, provide more accurate prognostic assessment, and enable real-time disease monitoring, all while minimizing invasiveness. Furthermore, this personalized approach to non-invasive brain tumor diagnosis could pave the way for precision medicine, tailored to the individual needs of each patient.

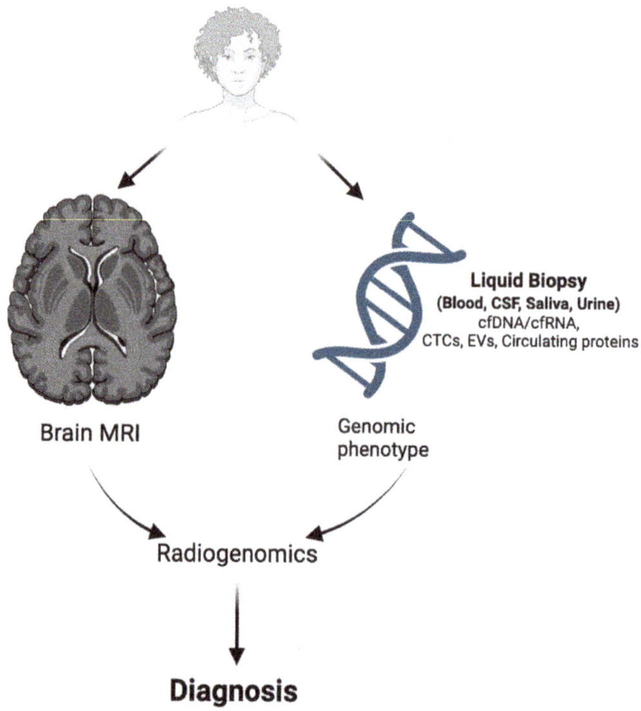


**Figure 2.** Impact of radiogenomics on diagnosis of brain tumors. Radiogenomics integrates large amounts of quantitative data obtained from medical images with individual genomic phenotypes from liquid biopsy analysis to enable efficient glioma diagnosis (created with BioRender, accessed 12 July 2023).

## 5. Conclusions: Future Perspectives

Significant advances have been made in the field of liquid biopsy in the last 20 years, moving from basic research to clinical applications in certain tumors such as lung, prostate and breast cancer, and melanoma. However, its integration into routine medical practices outside of academic hospitals remains limited. Despite various studies demonstrating promising results in various tumor types and clinical applications, including disease diagnosis, prognosis, monitoring, and therapeutic responses, the existence of standardized methods and useful genetic markers is still lacking. The process of biofluid collection, choice of biomarker, and detection strategy need to be extensively studied in future clinical trials in order to allow for the incorporation of LB into clinical practice.

Moreover, a series of factors may influence the utility, sensitivity, and specificity of different LB biomarkers, such as tumor subtype (GB vs. lower-grade gliomas for example), tumor size and location, extent of BBB disruption, and tumor stage. In more detail, the value of each biomarker will vary based on the different stages of the disease, such as during the initial diagnosis, patient surveillance with stable disease, active disease progression, or disease recurrence, so that biomarker combinations may need to be considered [155]. Moreover, the ideal site for biomarker collection needs to be specified for each patient and each tumor type, as well as the best method in cases of CSF collection, namely lumbar versus cisternal, as mentioned above. It is also necessary to find circulating biomarkers that are specific for each tumor type, as many are not entirely specific to GB and do not allow for the discrimination between different tumor types in patients with similar symptoms or subcategories of the same tumor [156].

Altogether, the utilization of liquid biopsy and radiogenomics presents a significant opportunity to achieve non-invasive brain tumor diagnoses and successfully identify their molecular alterations. If the abovementioned issues are overcome, analysis of tumor constituents in bodily fluids and utilization of imaging data to predict genetic abnormalities by using these techniques can provide critical insights into the nature and behavior of CNS tumors, thereby allowing for more effective treatment planning and monitoring.

**Author Contributions:** Conceptualization, C.P. and P.S.; methodology, P.S. and M.M.; software, P.S.; validation, C.P., M.M. and P.K.; formal analysis, P.S.; investigation, P.S.; resources, G.S.; data curation, M.M.; writing—original draft preparation, P.S.; writing—review and editing, M.M. and T.K.; visualization, P.S.; supervision, C.P.; project administration, C.P., G.S. and P.K.; funding acquisition, C.P. All authors have read and agreed to the published version of the manuscript.

**Funding:** This research received no external funding.

**Institutional Review Board Statement:** Not applicable.

**Informed Consent Statement:** Not applicable.

**Data Availability Statement:** Not applicable.

**Conflicts of Interest:** The authors declare no conflict of interest.

## References


1. Lapointe, S.; Perry, A.; Butowski, N.A. Primary Brain Tumours in Adults. *Lancet* **2018**, *392*, 432–446. [CrossRef] [PubMed]
2. Chen, S.; Zhang, Q.; Xu, D.; Li, Y.; Fan, Y.; Li, W.; Yin, X.; Zhang, Y.; Liu, J.; Li, X.; et al. Antitumor Effect of the Newcastle Disease Viral Hemagglutinin-Neuraminidase Gene Is Expressed through an Oncolytic Adenovirus Effect in Osteosarcoma Cells. *Anticancer Drugs* **2018**, *29*, 197–207. [CrossRef] [PubMed]
3. Ammirati, M.; Chotai, S.; Newton, H.; Lamki, T.; Wei, L.; Grecula, J. Hypofractionated Intensity Modulated Radiotherapy with Temozolomide in Newly Diagnosed Glioblastoma Multiforme. *J. Clin. Neurosci.* **2014**, *21*, 633–637. [CrossRef] [PubMed]
4. Mannas, J.P.; Lightner, D.D.; Defrates, S.R.; Pittman, T.; Villano, J.L. Long-Term Treatment with Temozolomide in Malignant Glioma. *J. Clin. Neurosci.* **2014**, *21*, 121–123. [CrossRef]
5. Cohen, J.D.; Li, L.; Wang, Y.; Thoburn, C.; Afsari, B.; Danilova, L.; Douville, C.; Javed, A.A.; Wong, F.; Mattox, A.; et al. Detection and Localization of Surgically Resectable Cancers with a Multi-Analyte Blood Test. *Science* **2018**, *359*, 926–930. [CrossRef]
6. Chen, M.; Zhao, H. Next-Generation Sequencing in Liquid Biopsy: Cancer Screening and Early Detection. *Hum. Genom.* **2019**, *13*, 34. [CrossRef]
7. Perakis, S.; Speicher, M.R. Emerging Concepts in Liquid Biopsies. *BMC Med.* **2017**, *15*, 75. [CrossRef]
8. Hirahata, T.; ul Quraish, R.; Ul Quraish, A.; Ul Quraish, S.; Naz, M.; Razzaq, M.A. Liquid Biopsy: A Distinctive Approach to the Diagnosis and Prognosis of Cancer. *Cancer Inf.* **2022**, *21*, 11769351221076062. [CrossRef]
9. Schilsky, R.L.; Longo, D.L. Closing the Gap in Cancer Genomic Testing. *N. Engl. J. Med.* **2022**, *387*, 2107–2110. [CrossRef]
10. Poulet, G.; Massias, J.; Taly, V. Liquid Biopsy: General Concepts. *Acta Cytol.* **2019**, *63*, 449–455. [CrossRef]
11. Eibl, R.H.; Schneemann, M. Liquid Biopsy and Primary Brain Tumors. *Cancers* **2021**, *13*, 5429. [CrossRef]
12. Lone, S.N.; Nisar, S.; Masoodi, T.; Singh, M.; Rizwan, A.; Hashem, S.; El-Rifai, W.; Bedognetti, D.; Batra, S.K.; Haris, M.; et al. Liquid Biopsy: A Step Closer to Transform Diagnosis, Prognosis and Future of Cancer Treatments. *Mol. Cancer* **2022**, *21*, 79. [CrossRef] [PubMed]
13. Bauman, M.M.J.; Bouchal, S.M.; Monie, D.D.; Aibaidula, A.; Singh, R.; Parney, I.F. Strategies, Considerations, and Recent Advancements in the Development of Liquid Biopsy for Glioblastoma: A Step towards Individualized Medicine in Glioblastoma. *Neurosurg. Focus* **2022**, *53*, E14. [CrossRef] [PubMed]
14. Hayes, D.F.; Bast, R.C.; Desch, C.E.; Fritsche, H., Jr.; Kemeny, N.E.; Jessup, J.M.; Locker, G.Y.; Macdonald, J.S.; Mennel, R.G.; Norton, L.; et al. Tumor Marker Utility Grading System: A Framework to Evaluate Clinical Utility of Tumor Markers. *JNCI J. Natl. Cancer Inst.* **1996**, *88*, 1456–1466. [CrossRef] [PubMed]
15. Febbo, P.G.; Ladanyi, M.; Aldape, K.D.; De Marzo, A.M.; Hammond, M.E.; Hayes, D.F.; Iafrate, A.J.; Kelley, R.K.; Marcucci, G.; Ogino, S.; et al. NCCN Task Force Report: Evaluating the Clinical Utility of Tumor Markers in Oncology. *J. Natl. Compr. Cancer Netw.* **2011**, *9* (Suppl. S5), S1–S32. [CrossRef] [PubMed]
16. Jelski, W.; Mroczko, B. Molecular and Circulating Biomarkers of Brain Tumors. *Int. J. Mol. Sci.* **2021**, *22*, 7039. [CrossRef]
17. Westphal, M.; Lamszus, K. Circulating Biomarkers for Gliomas. *Nat. Rev. Neurol.* **2015**, *11*, 556–566. [CrossRef]

18. Aarthy, R.; Mani, S.; Velusami, S.; Sundarsingh, S.; Rajkumar, T. Role of Circulating Cell-Free DNA in Cancers. *Mol. Diagn. Ther.* **2015**, *19*, 339–350. [CrossRef]
19. Rykova, E.Y.; Morozkin, E.S.; Ponomaryova, A.A.; Loseva, E.M.; Zaporozhchenko, I.A.; Cherdyntseva, N.V.; Vlassov, V.V.; Laktionov, P.P. Cell-Free and Cell-Bound Circulating Nucleic Acid Complexes: Mechanisms of Generation, Concentration and Content. *Expert Opin. Biol. Ther.* **2012**, *12* (Suppl. S1), S141–S153. [CrossRef]
20. Roth, C.; Pantel, K.; Müller, V.; Rack, B.; Kasimir-Bauer, S.; Janni, W.; Schwarzenbach, H. Apoptosis-Related Deregulation of Proteolytic Activities and High Serum Levels of Circulating Nucleosomes and DNA in Blood Correlate with Breast Cancer Progression. *BMC Cancer* **2011**, *11*, 4. [CrossRef]
21. Carpenter, E.L.; Bagley, S.J. Clinical Utility of Plasma Cell-Free DNA in Gliomas. *Neuro-Oncol. Adv.* **2022**, *4* (Suppl. S2), ii41–ii44. [CrossRef]
22. Bronkhorst, A.J.; Ungerer, V.; Holdenrieder, S. The Emerging Role of Cell-Free DNA as a Molecular Marker for Cancer Management. *Biomol. Detect. Quantif.* **2019**, *17*, 100087. [CrossRef] [PubMed]
23. Aucamp, J.; Bronkhorst, A.J.; Badenhorst, C.P.S.; Pretorius, P.J. The Diverse Origins of Circulating Cell-Free DNA in the Human Body: A Critical Re-Evaluation of the Literature. *Biol. Rev. Camb. Philos. Soc.* **2018**, *93*, 1649–1683. [CrossRef] [PubMed]
24. Mair, R.; Mouliere, F. Cell-Free DNA Technologies for the Analysis of Brain Cancer. *Br. J. Cancer* **2022**, *126*, 371–378. [CrossRef] [PubMed]
25. Sun, K.; Jiang, P.; Cheng, S.H.; Cheng, T.H.T.; Wong, J.; Wong, V.W.S.; Ng, S.S.M.; Ma, B.B.Y.; Leung, T.Y.; Chan, S.L.; et al. Orientation-Aware Plasma Cell-Free DNA Fragmentation Analysis in Open Chromatin Regions Informs Tissue of Origin. *Genome Res.* **2019**, *29*, 418–427. [CrossRef]
26. Mouliere, F.; Chandrananda, D.; Piskorz, A.M.; Moore, E.K.; Morris, J.; Ahlborn, L.B.; Mair, R.; Goranova, T.; Marass, F.; Heider, K.; et al. Enhanced Detection of Circulating Tumor DNA by Fragment Size Analysis. *Sci. Transl. Med.* **2018**, *10*, eaat4921. [CrossRef] [PubMed]
27. Euskirchen, P.; Bielle, F.; Labreche, K.; Kloosterman, W.P.; Rosenberg, S.; Daniau, M.; Schmitt, C.; Masliah-Planchon, J.; Bourdeaut, F.; Dehais, C.; et al. Same-Day Genomic and Epigenomic Diagnosis of Brain Tumors Using Real-Time Nanopore Sequencing. *Acta Neuropathol.* **2017**, *134*, 691–703. [CrossRef]
28. Wang, Y.; Springer, S.; Zhang, M.; McMahon, K.W.; Kinde, I.; Dobbyn, L.; Ptak, J.; Brem, H.; Chaichana, K.; Gallia, G.L.; et al. Detection of Tumor-Derived DNA in Cerebrospinal Fluid of Patients with Primary Tumors of the Brain and Spinal Cord. *Proc. Natl. Acad. Sci. USA* **2015**, *112*, 9704–9709. [CrossRef]
29. Helmerhorst, E.J.; Oppenheim, F.G. Saliva: A Dynamic Proteome. *J. Dent. Res.* **2007**, *86*, 680–693. [CrossRef]
30. Hyun, K.-A.; Gwak, H.; Lee, J.; Kwak, B.; Jung, H.-I. Salivary Exosome and Cell-Free DNA for Cancer Detection. *Micromachines* **2018**, *9*, 340. [CrossRef]
31. Yao, W.; Mei, C.; Nan, X.; Hui, L. Evaluation and Comparison of in Vitro Degradation Kinetics of DNA in Serum, Urine and Saliva: A Qualitative Study. *Gene* **2016**, *590*, 142–148. [CrossRef] [PubMed]
32. Nadano, D.; Yasuda, T.; Kishi, K. Measurement of Deoxyribonuclease I Activity in Human Tissues and Body Fluids by a Single Radial Enzyme-Diffusion Method. *Clin. Chem.* **1993**, *39*, 448–452. [CrossRef] [PubMed]
33. Bryzgunova, O.E.; Laktionov, P.P. Extracellular Nucleic Acids in Urine: Sources, Structure, Diagnostic Potential. *Acta Naturae* **2015**, *7*, 48–54. [CrossRef] [PubMed]
34. Li, P.; Ning, J.; Luo, X.; Du, H.; Zhang, Q.; Zhou, G.; Du, Q.; Ou, Z.; Wang, L.; Wang, Y. New method to preserve the original proportion and integrity of urinary cell-free DNA. *J. Clin. Lab. Anal.* **2018**, *33*, e22668. [CrossRef] [PubMed]
35. Lee, E.Y.; Lee, E.-J.; Yoon, H.; Lee, D.H.; Kim, K.H. Comparison of Four Commercial Kits for Isolation of Urinary Cell-Free DNA and Sample Storage Conditions. *Diagnostics* **2020**, *10*, 234. [CrossRef]
36. Senhaji, N.; Squalli Houssaini, A.; Lamrabet, S.; Louati, S.; Bennis, S. Molecular and Circulating Biomarkers in Patients with Glioblastoma. *Int. J. Mol. Sci.* **2022**, *23*, 7474. [CrossRef]
37. Derrien, T.; Johnson, R.; Bussotti, G.; Tanzer, A.; Djebali, S.; Tilgner, H.; Guernec, G.; Martin, D.; Merkel, A.; Knowles, D.G.; et al. The GENCODE v7 Catalog of Human Long Noncoding RNAs: Analysis of Their Gene Structure, Evolution, and Expression. *Genome Res.* **2012**, *22*, 1775–1789. [CrossRef]
38. Esteller, M. Non-Coding RNAs in Human Disease. *Nat. Rev. Genet.* **2011**, *12*, 861–874. [CrossRef]
39. Bhan, A.; Soleimani, M.; Mandal, S.S. Long Noncoding RNA and Cancer: A New Paradigm. *Cancer Res.* **2017**, *77*, 3965–3981. [CrossRef]
40. Umu, S.U.; Langseth, H.; Bucher-Johannessen, C.; Fromm, B.; Keller, A.; Meese, E.; Lauritzen, M.; Leithaug, M.; Lyle, R.; Rounge, T.B. A Comprehensive Profile of Circulating RNAs in Human Serum. *RNA Biol.* **2018**, *15*, 242–250. [CrossRef]
41. Lu, M. Circular RNA: Functions, Applications and Prospects. *ExRNA* **2020**, *2*, 1–7. [CrossRef]
42. Liu, S.J.; Dang, H.X.; Lim, D.A.; Feng, F.Y.; Maher, C.A. Long Noncoding RNAs in Cancer Metastasis. *Nat. Rev. Cancer* **2021**, *21*, 446–460. [CrossRef] [PubMed]
43. Boon, R.A.; Jaé, N.; Holdt, L.; Dimmeler, S. Long Noncoding RNAs: From Clinical Genetics to Therapeutic Targets? *J. Am. Coll. Cardiol.* **2016**, *67*, 1214–1226. [CrossRef] [PubMed]
44. Szilágyi, M.; Pös, O.; Márton, É.; Buglyó, G.; Soltész, B.; Keserű, J.; Penyige, A.; Szemes, T.; Nagy, B. Circulating Cell-Free Nucleic Acids: Main Characteristics and Clinical Application. *Int. J. Mol. Sci.* **2020**, *21*, 6827. [CrossRef] [PubMed]

45. Kalluri, R.; LeBleu, V.S. The Biology, Function, and Biomedical Applications of Exosomes. *Science* **2020**, *367*, eaau6977. [CrossRef] [PubMed]
46. Schey, K.L.; Luther, J.M.; Rose, K.L. Proteomics Characterization of Exosome Cargo. *Methods* **2015**, *87*, 75–82. [CrossRef] [PubMed]
47. Massagué, J.; Obenauf, A.C. Metastatic Colonization by Circulating Tumour Cells. *Nature* **2016**, *529*, 298–306. [CrossRef]
48. Sullivan, J.P.; Nahed, B.V.; Madden, M.W.; Oliveira, S.M.; Springer, S.; Bhere, D.; Chi, A.S.; Wakimoto, H.; Rothenberg, S.M.; Sequist, L.V.; et al. Brain Tumor Cells in Circulation Are Enriched for Mesenchymal Gene Expression. *Cancer Discov.* **2014**, *4*, 1299–1309. [CrossRef]
49. Perryman, L.; Erler, J.T. Brain Cancer Spreads. *Sci. Transl. Med.* **2014**, *6*, 247fs28. [CrossRef]
50. Pantel, K.; Denève, E.; Nocca, D.; Coffy, A.; Vendrell, J.-P.; Maudelonde, T.; Riethdorf, S.; Alix-Panabières, C. Circulating Epithelial Cells in Patients with Benign Colon Diseases. *Clin. Chem.* **2012**, *58*, 936–940. [CrossRef]
51. Keup, C.; Mach, P.; Aktas, B.; Tewes, M.; Kolberg, H.-C.; Hauch, S.; Sprenger-Haussels, M.; Kimmig, R.; Kasimir-Bauer, S. RNA Profiles of Circulating Tumor Cells and Extracellular Vesicles for Therapy Stratification of Metastatic Breast Cancer Patients. *Clin. Chem.* **2018**, *64*, 1054–1062. [CrossRef] [PubMed]
52. Gao, F.; Cui, Y.; Jiang, H.; Sui, D.; Wang, Y.; Jiang, Z.; Zhao, J.; Lin, S. Circulating Tumor Cell Is a Common Property of Brain Glioma and Promotes the Monitoring System. *Oncotarget* **2016**, *7*, 71330–71340. [CrossRef] [PubMed]
53. Krol, I.; Castro-Giner, F.; Maurer, M.; Gkountela, S.; Szczerba, B.M.; Scherrer, R.; Coleman, N.; Carreira, S.; Bachmann, F.; Anderson, S.; et al. Detection of Circulating Tumour Cell Clusters in Human Glioblastoma. *Br. J. Cancer* **2018**, *119*, 487–491. [CrossRef] [PubMed]
54. Théry, C. Exosomes: Secreted Vesicles and Intercellular Communications. *F1000 Biol. Rep.* **2011**, *3*, 15. [CrossRef] [PubMed]
55. Christianson, H.C.; Svensson, K.J.; van Kuppevelt, T.H.; Li, J.-P.; Belting, M. Cancer Cell Exosomes Depend on Cell-Surface Heparan Sulfate Proteoglycans for Their Internalization and Functional Activity. *Proc. Natl. Acad. Sci. USA* **2013**, *110*, 17380–17385. [CrossRef]
56. Cocucci, E.; Meldolesi, J. Ectosomes and Exosomes: Shedding the Confusion between Extracellular Vesicles. *Trends Cell Biol.* **2015**, *25*, 364–372. [CrossRef]
57. De Toro, J.; Herschlik, L.; Waldner, C.; Mongini, C. Emerging Roles of Exosomes in Normal and Pathological Conditions: New Insights for Diagnosis and Therapeutic Applications. *Front. Immunol.* **2015**, *6*, 203. [CrossRef]
58. Skouras, P.; Gargalionis, A.N.; Piperi, C. Exosomes as Novel Diagnostic Biomarkers and Therapeutic Tools in Gliomas. *Int. J. Mol. Sci.* **2023**, *24*, 10162. [CrossRef]
59. Romano, E.; Netti, P.A.; Torino, E. Exosomes in Gliomas: Biogenesis, Isolation, and Preliminary Applications in Nanomedicine. *Pharmaceuticals* **2020**, *13*, 319. [CrossRef]
60. André, F.; Schartz, N.E.C.; Chaput, N.; Flament, C.; Raposo, G.; Amigorena, S.; Angevin, E.; Zitvogel, L. Tumor-Derived Exosomes: A New Source of Tumor Rejection Antigens. *Vaccine* **2002**, *20* (Suppl. S4), A28–A31. [CrossRef]
61. Chaput, N.; Taïeb, J.; Schartz, N.E.C.; André, F.; Angevin, E.; Zitvogel, L. Exosome-Based Immunotherapy. *Cancer Immunol. Immunother.* **2004**, *53*, 234–239. [CrossRef] [PubMed]
62. Shi, M.; Sheng, L.; Stewart, T.; Zabetian, C.P.; Zhang, J. New Windows into the Brain: Central Nervous System-Derived Extracellular Vesicles in Blood. *Prog. Neurobiol.* **2019**, *175*, 96–106. [CrossRef] [PubMed]
63. Boisselier, B.; Gállego Pérez-Larraya, J.; Rossetto, M.; Labussière, M.; Ciccarino, P.; Marie, Y.; Delattre, J.-Y.; Sanson, M. Detection of IDH1 Mutation in the Plasma of Patients with Glioma. *Neurology* **2012**, *79*, 1693–1698. [CrossRef] [PubMed]
64. Lavon, I.; Refael, M.; Zelikovitch, B.; Shalom, E.; Siegal, T. Serum DNA Can Define Tumor-Specific Genetic and Epigenetic Markers in Gliomas of Various Grades. *Neuro Oncol.* **2010**, *12*, 173–180. [CrossRef] [PubMed]
65. Bettegowda, C.; Sausen, M.; Leary, R.J.; Kinde, I.; Wang, Y.; Agrawal, N.; Bartlett, B.R.; Wang, H.; Luber, B.; Alani, R.M.; et al. Detection of Circulating Tumor DNA in Early- and Late-Stage Human Malignancies. *Sci. Transl. Med.* **2014**, *6*, 224ra24. [CrossRef] [PubMed]
66. Balaña, C.; Ramirez, J.L.; Taron, M.; Roussos, Y.; Ariza, A.; Ballester, R.; Sarries, C.; Mendez, P.; Sanchez, J.J.; Rosell, R. O6-Methyl-Guanine-DNA Methyltransferase Methylation in Serum and Tumor DNA Predicts Response to 1,3-Bis(2-Chloroethyl)-1-Nitrosourea but Not to Temozolamide plus Cisplatin in Glioblastoma Multiforme. *Clin. Cancer Res.* **2003**, *9*, 1461–1468.
67. Mouliere, F.; Mair, R.; Chandrananda, D.; Marass, F.; Smith, C.G.; Su, J.; Morris, J.; Watts, C.; Brindle, K.M.; Rosenfeld, N. Detection of Cell-Free DNA Fragmentation and Copy Number Alterations in Cerebrospinal Fluid from Glioma Patients. *EMBO Mol. Med.* **2018**, *10*, e9323. [CrossRef]
68. Shi, G.; Cui, W.; Benchimol, M.; Liu, Y.-T.; Mattrey, R.F.; Mukthavaram, R.; Kesari, S.; Esener, S.C.; Simberg, D. Isolation of Rare Tumor Cells from Blood Cells with Buoyant Immuno-Microbubbles. *PLoS ONE* **2013**, *8*, e58017. [CrossRef]
69. Lynch, D.; Powter, B.; Po, J.W.; Cooper, A.; Garrett, C.; Koh, E.-S.; Sheridan, M.; van Gelder, J.; Darwish, B.; Mckechnie, S.; et al. Isolation of Circulating Tumor Cells from Glioblastoma Patients by Direct Immunomagnetic Targeting. *Appl. Sci.* **2020**, *10*, 3338. [CrossRef]
70. Macarthur, K.M.; Kao, G.D.; Chandrasekaran, S.; Alonso-Basanta, M.; Chapman, C.; Lustig, R.A.; Wileyto, E.P.; Hahn, S.M.; Dorsey, J.F. Detection of Brain Tumor Cells in the Peripheral Blood by a Telomerase Promoter-Based Assay. *Cancer Res.* **2014**, *74*, 2152–2159. [CrossRef]

71. Olioso, D.; Caccese, M.; Santangelo, A.; Lippi, G.; Zagonel, V.; Cabrini, G.; Lombardi, G.; Dechecchi, M.C. Serum Exosomal MicroRNA-21, 222 and 124-3p as Noninvasive Predictive Biomarkers in Newly Diagnosed High-Grade Gliomas: A Prospective Study. *Cancers* **2021**, *13*, 3006. [CrossRef] [PubMed]
72. Yue, X.; Lan, F.; Hu, M.; Pan, Q.; Wang, Q.; Wang, J. Downregulation of Serum MicroRNA-205 as a Potential Diagnostic and Prognostic Biomarker for Human Glioma. *J. Neurosurg.* **2016**, *124*, 122–128. [CrossRef]
73. Wang, X.; Cao, Q.; Shi, Y.; Wu, X.; Mi, Y.; Liu, K.; Kan, Q.; Fan, R.; Liu, Z.; Zhang, M. Identification of Low-Dose Radiation-Induced Exosomal Circ-METRN and MiR-4709-3p/GRB14/PDGFRα Pathway as a Key Regulatory Mechanism in Glioblastoma Progression and Radioresistance: Functional Validation and Clinical Theranostic Significance. *Int. J. Biol. Sci.* **2021**, *17*, 1061–1078. [CrossRef] [PubMed]
74. Jones, J.; Nguyen, H.; Drummond, K.; Morokoff, A. Circulating Biomarkers for Glioma: A Review. *Neurosurgery* **2021**, *88*, E221–E230. [CrossRef]
75. Faria, G.; Silva, E.; Da Fonseca, C.; Quirico-Santos, T. Circulating Cell-Free DNA as a Prognostic and Molecular Marker for Patients with Brain Tumors under Perillyl Alcohol-Based Therapy. *Int. J. Mol. Sci.* **2018**, *19*, 1610. [CrossRef] [PubMed]
76. Mathios, D.; Phallen, J. Advances in Molecular Biomarkers and Liquid Biopsy in Gliomas. *Neurooncol. Adv.* **2022**, *4* (Suppl. S2), ii15–ii21. [CrossRef]
77. Li, D.; Bonner, E.R.; Wierzbicki, K.; Panditharatna, E.; Huang, T.; Lulla, R.; Mueller, S.; Koschmann, C.; Nazarian, J.; Saratsis, A.M. Standardization of the Liquid Biopsy for Pediatric Diffuse Midline Glioma Using DdPCR. *Sci. Rep.* **2021**, *11*, 5098. [CrossRef]
78. Panditharatna, E.; Kilburn, L.B.; Aboian, M.S.; Kambhampati, M.; Gordish-Dressman, H.; Magge, S.N.; Gupta, N.; Myseros, J.S.; Hwang, E.I.; Kline, C.; et al. Clinically Relevant and Minimally Invasive Tumor Surveillance of Pediatric Diffuse Midline Gliomas Using Patient Derived Liquid Biopsy. *Clin. Cancer Res.* **2018**, *24*, 5850–5859. [CrossRef]
79. Lander, E.S.; Linton, L.M.; Birren, B.; Nusbaum, C.; Zody, M.C.; Baldwin, J.; Devon, K.; Dewar, K.; Doyle, M.; FitzHugh, W.; et al. International Human Genome Sequencing Consortium. Initial Sequencing and Analysis of the Human Genome. *Nature* **2001**, *409*, 860–921. [CrossRef]
80. Markouli, M.; Strepkos, D.; Basdra, E.K.; Papavassiliou, A.G.; Piperi, C. Prominent Role of Histone Modifications in the Regulation of Tumor Metastasis. *Int. J. Mol. Sci.* **2021**, *22*, 2778. [CrossRef]
81. Pfeifer, G.P. Defining Driver DNA Methylation Changes in Human Cancer. *Int. J. Mol. Sci.* **2018**, *19*, 1166. [CrossRef] [PubMed]
82. Coufal, N.G.; Garcia-Perez, J.L.; Peng, G.E.; Yeo, G.W.; Mu, Y.; Lovci, M.T.; Morell, M.; O'Shea, K.S.; Moran, J.V.; Gage, F.H. L1 Retrotransposition in Human Neural Progenitor Cells. *Nature* **2009**, *460*, 1127–1131. [CrossRef] [PubMed]
83. Baillie, J.K.; Barnett, M.W.; Upton, K.R.; Gerhardt, D.J.; Richmond, T.A.; De Sapio, F.; Brennan, P.M.; Rizzu, P.; Smith, S.; Fell, M.; et al. Somatic Retrotransposition Alters the Genetic Landscape of the Human Brain. *Nature* **2011**, *479*, 534–537. [CrossRef]
84. Zheng, S.; Houseman, E.A.; Morrison, Z.; Wrensch, M.R.; Patoka, J.S.; Ramos, C.; Haas-Kogan, D.A.; McBride, S.; Marsit, C.J.; Christensen, B.C.; et al. DNA Hypermethylation Profiles Associated with Glioma Subtypes and EZH2 and IGFBP2 MRNA Expression. *Neuro Oncol.* **2011**, *13*, 280–289. [CrossRef] [PubMed]
85. Ohka, F.; Natsume, A.; Motomura, K.; Kishida, Y.; Kondo, Y.; Abe, T.; Nakasu, Y.; Namba, H.; Wakai, K.; Fukui, T.; et al. The Global DNA Methylation Surrogate LINE-1 Methylation Is Correlated with MGMT Promoter Methylation and Is a Better Prognostic Factor for Glioma. *PLoS ONE* **2011**, *6*, e23332. [CrossRef] [PubMed]
86. Sabedot, T.S.; Malta, T.M.; Snyder, J.; Nelson, K.; Wells, M.; deCarvalho, A.C.; Mukherjee, A.; Chitale, D.A.; Mosella, M.S.; Sokolov, A.; et al. A Serum-Based DNA Methylation Assay Provides Accurate Detection of Glioma. *Neuro Oncol.* **2021**, *23*, 1494–1508. [CrossRef]
87. Nassiri, F.; Chakravarthy, A.; Feng, S.; Shen, S.Y.; Nejad, R.; Zuccato, J.A.; Voisin, M.R.; Patil, V.; Horbinski, C.; Aldape, K.; et al. Detection and Discrimination of Intracranial Tumors Using Plasma Cell-Free DNA Methylomes. *Nat. Med.* **2020**, *26*, 1044–1047. [CrossRef]
88. Martínez-Ricarte, F.; Mayor, R.; Martínez-Sáez, E.; Rubio-Pérez, C.; Pineda, E.; Cordero, E.; Cicuéndez, M.; Poca, M.A.; López-Bigas, N.; Ramon, Y.; et al. Molecular Diagnosis of Diffuse Gliomas through Sequencing of Cell-Free Circulating Tumor DNA from Cerebrospinal Fluid. *Clin. Cancer Res.* **2018**, *24*, 2812–2819. [CrossRef]
89. Muralidharan, K.; Yekula, A.; Small, J.L.; Rosh, Z.S.; Kang, K.M.; Wang, L.; Lau, S.; Zhang, H.; Lee, H.; Bettegowda, C.; et al. TERT Promoter Mutation Analysis for Blood-Based Diagnosis and Monitoring of Gliomas. *Clin. Cancer Res.* **2021**, *27*, 169–178. [CrossRef]
90. Fujioka, Y.; Hata, N.; Akagi, Y.; Kuga, D.; Hatae, R.; Sangatsuda, Y.; Michiwaki, Y.; Amemiya, T.; Takigawa, K.; Funakoshi, Y.; et al. Molecular Diagnosis of Diffuse Glioma Using a Chip-Based Digital PCR System to Analyze IDH, TERT, and $H_3$ Mutations in the Cerebrospinal Fluid. *J. Neurooncol.* **2021**, *152*, 47–54. [CrossRef]
91. Husain, A.; Mishra, S.; Hadi, R.; Sahu, A.; Kumari, S.; Rastogi, M.; Khurana, R.; Shukla, S.; Siddiqui, M.H.; Husain, N. Dynamics of Cell-Free DNA in Predicting Response in Adult Diffuse Glioma on Chemoradiotherapy. *Cancer Genet.* **2022**, *268–269*, 55–63. [CrossRef] [PubMed]
92. Palande, V.; Siegal, T.; Detroja, R.; Gorohovski, A.; Glass, R.; Flueh, C.; Kanner, A.A.; Laviv, Y.; Har-Nof, S.; Levy-Barda, A.; et al. Detection of Gene Mutations and Gene–Gene Fusions in Circulating Cell-Free DNA of Glioblastoma Patients: An Avenue for Clinically Relevant Diagnostic Analysis. *Mol. Oncol.* **2022**, *16*, 2098–2114. [CrossRef] [PubMed]
93. Dong, L.; Li, Y.; Han, C.; Wang, X.; She, L.; Zhang, H. MiRNA Microarray Reveals Specific Expression in the Peripheral Blood of Glioblastoma Patients. *Int. J. Oncol.* **2014**, *45*, 746–756. [CrossRef] [PubMed]

94. Wang, Q.; Li, P.; Li, A.; Jiang, W.; Wang, H.; Wang, J.; Xie, K. Plasma Specific MiRNAs as Predictive Biomarkers for Diagnosis and Prognosis of Glioma. *J. Exp. Clin. Cancer Res.* **2012**, *31*, 97. [CrossRef] [PubMed]
95. Wu, L.; Li, G.; Feng, D.; Qin, H.; Gong, L.; Zhang, J.; Zhang, Z. MicroRNA-21 Expression Is Associated with Overall Survival in Patients with Glioma. *Diagn. Pathol.* **2013**, *8*, 200. [CrossRef] [PubMed]
96. Ilhan-Mutlu, A.; Wagner, L.; Wöhrer, A.; Furtner, J.; Widhalm, G.; Marosi, C.; Preusser, M. Plasma MicroRNA-21 Concentration May Be a Useful Biomarker in Glioblastoma Patients. *Cancer Investig.* **2012**, *30*, 615–621. [CrossRef] [PubMed]
97. Mazor, G.; Levin, L.; Picard, D.; Ahmadov, U.; Carén, H.; Borkhardt, A.; Reifenberger, G.; Leprivier, G.; Remke, M.; Rotblat, B. The LncRNA TP73-AS1 Is Linked to Aggressiveness in Glioblastoma and Promotes Temozolomide Resistance in Glioblastoma Cancer Stem Cells. *Cell Death Dis.* **2019**, *10*, 1–14. [CrossRef]
98. Shen, J.; Hodges, T.R.; Song, R.; Gong, Y.; Calin, G.A.; Heimberger, A.B.; Zhao, H. Serum HOTAIR and GAS5 Levels as Predictors of Survival in Patients with Glioblastoma. *Mol. Carcinog.* **2018**, *57*, 137–141. [CrossRef]
99. Yuan, Z.; Yang, Z.; Li, W.; Wu, A.; Su, Z.; Jiang, B. Exosome-Mediated Transfer of Long Noncoding RNA HOTAIR Regulates Temozolomide Resistance by MiR-519a-3p/RRM1 Axis in Glioblastoma. *Cancer Biother. Radiopharm.* **2020**. [CrossRef]
100. Naryzhny, S.; Ronzhina, N.; Zorina, E.; Kabachenko, F.; Zavialova, M.; Zgoda, V.; Klopov, N.; Legina, O.; Pantina, R. Evaluation of Haptoglobin and Its Proteoforms as Glioblastoma Markers. *Int. J. Mol. Sci.* **2021**, *22*, 6533. [CrossRef]
101. Gautam, P.; Nair, S.C.; Gupta, M.K.; Sharma, R.; Polisetty, R.V.; Uppin, M.S.; Sundaram, C.; Puligopum, A.K.; Ankathi, P.; Purohit, A.K.; et al. Proteins with altered levels in plasma from glioblastoma patients as revealed by iTRAQ-based quantitative proteomic analysis. *PLoS ONE* **2012**, *7*, e46153. [CrossRef] [PubMed]
102. Bellia, F.; Vecchio, G.; Rizzarelli, E. Carnosinases, their substrates and diseases. *Molecules* **2014**, *19*, 2299–2329. [CrossRef] [PubMed]
103. Gállego Pérez-Larraya, J.; Paris, S.; Idbaih, A.; Dehais, C.; Laigle-Donadey, F.; Navarro, S.; Capelle, L.; Mokhtari, K.; Marie, Y.; Sanson, M.; et al. Diagnostic and Prognostic Value of Preoperative Combined GFAP, IGFBP-2, and YKL-40 Plasma Levels in Patients with Glioblastoma. *Cancer* **2014**, *120*, 3972–3980. [CrossRef]
104. Ochieng, J.; Nangami, G.; Sakwe, A.; Moye, C.; Alvarez, J.; Whalen, D.; Thomas, P.; Lammers, P. Impact of Fetuin-A (AHSG) on Tumor Progression and Type 2 Diabetes. *Int. J. Mol. Sci.* **2018**, *19*, 2211. [CrossRef] [PubMed]
105. Naryzhny, S.; Volnitskiy, A.; Kopylov, A.; Zorina, E.; Kamyshinsky, R.; Bairamukov, V.; Garaeva, L.; Shlikht, A.; Shtam, T. Proteome of Glioblastoma-Derived Exosomes as a Source of Biomarkers. *Biomedicines* **2020**, *8*, 216. [CrossRef]
106. Miyauchi, E.; Furuta, T.; Ohtsuki, S.; Tachikawa, M.; Uchida, Y.; Sabit, H.; Obuchi, W.; Baba, T.; Watanabe, M.; Terasaki, T.; et al. Identification of Blood Biomarkers in Glioblastoma by SWATH Mass Spectrometry and Quantitative Targeted Absolute Proteomics. *PLoS ONE* **2018**, *13*, e0193799. [CrossRef]
107. Qin, G.; Li, X.; Chen, Z.; Liao, G.; Su, Y.; Chen, Y.; Zhang, W. Prognostic Value of YKL-40 in Patients with Glioblastoma: A Systematic Review and Meta-Analysis. *Mol. Neurobiol.* **2017**, *54*, 3264–3270. [CrossRef]
108. Müller, C.; Holtschmidt, J.; Auer, M.; Heitzer, E.; Lamszus, K.; Schulte, A.; Matschke, J.; Langer-Freitag, S.; Gasch, C.; Stoupiec, M.; et al. Hematogenous Dissemination of Glioblastoma Multiforme. *Sci. Transl. Med.* **2014**, *6*, 247ra101. [CrossRef]
109. Sastre, J.; Maestro, M.L.; Puente, J.; Veganzones, S.; Alfonso, R.; Rafael, S.; García-Saenz, J.A.; Vidaurreta, M.; Martín, M.; Arroyo, M.; et al. Circulating Tumor Cells in Colorectal Cancer: Correlation with Clinical and Pathological Variables. *Ann. Oncol.* **2008**, *19*, 935–938. [CrossRef]
110. Kichkailo, A.S.; Narodov, A.A.; Komarova, M.A.; Zamay, T.N.; Zamay, G.S.; Kolovskaya, O.S.; Erakhtin, E.E.; Glazyrin, Y.E.; Veprintsev, D.V.; Moryachkov, R.V.; et al. Development of DNA Aptamers for Visualization of Glial Brain Tumors and Detection of Circulating Tumor Cells. *Mol. Ther. Nucleic Acids* **2023**, *32*, 267–288. [CrossRef]
111. Ni, X.; Castanares, M.; Mukherjee, A.; Lupold, S.E. Nucleic Acid Aptamers: Clinical Applications and Promising New Horizons. *Curr. Med. Chem.* **2011**, *18*, 4206–4214. [CrossRef] [PubMed]
112. Song, Y.; Shi, Y.; Huang, M.; Wang, W.; Wang, Y.; Cheng, J.; Lei, Z.; Zhu, Z.; Yang, C. Bioinspired Engineering of a Multivalent Aptamer-Functionalized Nanointerface to Enhance the Capture and Release of Circulating Tumor Cells. *Angew. Chem. Int. Ed.* **2019**, *58*, 2236–2240. [CrossRef] [PubMed]
113. Gourlay, J.; Morokoff, A.P.; Luwor, R.B.; Zhu, H.-J.; Kaye, A.H.; Stylli, S.S. The Emergent Role of Exosomes in Glioma. *J. Clin. Neurosci.* **2017**, *35*, 13–23. [CrossRef] [PubMed]
114. Mallawaaratchy, D.M.; Hallal, S.; Russell, B.; Ly, L.; Ebrahimkhani, S.; Wei, H.; Christopherson, R.I.; Buckland, M.E.; Kaufman, K.L. Comprehensive Proteome Profiling of Glioblastoma-Derived Extracellular Vesicles Identifies Markers for More Aggressive Disease. *J. Neurooncol.* **2017**, *131*, 233–244. [CrossRef] [PubMed]
115. H Rashed, M.; Bayraktar, E.; K Helal, G.; Abd-Ellah, M.F.; Amero, P.; Chavez-Reyes, A.; Rodriguez-Aguayo, C. Exosomes: From Garbage Bins to Promising Therapeutic Targets. *Int. J. Mol. Sci.* **2017**, *18*, 538. [CrossRef]
116. Klekner, Á.; Szivos, L.; Virga, J.; Árkosy, P.; Bognár, L.; Birkó, Z.; Nagy, B. Significance of Liquid Biopsy in Glioblastoma—A Review. *J. Biotechnol.* **2019**, *298*, 82–87. [CrossRef]
117. Shao, H.; Chung, J.; Balaj, L.; Charest, A.; Bigner, D.D.; Carter, B.S.; Hochberg, F.H.; Breakefield, X.O.; Weissleder, R.; Lee, H. Protein Typing of Circulating Microvesicles Allows Real-Time Monitoring of Glioblastoma Therapy. *Nat. Med.* **2012**, *18*, 1835–1840. [CrossRef]
118. Whitehead, C.A.; Kaye, A.H.; Drummond, K.J.; Widodo, S.S.; Mantamadiotis, T.; Vella, L.J.; Stylli, S.S. Extracellular Vesicles and Their Role in Glioblastoma. *Crit. Rev. Clin. Lab. Sci.* **2019**, *57*, 227–252. [CrossRef]

119. Gatto, L.; Franceschi, E.; Di Nunno, V.; Tosoni, A.; Lodi, R.; Brandes, A.A. Liquid Biopsy in Glioblastoma Management: From Current Research to Future Perspectives. *Oncology* **2021**, *26*, 865–878. [CrossRef]
120. Quezada, C.; Torres, Á.; Niechi, I.; Uribe, D.; Contreras-Duarte, S.; Toledo, F.; San Martín, R.; Gutiérrez, J.; Sobrevia, L. Role of Extracellular Vesicles in Glioma Progression. *Mol. Asp. Med.* **2018**, *60*, 38–51. [CrossRef]
121. Hagiwara, K.; Ochiya, T.; Kosaka, N. A Paradigm Shift for Extracellular Vesicles as Small RNA Carriers: From Cellular Waste Elimination to Therapeutic Applications. *Drug Deliv. Transl. Res.* **2014**, *4*, 31–37. [CrossRef]
122. Jaiswal, R.; Sedger, L.M. Intercellular Vesicular Transfer by Exosomes, Microparticles and Oncosomes—Implications for Cancer Biology and Treatments. *Front. Oncol.* **2019**, *9*. [CrossRef] [PubMed]
123. Skog, J.; Würdinger, T.; van Rijn, S.; Meijer, D.H.; Gainche, L.; Sena-Esteves, M.; Curry, W.T.; Carter, B.S.; Krichevsky, A.M.; Breakefield, X.O. Glioblastoma Microvesicles Transport RNA and Proteins That Promote Tumour Growth and Provide Diagnostic Biomarkers. *Nat. Cell Biol.* **2008**, *10*, 1470–1476. [CrossRef] [PubMed]
124. Putz, U.; Howitt, J.; Doan, A.; Goh, C.-P.; Low, L.-H.; Silke, J.; Tan, S.-S. The Tumor Suppressor PTEN Is Exported in Exosomes and Has Phosphatase Activity in Recipient Cells. *Sci. Signal.* **2012**, *5*, ra70. [CrossRef] [PubMed]
125. Shi, R.; Wang, P.-Y.; Li, X.-Y.; Chen, J.-X.; Li, Y.; Zhang, X.-Z.; Zhang, C.-G.; Jiang, T.; Li, W.-B.; Ding, W.; et al. Exosomal Levels of MiRNA-21 from Cerebrospinal Fluids Associated with Poor Prognosis and Tumor Recurrence of Glioma Patients. *Oncotarget* **2015**, *6*, 26971–26981. [CrossRef]
126. Ivo D'Urso, P.; Fernando D'Urso, O.; Damiano Gianfreda, C.; Mezzolla, V.; Storelli, C.; Marsigliante, S. MiR-15b and MiR-21 as Circulating Biomarkers for Diagnosis of Glioma. *Curr. Genom.* **2015**, *16*, 304–311. [CrossRef] [PubMed]
127. Akers, J.C.; Ramakrishnan, V.; Kim, R.; Phillips, S.; Kaimal, V.; Mao, Y.; Hua, W.; Yang, I.; Fu, C.-C.; Nolan, J.; et al. MiRNA Contents of Cerebrospinal Fluid Extracellular Vesicles in Glioblastoma Patients. *J. Neurooncol.* **2015**, *123*, 205–216. [CrossRef] [PubMed]
128. Zeng, A.; Wei, Z.; Yan, W.; Yin, J.; Huang, X.; Zhou, X.; Li, R.; Shen, F.; Wu, W.; Wang, X.; et al. Exosomal Transfer of MiR-151a Enhances Chemosensitivity to Temozolomide in Drug-Resistant Glioblastoma. *Cancer Lett.* **2018**, *436*, 10–21. [CrossRef] [PubMed]
129. Figueroa, J.M.; Skog, J.; Akers, J.; Li, H.; Komotar, R.; Jensen, R.; Ringel, F.; Yang, I.; Kalkanis, S.; Thompson, R.; et al. Detection of Wild-Type EGFR Amplification and EGFRvIII Mutation in CSF-Derived Extracellular Vesicles of Glioblastoma Patients. *Neuro Oncol.* **2017**, *19*, 1494–1502. [CrossRef]
130. Huang, K.; Fang, C.; Yi, K.; Liu, X.; Qi, H.; Tan, Y.; Zhou, J.; Li, Y.; Liu, M.; Zhang, Y.; et al. The Role of PTRF/Cavin1 as a Biomarker in Both Glioma and Serum Exosomes. *Theranostics* **2018**, *8*, 1540–1557. [CrossRef] [PubMed]
131. Shao, H.; Chung, J.; Lee, K.; Balaj, L.; Min, C.; Carter, B.S.; Hochberg, F.H.; Breakefield, X.O.; Lee, H.; Weissleder, R. Chip-Based Analysis of Exosomal MRNA Mediating Drug Resistance in Glioblastoma. *Nat. Commun.* **2015**, *6*, 6999. [CrossRef] [PubMed]
132. Manterola, L.; Guruceaga, E.; Pérez-Larraya, J.G.; González-Huarriz, M.; Jauregui, P.; Tejada, S.; Diez-Valle, R.; Segura, V.; Samprón, N.; Barrena, C.; et al. A Small Noncoding RNA Signature Found in Exosomes of GBM Patient Serum as a Diagnostic Tool. *Neuro Oncol.* **2014**, *16*, 520–527. [CrossRef] [PubMed]
133. Lan, F.; Qing, Q.; Pan, Q.; Hu, M.; Yu, H.; Yue, X. Serum Exosomal MiR-301a as a Potential Diagnostic and Prognostic Biomarker for Human Glioma. *Cell. Oncol.* **2018**, *41*, 25–33. [CrossRef] [PubMed]
134. Zhi, F.; Shao, N.; Wang, R.; Deng, D.; Xue, L.; Wang, Q.; Zhang, Y.; Shi, Y.; Xia, X.; Wang, S.; et al. Identification of 9 Serum MicroRNAs as Potential Noninvasive Biomarkers of Human Astrocytoma. *Neuro Oncol.* **2015**, *17*, 383–391. [CrossRef]
135. Tzaridis, T.; Reiners, K.S.; Weller, J.; Bachurski, D.; Schäfer, N.; Schaub, C.; Hallek, M.; Scheffler, B.; Glas, M.; Herrlinger, U.; et al. Analysis of Serum MiRNA in Glioblastoma Patients: CD44-Based Enrichment of Extracellular Vesicles Enhances Specificity for the Prognostic Signature. *Int. J. Mol. Sci.* **2020**, *21*, 7211. [CrossRef]
136. Regazzo, G.; Terrenato, I.; Spagnuolo, M.; Carosi, M.; Cognetti, G.; Cicchillitti, L.; Sperati, F.; Villani, V.; Carapella, C.; Piaggio, G.; et al. A Restricted Signature of Serum MiRNAs Distinguishes Glioblastoma from Lower Grade Gliomas. *J. Exp. Clin. Cancer Res.* **2016**, *35*, 124. [CrossRef]
137. Zhang, C.; Zhang, J.; Hao, J.; Shi, Z.; Wang, Y.; Han, L.; Yu, S.; You, Y.; Jiang, T.; Wang, J.; et al. High Level of MiR-221/222 Confers Increased Cell Invasion and Poor Prognosis in Glioma. *J. Transl. Med.* **2012**, *10*, 119. [CrossRef]
138. Kucharzewska, P.; Christianson, H.C.; Welch, J.E.; Svensson, K.J.; Fredlund, E.; Ringnér, M.; Mörgelin, M.; Bourseau-Guilmain, E.; Bengzon, J.; Belting, M. Exosomes Reflect the Hypoxic Status of Glioma Cells and Mediate Hypoxia-Dependent Activation of Vascular Cells during Tumor Development. *Proc. Natl. Acad. Sci. USA* **2013**, *110*, 7312–7317. [CrossRef]
139. Zhang, X.; Liu, C.; Pei, Y.; Song, W.; Zhang, S. Preparation of a Novel Raman Probe and Its Application in the Detection of Circulating Tumor Cells and Exosomes. *ACS Appl. Mater. Interfaces* **2019**, *11*, 28671–28680. [CrossRef]
140. Shui, L.; Ren, H.; Yang, X.; Li, J.; Chen, Z.; Yi, C.; Zhu, H.; Shui, P. The Era of Radiogenomics in Precision Medicine: An Emerging Approach to Support Diagnosis, Treatment Decisions, and Prognostication in Oncology. *Front. Oncol.* **2021**, *10*, 570465. [CrossRef]
141. Gore, S.; Chougule, T.; Jagtap, J.; Saini, J.; Ingalhalikar, M. A Review of Radiomics and Deep Predictive Modeling in Glioma Characterization. *Acad. Radiol.* **2021**, *28*, 1599–1621. [CrossRef] [PubMed]
142. Jian, A.; Jang, K.; Manuguerra, M.; Liu, S.; Magnussen, J.; Di Ieva, A. Machine Learning for the Prediction of Molecular Markers in Glioma on Magnetic Resonance Imaging: A Systematic Review and Meta-Analysis. *Neurosurgery* **2021**, *89*, 31–44. [CrossRef] [PubMed]
143. Jain, R.; Chi, A.S. Radiogenomics Identifying Important Biological Pathways in Gliomas. *Neuro Oncol.* **2021**, *23*, 177–178. [CrossRef] [PubMed]

144. Matsutani, A.; Udagawa, C.; Matsunaga, Y.; Nakamura, S.; Zembutsu, H. Liquid Biopsy for the Detection of Clinical Biomarkers in Early Breast Cancer: New Insights and Challenges. *Pharmacogenomics* **2020**, *21*, 359–367. [CrossRef] [PubMed]
145. Lafata, K.J.; Corradetti, M.N.; Gao, J.; Jacobs, C.D.; Weng, J.; Chang, Y.; Wang, C.; Hatch, A.; Xanthopoulos, E.; Jones, G.; et al. Radiogenomic Analysis of Locally Advanced Lung Cancer Based on CT Imaging and Intratreatment Changes in Cell-Free DNA. *Radiol. Imaging Cancer* **2021**, *3*, e200157. [CrossRef]
146. Li, Y.; Liu, X.; Qian, Z.; Sun, Z.; Xu, K.; Wang, K.; Fan, X.; Zhang, Z.; Li, S.; Wang, Y.; et al. Genotype Prediction of ATRX Mutation in Lower-Grade Gliomas Using an MRI Radiomics Signature. *Eur. Radiol.* **2018**, *28*, 2960–2968. [CrossRef]
147. Calabrese, E.; Villanueva-Meyer, J.E.; Cha, S. A Fully Automated Artificial Intelligence Method for Non-Invasive, Imaging-Based Identification of Genetic Alterations in Glioblastomas. *Sci. Rep.* **2020**, *10*, 11852. [CrossRef]
148. Li, Q.; Dong, F.; Jiang, B.; Zhang, M. Exploring MRI Characteristics of Brain Diffuse Midline Gliomas With the $H_3$ K27M Mutation Using Radiomics. *Front. Oncol.* **2021**, *11*, 646267. [CrossRef]
149. Tian, H.; Wu, H.; Wu, G.; Xu, G. Noninvasive Prediction of TERT Promoter Mutations in High-Grade Glioma by Radiomics Analysis Based on Multiparameter MRI. *BioMed Res. Int.* **2020**, *2020*, 3872314. [CrossRef]
150. Moon, W.-J.; Choi, J.W.; Roh, H.G.; Lim, S.D.; Koh, Y.-C. Imaging Parameters of High Grade Gliomas in Relation to the MGMT Promoter Methylation Status: The CT, Diffusion Tensor Imaging, and Perfusion MR Imaging. *Neuroradiology* **2012**, *54*, 555–563. [CrossRef] [PubMed]
151. Louis, D.N.; Perry, A.; Wesseling, P.; Brat, D.J.; Cree, I.A.; Figarella-Branger, D.; Hawkins, C.; Ng, H.K.; Pfister, S.M.; Reifenberger, G.; et al. The 2021 WHO Classification of Tumors of the Central Nervous System: A Summary. *Neuro Oncol.* **2021**, *23*, 1231–1251. [CrossRef] [PubMed]
152. Brown, R.; Zlatescu, M.; Sijben, A.; Roldan, G.; Easaw, J.; Forsyth, P.; Parney, I.; Sevick, R.; Yan, E.; Demetrick, D.; et al. The Use of Magnetic Resonance Imaging to Noninvasively Detect Genetic Signatures in Oligodendroglioma. *Clin. Cancer Res.* **2008**, *14*, 2357–2362. [CrossRef] [PubMed]
153. Kong, Z.; Jiang, C.; Zhang, Y.; Liu, S.; Liu, D.; Liu, Z.; Chen, W.; Liu, P.; Yang, T.; Lyu, Y.; et al. Thin-Slice Magnetic Resonance Imaging-Based Radiomics Signature Predicts Chromosomal 1p/19q Co-Deletion Status in Grade II and III Gliomas. *Front. Neurol.* **2020**, *11*, 551771. [CrossRef] [PubMed]
154. Casale, R.; Lavrova, E.; Sanduleanu, S.; Woodruff, H.C.; Lambin, P. Development and External Validation of a Non-Invasive Molecular Status Predictor of Chromosome 1p/19q Co-Deletion Based on MRI Radiomics Analysis of Low Grade Glioma Patients. *Eur. J. Radiol.* **2021**, *139*, 109678. [CrossRef] [PubMed]
155. Soffietti, R.; Bettegowda, C.; Mellinghoff, I.K.; Warren, K.E.; Ahluwalia, M.S.; De Groot, J.F.; Galanis, E.; Gilbert, M.R.; Jaeckle, K.A.; Le Rhun, E.; et al. Liquid Biopsy in Gliomas: A RANO Review and Proposals for Clinical Applications. *Neuro Oncol.* **2022**, *24*, 855–871. [CrossRef]
156. Ronvaux, L.; Riva, M.; Coosemans, A.; Herzog, M.; Rommelaere, G.; Donis, N.; D'Hondt, L.; Douxfils, J. Liquid Biopsy in Glioblastoma. *Cancers* **2022**, *14*, 3394. [CrossRef]

biomedicines

MDPI

*Systematic Review*

# From Voxel to Gene: A Scoping Review on MRI Radiogenomics' Artificial Intelligence Predictions in Adult Gliomas and Glioblastomas—The Promise of Virtual Biopsy?

**Xavier Maximin Le Guillou Horn [1,2,*], François Lecellier [3], Clement Giraud [1,4], Mathieu Naudin [4], Pierre Fayolle [1,4], Céline Thomarat [4], Christine Fernandez-Maloigne [5] and Rémy Guillevin [4,6]**

1 Laboratoire de Mathématique Appliquées LMA, Labcom i3M, Université de Poitiers, CNRS UMR 7348, F-86000 Poitiers, France; pierre.fayolle@univ-poitiers.fr (P.F.)
2 Service de Génétique Médicale, CHU de Poitiers, F-86000 Poitiers, France
3 Laboratoire XLIM, Université de Poitiers, CNRS UMR 7252, F-86000 Poitiers, France
4 Plateforme Ultra-Haut Champ 3T-7T, Laboratoire de Mathématique Appliquées LMA, Labcom i3M, CHU de Poitiers, Université de Poitiers, CNRS UMR 7348, F-86000 Poitiers, France; remy.guillevin@chu-poitiers.fr (R.G.)
5 Laboratoire XLIM, Labcom i3M, Université de Poitiers, CNRS UMR 7252, F-86000 Poitiers, France; christine.fernandez@univ-poitiers.fr
6 Service de Radiologie, CHU de Poitiers, F-86000 Poitiers, France
* Correspondence: xavier.le-guillou@chu-poitiers.fr

**Abstract:** Background: Gliomas, including the most severe form known as glioblastomas, are primary brain tumors arising from glial cells, with significant impact on adults, particularly men aged 45 to 70. Recent advancements in the WHO (World Health Organization) classification now correlate genetic markers with glioma phenotypes, enhancing diagnostic precision and therapeutic strategies. Aims and Methods: This scoping review aims to evaluate the current state of deep learning (DL) applications in the genetic characterization of adult gliomas, addressing the potential of these technologies for a reliable virtual biopsy. Results: We reviewed 17 studies, analyzing the evolution of DL algorithms from fully convolutional networks to more advanced architectures (ResNet and DenseNet). The methods involved various validation techniques, including k-fold cross-validation and external dataset validation. Conclusions: Our findings highlight significant variability in reported performance, largely due to small, homogeneous datasets and inconsistent validation methods. Despite promising results, particularly in predicting individual genetic traits, the lack of robust external validation limits the generalizability of these models. Future efforts should focus on developing larger, more diverse datasets and integrating multidisciplinary collaboration to enhance model reliability. This review underscores the potential of DL in advancing glioma characterization, paving the way for more precise, non-invasive diagnostic tools. The development of a robust algorithm capable of predicting the somatic genetics of gliomas or glioblastomas could accelerate the diagnostic process and inform therapeutic decisions more quickly, while maintaining the same level of accuracy as the traditional diagnostic pathway, which involves invasive tumor biopsies.

**Keywords:** adult gliomas; adult glioblastomas; MRI; deep learning; radiogenomics; virtual biopsy; scoping review

## 1. Introduction

The brain is primarily composed of neurons, which are responsible for processing information, and glial cells, which provide structural and functional support. Glial cells account for at least 50% to 90% of the brain's composition [1]. Among them, astrocytes play a critical role in supplying nutrients to neurons, managing interneuronal connections, and regulating processes such as memory, movement, and odor processing [2].

Gliomas are tumors that develop from these glial cells in the brain or spinal cord. Most brain tumors originate in the cerebral hemispheres, particularly in the white matter, which contains a dense network of axons (projections from neurons) surrounded by glial cells. However, gliomas can occur throughout the central nervous system. They are the most common type of brain tumor in children, adolescents, and adults, and are the most prevalent form of primary (i.e., non-metastatic) brain tumor in adults. The most severe form of glioma is glioblastoma, which is also the most common central nervous system tumor in the U.S. population, accounting for 14.2% of all tumors and 50.1% of all malignant tumors. In 2022, there were 13,272 new cases of glioblastoma in the United States, primarily affecting adults and occurring more frequently in men than in women [3].

Over time, the World Health Organization (WHO) classifications have increasingly incorporated correlations between a tumor's genetic profile and its phenotype. This effort culminated in the 2021 classification, which clearly distinguishes between childhood and adult glial tumors [4]. Adult diffuse gliomas are now categorized into three classes: *IDH1*- and *IDH2*-mutated oligodendrogliomas, which exhibit a 1p/19q codeletion (loss of the p arm of chromosome 1 and the q arm of chromosome 19) and are associated with longer survival, representing low-grade gliomas; *IDH1*- and *IDH2*-mutated diffuse astrocytomas, which have a variable prognosis, ranging from low-grade to high-grade gliomas; and glioblastomas, which lack *IDH1* or *IDH2* mutations but exhibit an activating mutation in the *TERT* promoter (a telomere-building protein), *EGFR* amplification, and specific karyotypic features. These glioblastomas are classified as high-grade gliomas and are associated with the worst prognosis. Additionally, methylation of the *MGMT* gene, which encodes the DNA repair enzyme O-6-methylguanine-DNA methyltransferase, is an important factor in determining the efficacy of temozolomide chemotherapy. If the *MGMT* gene is unmethylated, the cancer cells are capable of repairing the DNA damage induced by temozolomide, reducing the drug's effectiveness [5,6].

Magnetic resonance imaging (MRI) is considered the gold standard for detecting brain tumors, whether discovered incidentally or as a result of neurological symptoms. MRI can assess the extent of the disease and determine the feasibility of surgical resection. It also guides stereotactic biopsy, the only method capable of providing histological and molecular analyses to confirm the diagnosis and characterize the tumor. This histo-molecular characterization is critical for optimal, patient-specific therapeutic management. However, biopsy is not feasible for all patients, and there is a risk that the sampled area may not represent the most severe or prognostically relevant region of the tumor. Additionally, biopsy is an invasive procedure and carries inherent risks for the patient. Since 2012, the classification of non-medical images has been revolutionized by deep learning (DL) algorithms, particularly those utilizing convolutional neural networks (CNNs) [7]. This technology offers significant potential for application in the genetic characterization of adult glial tumors, presenting a compelling opportunity for non-invasive diagnostics. The development of a robust algorithm capable of predicting the somatic genetics of gliomas or glioblastomas from MRI could accelerate the diagnostic process and inform therapeutic decisions more quickly, while maintaining the same level of accuracy as the traditional diagnostic pathway, which involves invasive tumor biopsies. Furthermore, such an algorithm could provide valuable genetic insights for patients' ineligible for biopsy, offering a non-invasive alternative for diagnosis and treatment planning.

The aim of our scoping review is to provide a comprehensive, state-of-the-art overview of the use of DL algorithms in the genetic characterization of adult glial tumors. This review will cover reported performance, potential limitations, datasets used (including those available via open access), and future directions needed to advance the field toward achieving a reliable and reproducible capacity for virtual biopsy.

## 2. Materials and Methods

This scoping review was conducted with reference to the Preferred Reporting Items for Systematic Reviews and Meta-Analyses (PRISMA) guidelines, commonly used for systematic reviews [8].

### *2.1. PICOs (Inclusion Criteria)*

We defined the PICOs as follows. Population: adults with gliomas or glioblastomas. Intervention: the development and use of AI models to predict somatic genetics from MRI. Comparison: not applicable. Outcomes: the ability to predict genetic features, evaluated using ROC curves. Our focus was on artificial intelligence models utilizing deep learning (DL) technology, rather than traditional machine learning (ML) approaches.

### *2.2. Search Strategy*

For our research strategy, we selected three databases: PubMed, Embase, and the Cochrane Library. These databases were chosen for their focus on medical articles relevant to our topic and the management of patients with gliomas. Although Google Scholar was considered, it was not used due to the challenges associated with obtaining complete bibliographic data through scraping methods. IEEE Xplore was also not included, despite its potential relevance due to its coverage of computer science articles that may involve the development of deep learning algorithms for genetic feature prediction. Notably, IEEE Xplore references articles from the IEEE (the Institute of Electrical and Electronics Engineers) and its partners that are often indexed in PubMed.

We established the following criteria for accessing articles: if an article could not be accessed through OpenAccess or Shibboleth library access provided by our institution, it was excluded. Preprints and articles written in languages other than English were also excluded. For PubMed searches, we excluded all references that did not have full-text availability. During the review writing process, we planned to re-query one database for additional relevant articles, choosing PubMed for this second query for simplicity. (query of each databases in Supplemental Material S1).

Any article known to us but not included in the search results was included in our review at our discretion. Similarly, if an article cited by one of the query results met the PICOs criteria but was not included in the initial search results, it was also included in our review.

### *2.3. Data Extraction*

Search results were compiled using Zotero, an open-access bibliographic tool, which automatically excluded duplicates and irrelevant records. For data extraction, we used a tabular file format (".csv").

## 3. Results

The flowchart of the review process is shown in Figure 1. The initial search was conducted across three databases on 30 December 2023, and an additional search on PubMed was performed on 1 June 2024, during the writing of the review. The results of the 17 articles selected for this review are summarized in Table 1. Among these, three articles employed algorithmic methodologies using DL strategies for the segmentation and extraction of radiomic features, which were then processed by ML algorithms such as Support Vector Machine (SVM), k-Nearest Neighbors (k-NN), or Random Forest: S. Rathore et al. [9], S. Kihira et al. [10], and S. Qureshi et al. [11].

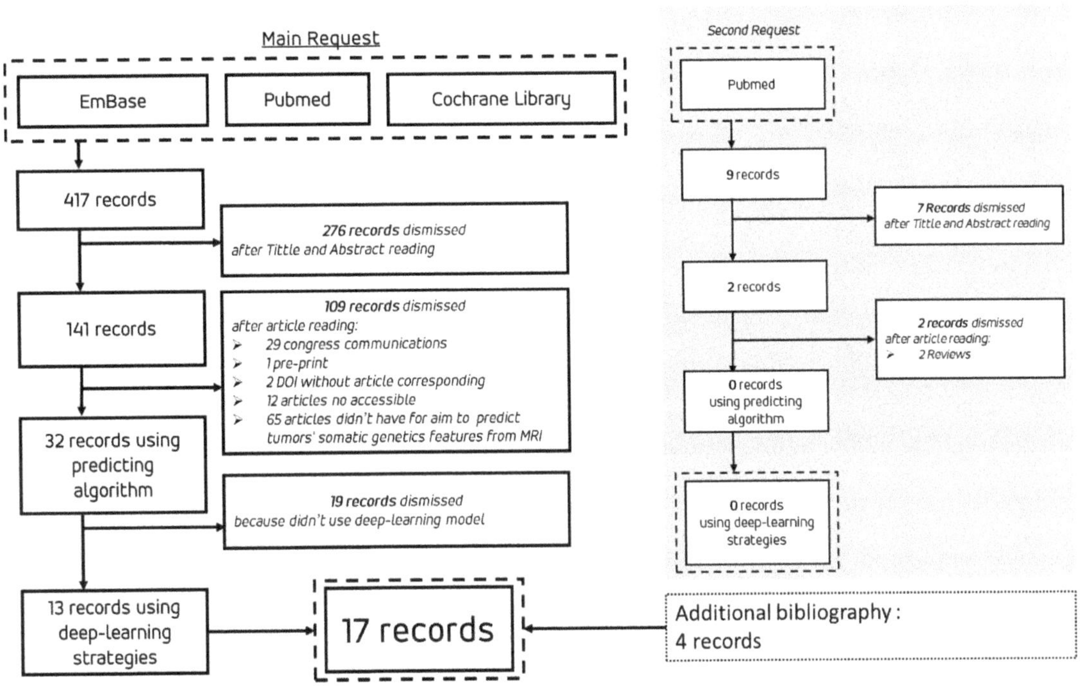


**Figure 1.** Flowchart of the review process from initial search to selected records.

**Table 1.** Summary of the 17 articles sorted by year of publication. SJR: Scimago Journal Ranking; TCIA: The Cancer Imaging Archive; TCGA: The Cancer Genome Atlas; BrATS: Brain Automated Tumors Segmentation; SNUH set: Seoul National University Hospital dataset.

| Authors | Year | Journal | SJR Pub. Year | Title | Number of Patients | Gliomas/ Glioblastomas/ Both | MRI Modalities | Dataset | Algorithms |
|---|---|---|---|---|---|---|---|---|---|
| I. Levner et al. [12] | 2009 | Medical Image Computing and Computer-Assisted Intervention | 0.297 | Predicting MGMT Methylation Status of Glioblastomas from MRI Texture | 59 | Glioblastomas | T1-Gd, T2, T2-FLAIR | Local | CNN (2 layers) |
| P. Eichinger et al. [13] | 2017 | Scientific Reports | 1.533 | Diffusion tensor image features predict *IDH* genotype in newly diagnosed WHO grade II/III gliomas | 79 | Gliomas | T2-FLAIR | TCIA | N-net |
| P. Chang et al. [14] | 2018 | AJNR Am J Neuroradiol | 1.543 | Deep-Learning Convolutional Neural Networks Accurately Classify Genetic Mutations in Gliomas | 259 | Gliomas | T1w, T1-Gd, T2w, T2-FLAIR | TCIA, TCGA | CNN |
| S. Liang et al. [15] | 2018 | Genes | 1.592 | Multimodal 3D DenseNet for *IDH* Genotype Prediction in Gliomas | 167 | Both | T1w, T1-Gd, T2w, T2-FLAIR | BrATS-2017, TCGA | M3D-DenseNet |
| M. Hedye-hzadeh et al. [16] | 2020 | Journal of Digital Imaging | 1.055 | A Comparison of the Efficiency of Using a Deep CNN Approach with Other Common Regression Methods for the Prediction of EGFR Expression in Glioblastoma Patients | 166 | Glioblastomas | T1w, T1-Gd, T2w, T2-FLAIR | TCIA, TCGA | CNN |

**Table 1.** *Cont.*

| Authors | Year | Journal | SJR Pub. Year | Title | Number of Patients | Gliomas/ Glioblastomas/ Both | MRI Modalities | Dataset | Algorithms |
|---|---|---|---|---|---|---|---|---|---|
| Y. Matsui et al. [17] | 2020 | Journal of Neuro-Oncology | 1.256 | Prediction of lower-grade glioma molecular subtypes using deep learning | 217 | Gliomas | T1w, T2w, T2-FLAIR, Spectrometry, PET scan | Local | ResNet into CNN |
| B Kocak et al. [18] | 2020 | European Radiology | 1.606 | Radiogenomics of lower-grade gliomas: Machine Learning-based MRI texture analysis for predicting 1p/19q codeletion status | 107 | Gliomas | T1w, T2w | TCIA | CNN against ML algorithms |
| S Rathore et al. [9] * | 2020 | Neuro-Oncology Advances | 1.052 | Multi-institutional non-invasive in vivo characterization of *IDH*, 1p/19q, and EGFRvIII in glioma using neuro-Cancer Imaging Phenomics Toolkit (neuro-CaPTk) | 473 | Both | T1w, T1-Gd, T2w, T2-FLAIR, DSC, DCE | Local, TCIA, TCGA | Neuro-CaPTK (Cancer Imaging Phenomics Toolkit) |
| C. G. B. Yogananda et al. [19] | 2021 | AJNR Am J Neuroradiol | 1.34 | MRI-Based Deep-Learning Method for Determining Glioma MGMT Promoter Methylation Status | 247 | Gliomas | T2w | TCIA, TCGA | 3D-dense-Unets |
| Y. S. Choi et al. [20] | 2021 | Neuro-Oncology | 3.097 | Fully automated hybrid approach to predict the *IDH* mutation status of gliomas via deep learning and radiomics | 856 | Both | T1w, T2w, T2-FLAIR | Local, SNUH set, TCIA | CNN |
| I. Hrapşa et al. [21] | 2022 | Medicina | 0.59 | External Validation of a Convolutional Neural Network for *IDH* Mutation Prediction | 21 | Glioblastomas | T1w, T2w, T2-FLAIR | Local, TCIA, TCGA | CHOI et al.'s CNN [20] |
| E. Calabrese et al. [22] | 2022 | Neuro-Oncology Advances | 1.052 | Combining radiomics and deep convolutional neural network features from preoperative MRI for predicting clinically relevant genetic biomarkers in glioblastoma | 400 | Glioblastomas | T1w, T2w, T2-FLAIR, SWI, DWI, ASL, MD, AD, RD | Local | CNN Limb |
| B.-H. Kim et al. [23] | 2022 | Cancers | 1.312 | Validation of MRI-Based Models to Predict MGMT Promoter Methylation in Gliomas: BraTS 2021 Radiogenomics Challenge | 400 (+585) | Both | T1w, T1-Gd, T2w, T2-FLAIR | Local, SNUH set, BrATS 2021 | Efficient-Net, squeeze-and-excitation networks, SEResNet, SEResNeXt, DenseNet |
| S. Kihira et al. [10] * | 2022 | Cancers | 1.312 | U-Net Based Segmentation and Characterization of Gliomas | 208 | Both | T2-FLAIR | Local | DenseNet121 |
| H. Sakly et al. [24] | 2023 | Cancer Control: Journal of the Moffitt Cancer Center | 0.698 | Brain Tumor Radiogenomic Classification of O6-Methylguanine-DNA Methyltransferase Promoter Methylation in Malignant Gliomas-Based Transfer Learning | 585 | Glioblastomas | T1w, T1-Gd, T2w, T2-FLAIR | BrATS 2021 | Alexnet, Googlenet, Resnet, ImageNet, VGG, DenseNet, Xception, InceptionV3Squeezenet |

**Table 1.** *Cont.*

| Authors | Year | Journal | SJR Pub. Year | Title | Number of Patients | Gliomas/ Glioblastomas/ Both | MRI Modalities | Dataset | Algorithms |
|---|---|---|---|---|---|---|---|---|---|
| S. A. Qureshi et al. [11] * | 2023 | Scientific Reports | 0.9 | Radiogenomic classification for MGMT promoter methylation status using multi-omics-fused feature space for least invasive diagnosis through mpMRI scans | 585 | Glioblastomas | T1w, T1-Gd, T2w | BrATS 2021 | CNN for segmentation and extraction feature but SVM or k-NN for classification |
| N. Saeed et al. [25] | 2023 | Medical Image Analysis | 4.112 | MGMT promoter methylation status prediction using MRI scans. An extensive experimental evaluation of deep learning models | 585 | Glioblastomas | T1w, T1-Gd, T2w, T2-FLAIR | BrATS 2021 | ResNet, DenseNet, EfficientNEt |

* Articles employing deep learning strategies did not advance to the genetic characterization of tumors.

### 3.1. Bibliographical and Descriptive Data on Publications

The 17 publications span from 2009 to 2023 (Figure 2a). For each publication, we collected the Scimago Journal Rank (SJR) score for the year of publication, with a median of 1.312, a minimum of 0.297, and a maximum of 4.112. For comparison, the SJR scores in 2023 for major journals were as follows: Nature—18.509, Cell—24.342, and The New England Journal of Medicine—20.544. Among medical imaging journals, Medical Image Analysis had the highest SJR in 2023, with a score of 4.112.

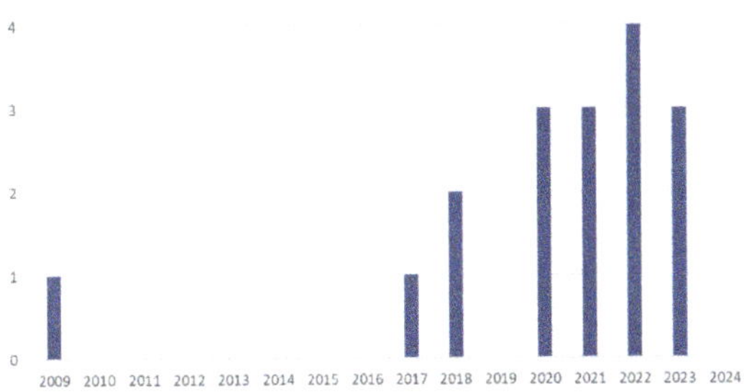


(a)

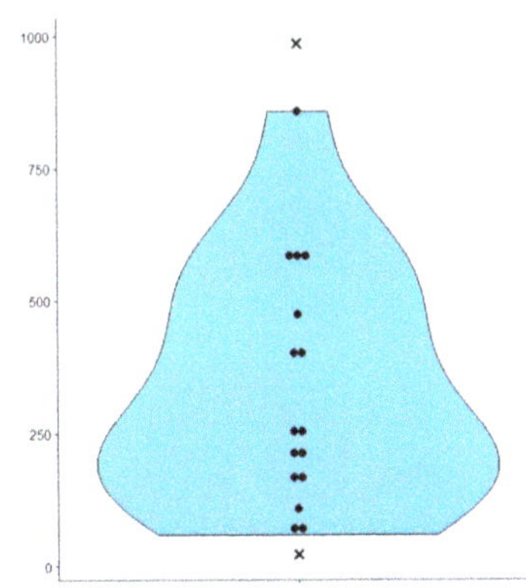


(b)

**Figure 2.** (**a**) Publications per year; (**b**) violin plot showing the number of patients used in algorithm training and testing. The cross points correspond to 21 patients (I. Hrapșa et al. [21]) and 985 patients (B.-H. Kim et al. [23]). $x$-axis: no unit, $y$-axis: number of patient.

The number of patients used to develop the algorithms ranged from 21 to 985 (Figure 2b). However, I. Hrapșa et al. [21] presented replication work involving 21 patients without retraining, based on the work of Y.S. Choi et al. [20]. In contrast, B.-H. Kim et al. [23] used up to 985 patients, which were combined from the Seoul National University Hospital (SNUH) dataset (400 patients) and the MICCAI BrATS 2021 dataset (585 patients). The algorithm described in their study was initially trained on the 400 patients from the SNUH dataset.

Among the seventeen articles, seven used data from patients with glioblastomas, five used data from patients with gliomas, and five used data from patients with both glioblastomas and gliomas.

The MRI sequences employed in these studies were primarily T2-weighted (T2w) and T2 Fluid-Attenuated Inversion Recovery (T2-FLAIR) (Table 2).

**Table 2.** MRI sequences used in articles employing DL algorithms. T1w: T1-weighted; T1-Gd: T1-weighted post-gadolinium contrast; T2w: T2-weighted; T2-FLAIR: T2 Fluid-Attenuated Inversion Recovery.

| MRI Sequence | Number | Percent |
|---|---|---|
| T1w | 13 | 76% |
| T1-Gd | 9 | 53% |
| T2w | 14 | 82% |
| T2-FLAIR | 14 | 82% |
| Spectrometry | 1 | 6% |
| Other | 3 | 18% |

### 3.2. *Deep-Learning Strategy*

All the articles utilized algorithms that fit the definition of deep learning. Among these, three employed DL algorithms, but the task of classifying tumors based on their genetic characteristics was handled by a separate machine learning (ML) algorithm [2–4].

Of the 17 papers, 10 used CNNs with fully convolutional architectures. These CNNs were applied either directly to the images, after automatic segmentation by a deep learning algorithm with a specific architecture, or following manual segmentation.

The specific architectures mentioned included DenseNet, ResNet, and U-Net. One of the studies explored transfer learning by adapting non-medical image classification models for use in medical image classification (H. Sakly et al. 2023 [24]).

### 3.3. *Tumor Genetics' Explored*

The articles primarily focused on predicting genetic or epigenetic abnormalities, including the presence of *IDH1/2* mutations, methylation of the *MGMT* promoter, over-expression of *EGFR*, and the 1p/19q co-deletion (Table 3). Additionally, one study investigated the prediction of tumor variations affecting *PTEN*, *ATRX*, *TERT*, *CDKN2A/B*, *TP53*, and chromosomal rearrangements such as Trisomy 7 and Monosomy 10 [9].

**Table 3.** Genetic and epigenetic abnormalities explored in the studies.

| Genetic Features | Number | Percent |
|---|---|---|
| *IDH1/2* mutation | 9 | 53% |
| *MGMT* methylation | 9 | 53% |
| *EGFR* expression | 3 | 18% |
| 1p19q codeletion | 4 | 24% |
| Other | 1 | 6% |

#### 3.3.1. IDH Mutation Prediction

Based on the Area Under the Receiver Operating Characteristic (ROC) Curve (AUC), the algorithm proposed by Y.S. Choi et al. [20] achieved the best performance with an AUC of 0.96 (95% CI: 0.93–0.99) (Table 4). However, when this algorithm was replicated on external data by I. Hrapșa et al. [21], the AUC was 0.74 (95% CI: 0.53–0.91), with a sensitivity of 78% and a specificity of 75%. The first study included patients with both gliomas and glioblastomas, while the second focused solely on glioblastoma patients. One of the highest performances was reported by P. Eichinger et al. [13], who achieved an AUC of 0.952 using 79 patients, compared to the 856 patients in the study by Y.S. Choi et al. [20].

**Table 4.** Prediction of *IDH* mutation presence. Includes AUC (Area Under the ROC Curve); 95% confidence intervals (CI) are shown in brackets where available. NA: Not Available.

| | Tumors Type | AUC (95% CI) | Accuracy (%) | Sensitivity (%) | Specificity (%) |
|---|---|---|---|---|---|
| P. Eichinger et al. [13] | Gliomas | 0.952 | 0.95 | NA | NA |
| P. Chang et al. [14] | Gliomas | 0.91 (0.89–0.92) | NA | NA | NA |
| S. Liang et al. [15] | Both | 0.857 | 84.6 | 78.5 | 88.0 |
| Y. Matsui et al. [17] | Gliomas | NA | 82.9 | NA | NA |
| S. Rathore et al. [9] | Both | 0.87 | 82.5 | 70.43 | 88.32 |
| Y.S. Choi et al. [20] | Both | 0.96 (0.93–0.99) | 93.8 | NA | NA |
| I. Hrapșa et al. [21] | Glioblastomas | 0.74 (0.53–0.91) | 76 | 78 | 75 |
| E. Calabrese et al. [22] | Glioblastomas | 0.96 (0.88–1) | 84 | 100 | 83 |
| S. Kihira et al. [10] | Both | 0.93 (0.90–0.97) | NA | 0.98 | 0.32 |

### 3.3.2. MGMT Promoter Methylation Prediction

The highest performance in predicting *MGMT* promoter methylation was achieved by the algorithm of S.A. Qureshi et al. [11], with an AUC of 0.96 (95% CI: 0.94–0.98) (Table 5). This performance was achieved using a deep learning approach for segmentation and radiomics extraction, followed by classification with a Random Forest algorithm, which is a machine learning strategy.

**Table 5.** Prediction of *MGMT* promoter methylation. AUC (Area Under the ROC Curve), 95% confidence intervals (CI) are shown in brackets where available. NA: Not Available.

| | Tumor Type | AUC (95% CI) | Accuracy (%) | Sensitivity (%) |
|---|---|---|---|---|
| I. Levner et al. [12] | Glioblastomas | NA | 87.7 | 85.4 |
| P. Chang et al. [14] | Gliomas | 0.81 (0.76–0.84) | NA | NA |
| C. G. B. Yogananda et al. [19] | Gliomas | 0.58 (0.4182–0.7422) [1] | 65.95 | NA |
| E. Calabrese et al. [22] | Glioblastomas | 0.73 (0.65–0.81) [2] | 68 | 72 |
| B.-H. Kim et al. [23] | Both | 0.517 (0.459–0.645) | 51.9 | NA |
| S. Kihira et al. [10] | Both | 0.62 (0.54–0.71) | NA | 0.45 |
| H. Sakly et al. [24] [3] | Glioblastomas | NA | NA | NA |
| S. A. Qureshi et al. [11] | Glioblastomas | 0.96 (0.94–0.98) [4] | 96.94 | 96.31 |
| N. Saeed et al. [25] | Glioblastomas | 0.631 (0.629–0.633) | NA | NA |

[1] Results of C. G. B. Yogananda et al. include erratum: https://doi.org/10.3174/ajnr.A7715. [2] Combined method using CCN and Random Forest algorithm on radiomics feature achieved an AUC of 0.77 (95% CI: 0.63–0.91). [3] Did not present results on one or more validation datasets. [4] Achieved best performance but utilized a final step of ML classifier.

The second-best performance was achieved by P. Chang et al., with an AUC of 0.81 (95% CI: 0.76–0.84) [14]. Two studies reported AUCs with confidence intervals around 0.5, suggesting that their performance may be no better than chance [19,23]. I. Levner's 2009 study [12],which utilized a small neural network based on MRI texture analysis, achieved an accuracy of 87.7% for predicting *MGMT* promoter methylation. This study, while innovative, is primarily a proof of concept due to its very small dataset. H. Sakly et al. [24] explored transfer learning with image classification algorithms.

Additionally, the work by Saeed et al. [25] did not present the highest performance (Table 5), but was notable for comparing several deep learning algorithm architectures. Their study focused on understanding discrepancies in results from similar algorithms, rather than solely developing the most robust model.

### 3.3.3. EGFR Amplification Prediction

In our review, only three articles attempted to predict *EGFR* amplification (overexpression) (Table 6). Among these, M. Hedyehzadeh et al. did not present results that could be directly compared, but their deep learning strategy was able to statistically detect *EGFR* amplification in glioblastomas [16].

**Table 6.** Prediction for *EGFR* amplification. AUC (Area under the ROC curve) and 95% confidence interval (CI) are shown in brackets where available. NA: Not Available.

| | Tumor Type | AUC (95% CI) | Accuracy (%) | Sensitivity (%) | Specificity (%) |
|---|---|---|---|---|---|
| M. Hedyehzadeh et al. [16] [1] | Glioblastomas | NA | NA | NA | NA |
| S. Rathore et al. [9] | Both | 0.80 [2] | 86.74 | 84.91 | 87.5 |
| E. Calabrese et al. [22] | Glioblastomas | 0.72 (0.64–0.80) [3] | 66 | 68 | 66 |

[1] No metrics could be used as comparison. [2] Trained on 248 patients, with the maximum of 473 patients available in the study. [3] Appeared less performant than the Random Forest classifier on radiomics (AUC = 0.77 [95% CI: 0.67–0.87]) and combined classifier (AUC = 0.80 [95% CI: 0.74–0.86]).

Of the remaining two studies, S. Rathore et al. achieved slightly better results, although their final classifier was a Random Forest algorithm rather than a deep learning approach [9]. Similarly, E. Calabrese et al. reported better performance with their Random Forest classifier compared to their CNN classifier [22].

#### 3.3.4. Chromosome 1p19q Co-Deletion Prediction

The algorithm with the best performance for predicting 1p/19q co-deletion status was proposed by P. Chang et al. [14] (Table 7). Among the four articles addressing this prediction, B. Kocak et al. [18] tested a CNN and various ML algorithms (k-NN, Random Forest, SVM, etc.) on the same dataset, with the CNN yielding better results than the other methods.

**Table 7.** Prediction for co-deletion of chromosomes 1p and 19q. AUC (Area Under the ROC Curve) and 95% confidence interval (CI) are shown in brackets where available. NA: Not Available.

| | Tumor Type | AUC (95% CI) | Accuracy (%) | Sensitivity (%) | Specificity (%) |
|---|---|---|---|---|---|
| P. Chang et al. [14] | Gliomas | 0.88 (0.85–0.90) | NA | NA | NA |
| Y. Matsui et al. [17] | Gliomas [1] | NA | 75.1 | NA | NA |
| B. Kocak et al. [18] | Gliomas | 0.869 (0.751–0.987) [2] | 83.8 | 87.5 | 75.8 |
| S. Rathore et al. [9] | Both | 0.79 [3] | 75.15 | 81.49 | 73.96 |

[1] Presented patients with *IDH* wild-type gliomas, which have been classified as glioblastomas since 2021. [2] Both the CNN and ML algorithms trained on the same dataset. [3] Trained on 192 patients, with a maximum of 473 patients available in the study.

## 4. Discussion

### *4.1. Evolution of Publications*

A chronological analysis of the publications reveals a significant gap between the initial study in 2009 and the subsequent surge in research starting from 2017. Several factors may account for this hiatus. In 2009, glioma and glioblastoma classification was primarily based on histological criteria, with limited integration of genetic characteristics into therapeutic decision-making. At that time, the methylation status of the *MGMT* promoter was the primary genetic factor influencing chemotherapy choices.

I. Levner et al. [12] conducted a proof-of-concept study in 2009 using a two-layer CNN to predict MGMT promoter methylation with 87% accuracy on a training dataset. It was not until 2017 that more sophisticated research emerged, incorporating advanced CNN architectures such as ResNet and DenseNet. This shift reflects advancements in DL technologies and the increasing importance of genetic factors in glioma classification and treatment.

### *4.2. Deep Learning Algorithms*

The neural network architectures employed in the studies reviewed have evolved significantly over time. Early studies utilized Fully Convolutional Neural Networks (FCNNs). Later, more advanced architectures such as ResNet, DenseNet, and U-Net were introduced.

Some researchers also explored transfer learning using models like AlexNet, GoogleNet, and InceptionV2 [24].

The use of DL algorithms for the genetic characterization of glial tumors from medical imaging is well justified by the extensive application of these algorithms in computer vision tasks. Deep learning has become a cornerstone in the classification and analysis of common objects, providing highly detailed and nuanced representations through distributed parameterization. These DL algorithms can capture complex patterns in imaging data that traditional methods might miss. While some studies indicate that traditional ML methods may outperform DL strategies in predicting specific genetic characteristics (S. Rathore et al. [9] or S. A. Qureshi et al. [11]), the key question is not whether ML is superior to DL, but rather under which conditions and architectural configurations DL algorithms will fully surpass ML approaches for this task. The focus should be on identifying the optimal DL architectures and settings that can achieve superior performance in genetic characterization, rather than comparing DL and ML as broad categories.

However, there was considerable variation in the validation methods used across studies. While some studies employed a classic train–test split, others used k-fold cross-validation with varying k-values (3, 4, or 5) or Leave-One-Out Cross-Validation (LOOCV) for small sample sizes. External validation on independent datasets was notably infrequent, which could be a significant limitation.

Notably, Y. S. Choi et al. conducted an external validation of both the TCIA and SNUH datasets [20]. Their study revealed significantly poorer results on the TCIA dataset compared to the SNUH dataset, which showed comparable results. This disparity may be attributed to differences in patient management and imaging protocols between Seoul and the institutions contributing to the TCIA dataset. The consistency in management and imaging modalities, as well as histological and molecular studies, within the SNUH dataset could explain the better performance on this dataset.

### 4.3. Performance and Reproducibility

Algorithms with very high AUCs (close to 1) raise questions about their generalizability. A publication by N. Saeed et al. [25] explored this issue, highlighting that the limited size of datasets and inadequate validation methods can artificially inflate performance. Most studies are based on a single dataset, thus limiting the reproducibility of results.

The reproducibility of these algorithms, based on this literature review, is largely questionable. Many studies report exceptional performance metrics, however, these results often derive from specific datasets with limited external validation. The lack of robust testing across diverse datasets and clinical settings raises concerns about the reliability of these algorithms in real-world healthcare applications. For these deep-learning models to be applicable in clinical practice, it is imperative to enhance their robustness and generalizability. Future research must focus on improving the validation processes and expanding datasets to ensure that these algorithms can consistently perform well across various populations and imaging conditions. Only through such advancements can we hope to make these algorithms reliably exploitable in healthcare settings.

Although some attempts have been made to predict multiple genetic traits within a single model, results are often presented individually for each trait. This suggests a juxtaposition of classifiers rather than an integrated model capable of distributed prediction in a multidimensional space.

### 4.4. Limitations and Challenges

The main limitations identified include the limited size and homogeneity of datasets. With fewer than 1000 patients, these datasets may not capture the full variability needed for robust model training, particularly given that a single DICOM MRI scan can yield up to 1000 radiomics parameters. Additionally, the frequent lack of well-defined external validation datasets undermines the assessment of algorithm reliability and reproducibility, which are crucial for ensuring model generalizability.

An important area for improvement is the integration of an explanatory dimension into algorithm development. This would facilitate a better understanding of algorithm limitations and biases, which is crucial for ensuring medical liability. Without such transparency, the deployment of these algorithms in clinical settings could face significant legal and regulatory challenges, potentially hindering their future use in healthcare.

Furthermore, addressing the genetic heterogeneity of tumors presents a significant challenge. The current datasets are inadequate for this purpose; therefore, there is a need to develop datasets incorporating three-dimensional information from biopsy areas, potentially in the form of a three-dimensional biopsy probability field.

Another challenge is the variability between datasets, as exemplified by differences in results between the TCIA and SNUH datasets reported by Y. S. Choi et al. [20]. The lack of standardized guidelines for MRI assessment of brain tumors—such as sequence types and field strength (3 T versus 1.5 T)—and the variability in genetic data acquisition and analysis due to factors like neurosurgical expertise, histological study, and molecular biology methods further complicate the situation.

To advance this field, it is crucial to foster interdisciplinary collaboration among mathematicians, radiologists, pathologists, geneticists, and neurosurgeons to effectively address clinical demands and overcome these challenges.

## 5. Conclusions

Our review of 17 articles on the application of deep learning (DL) algorithms in the radiogenomic characterization of gliomas and glioblastomas underscores both the advancements and the persistent challenges in this emerging field. Since the early studies in 2009, and particularly since 2017, significant progress has been made. However, numerous obstacles remain to ensure the effectiveness and generalization of these models for a potential virtual biopsy.

The increased interest in this field since 2017 is attributed to technological advancements and a growing recognition of the importance of genetic markers in the classification, prognosis, and treatment of gliomas and glioblastomas, as reflected in the WHO 2021 classification. However, the methodologies for model validation and the limited size of datasets continue to be major concerns. The seemingly high performance of some algorithms must be approached with caution due to the frequent lack of external validation, raising questions about their generalizability.

Efforts to integrate the prediction of multiple genetic traits into a single model are still in the early stages. Most studies focus on predicting individual traits, indicating a reliance on separate classifiers rather than a comprehensive, multidimensional approach. To address this, it is essential to develop larger and more diverse datasets, incorporating clinical history, ethnic diversity, tumor characteristics, and molecular and cytogenetic features. ITo account for tumor heterogeneity, and incorporating three-dimensional information on the areas of probability of biopsy could further enhance our understanding of genetic variation within tumors and improve the assessment of tissue characteristics at tumor resection margins.

The limitations of current datasets and validation methods underscore the need for interdisciplinary collaboration. Integrating expertise from mathematicians, radiologists, pathologists, geneticists, and neurosurgeons is crucial for developing models that are both reliable and clinically relevant. Additionally, incorporating an explanatory dimension into algorithms is vital for understanding and addressing their biases and limitations, which is essential for overcoming potential legal and regulatory hurdles.

*Key Recommendations for Future Research*

- External Validation: Ensure robust validation on external datasets to assess generalizability and avoid overfitting on internal datasets, which may artificially inflate performance.

- Improved Validation Methods: Apply advanced validation techniques such as k-fold cross-validation with sufficiently large k-values and Leave-One-Out Cross-Validation (LOOCV), especially for small datasets, to improve reliability.
- Dataset Size and Diversity: Use larger and more diverse datasets, capturing clinical, genetic, and demographic variability (e.g., tumor types, patient populations, and ethnicities) to ensure broad applicability of the algorithms.
- Integrated Multi-Genetic Trait Models: Focus on developing integrated models capable of predicting multiple genetic traits simultaneously, rather than separate classifiers, to better reflect the complexity of gliomas.
- Integration of Tumor Heterogeneity: Develop models that take account of tumor heterogeneity to improve the understanding of tumor complexity.
- Explanatory and Interpretable AI: Ensure that deep learning models include interpretable components to allow clinicians to understand algorithm predictions, thus enhancing their trust in AI tools and ensuring accountability in clinical settings.
- Standardized MRI Acquisition Protocols: Establish standardized protocols for MRI acquisition (e.g., field strength and sequence types) to reduce variability between datasets and improve model reproducibility.
- Cross-Disciplinary Collaboration: Promote interdisciplinary collaboration among data scientists, radiologists, pathologists, geneticists, and neurosurgeons to design clinically relevant models that align with real-world clinical workflows.
- Ethical and Legal Frameworks: Address ethical and regulatory considerations, ensuring that the developed models comply with standards for medical liability, data privacy, and patient safety, especially given the potential future deployment of AI in healthcare.

Finally, although DL holds promise for advancing the radiogenomic characterization of gliomas and glioblastomas, ongoing improvements in datasets, validation methods, and interdisciplinary co-operation are necessary to fully realize the clinical potential of these tools. Such advancements will be instrumental in achieving objective early therapeutic decision-making and aligning with the principles of personalized, preventive, predictive, participative, and pertinent (5P) medicine.

**Supplementary Materials:** The following supporting information can be downloaded at: https://www.mdpi.com/article/10.3390/biomedicines12092156/s1, S1: Requests of three bases.

**Author Contributions:** Conceptualization, X.M.L.G.H. and F.L.; methodology M.N. and C.G.; software, P.F. and C.T.; formal analysis, X.M.L.G.H. and F.L.; writing—original draft preparation, X.M.L.G.H.; writing—review and editing, C.F.-M. and R.G.; visualization, C.G.; supervision, R.G. All authors have read and agreed to the published version of the manuscript.

**Funding:** This research received no external funding.

**Institutional Review Board Statement:** Not applicable.

**Informed Consent Statement:** Not applicable.

**Data Availability Statement:** No new data were created or analyzed in this study. Data sharing is not applicable to this article.

**Acknowledgments:** During the preparation of this work the authors used ChatGPT (model GPT3.5) by OpenAI (sanFranscico, USA) in order to improve the English level. After using this tool/service, the authors reviewed and edited the content as needed and take full responsibility for the content of the publication. No AI tools were used for the data analysis and/or synthesis from this literature review. After a first turn review process and due to a limited English level, a version completely proofread and corrected by a professional human translator from our institution (CHU de Poitiers) was returned. We would like to thank Harsham for this task.

**Conflicts of Interest:** The authors declare no conflicts of interest.

## References


1. Herculano-Houzel, S. The human brain in numbers: A linearly scaled-up primate brain. *Front. Hum. Neurosci.* **2009**, *3*, 31. [CrossRef] [PubMed]
2. Parpura, V.; Heneka, M.T.; Montana, V.; Oliet, S.H.R.; Schousboe, A.; Haydon, P.G.; Stout, R.F., Jr.; Spray, D.C.; Reichenbach, A.; Pannicke, T.; et al. Glial cells in (patho)physiology. *J. Neurochem.* **2012**, *121*, 4–27. [CrossRef] [PubMed]
3. Ostrom, Q.T.; Price, M.; Neff, C.; Cioffi, G.; Waite, K.A.; Kruchko, C.; Barnholtz-Sloan, J.S. CBTRUS Statistical Report: Primary Brain and Other Central Nervous System Tumors Diagnosed in the United States in 2015–2019. *Neuro-Oncology* **2022**, *24*, v1–v95. [CrossRef]
4. Louis, D.N.; Perry, A.; Wesseling, P.; Brat, D.J.; Cree, I.A.; Figarella-Branger, D.; Hawkins, C.; Ng, H.K.; Pfister, S.M.; Reifenberger, G.; et al. The 2021 WHO Classification of Tumors of the Central Nervous System: A summary. *Neuro-Oncology* **2021**, *23*, 1231–1251. [CrossRef]
5. Hegi, M.E.; Diserens, A.-C.; Gorlia, T.; Hamou, M.-F.; De Tribolet, N.; Weller, M.; Kros, J.M.; Hainfellner, J.A.; Mason, W.; Mariani, L.; et al. MGMT Gene Silencing and Benefit from Temozolomide in Glioblastoma. *N. Engl. J. Med.* **2005**, *352*, 997–1003. [CrossRef]
6. Stupp, R.; Mason, W.P.; van den Bent, M.J.; Weller, M.; Fisher, B.; Taphoorn, M.J.B.; Belanger, K.; Brandes, A.A.; Marosi, C.; Bogdahn, U.; et al. Radiotherapy plus Concomitant and Adjuvant Temozolomide for Glioblastoma. *N. Engl. J. Med.* **2005**, *352*, 987–996. [CrossRef] [PubMed]
7. Krizhevsky, A.; Sutskever, I.; Hinton, G.E. Imagenet classification with deep convolutional neural networks. *Commun. ACM* **2017**, *60*, 84–90. [CrossRef]
8. Moher, D.; Liberati, A.; Tetzlaff, J.; Altman, D.G.; The PRISMA Group. Preferred reporting items for systematic reviews and meta-analyses: The PRISMA statement. *J. Clin. Epidemiol.* **2009**, *62*, 1006–1012. [CrossRef]
9. Rathore, S.; Mohan, S.; Bakas, S.; Sako, C.; Badve, C.; Pati, S.; Singh, A.; Bounias, D.; Ngo, P.; Akbari, H.; et al. Multi-institutional noninvasive in vivo characterization of *IDH*, 1p/19q, and EGFRvIII in glioma using neuro-Cancer Imaging Phenomics Toolkit (neuro-CaPTk). *Neuro-Oncol. Adv.* **2020**, *2*, iv22–iv34. [CrossRef]
10. Kihira, S.; Mei, X.; Mahmoudi, K.; Liu, Z.; Dogra, S.; Belani, P.; Tsankova, N.; Hormigo, A.; Fayad, Z.A.; Doshi, A.; et al. U-Net Based Segmentation and Characterization of Gliomas. *Cancers* **2022**, *14*, 4457. [CrossRef]
11. Qureshi, S.A.; Hussain, L.; Ibrar, U.; Alabdulkreem, E.; Nour, M.K.; Alqahtani, M.S.; Nafie, F.M.; Mohamed, A.; Mohammed, G.P.; Duong, T.Q. Radiogenomic classification for MGMT promoter methylation status using multi-omics fused feature space for least invasive diagnosis through mpMRI scans. *Sci. Rep.* **2023**, *13*, 3291. [CrossRef]
12. Levner, I.; Drabycz, S.; Roldan, G.; De Robles, P.; Cairncross, J.G.; Mitchell, R. Predicting MGMT methylation status of glioblastomas from MRI texture. *Med. Image Comput. Comput. Assist. Interv.* **2009**, *12*, 522–530. [CrossRef] [PubMed]
13. Eichinger, P.; Alberts, E.; Delbridge, C.; Trebeschi, S.; Valentinitsch, A.; Bette, S.; Huber, T.; Gempt, J.; Meyer, B.; Schlegel, J.; et al. Diffusion tensor image features predict IDH genotype in newly diagnosed WHO grade II/III gliomas. *Sci. Rep.* **2017**, *7*, 13396. [CrossRef]
14. Chang, P.; Grinband, J.; Weinberg, B.D.; Bardis, M.; Khy, M.; Cadena, G.; Su, M.-Y.; Cha, S.; Filippi, C.G.; Bota, D.; et al. Deep-Learning Convolutional Neural Networks Accurately Classify Genetic Mutations in Gliomas. *Am. J. Neuroradiol.* **2018**, *39*, 1201–1207. [CrossRef] [PubMed]
15. Liang, S.; Zhang, R.; Liang, D.; Song, T.; Ai, T.; Xia, C.; Xia, L.; Wang, Y. Multimodal 3D densenet for *IDH* genotype prediction in gliomas. *Genes* **2018**, *9*, 382. [CrossRef] [PubMed]
16. Hedyehzadeh, M.; Maghooli, K.; MomenGharibvand, M.; Pistorius, S. A Comparison of the Efficiency of Using a Deep CNN Approach with Other Common Regression Methods for the Prediction of EGFR Expression in Glioblastoma Patients. *J. Digit. Imaging* **2020**, *33*, 391–398. [CrossRef]
17. Matsui, Y.; Maruyama, T.; Nitta, M.; Saito, T.; Tsuzuki, S.; Tamura, M.; Kusuda, K.; Fukuya, Y.; Asano, H.; Kawamata, T.; et al. Prediction of lower-grade glioma molecular subtypes using deep learning. *J. Neuro-Oncol.* **2020**, *146*, 321–327. [CrossRef]
18. Kocak, B.; Durmaz, E.S.; Ates, E.; Sel, I.; Gunes, S.T.; Kaya, O.K.; Zeynalova, A.; Kilickesmez, O. Radiogenomics of lower-grade gliomas: Machine learning–based MRI texture analysis for predicting 1p/19q codeletion status. *Eur. Radiol.* **2020**, *30*, 877–886. [CrossRef]
19. Yogananda, C.; Shah, B.; Nalawade, S.; Murugesan, G.; Yu, F.; Pinho, M.; Wagner, B.; Mickey, B.; Patel, T.; Fei, B.; et al. MRI-Based Deep-Learning Method for Determining Glioma *MGMT* Promoter Methylation Status. *Am. J. Neuroradiol.* **2021**, *42*, 845–852. [CrossRef]
20. Choi, Y.S.; Bae, S.; Chang, J.H.; Kang, S.-G.; Kim, S.H.; Kim, J.; Rim, T.H.; Choi, S.H.; Jain, R.; Lee, S.-K. Fully automated hybrid approach to predict the *IDH* mutation status of gliomas via deep learning and radiomics. *Neuro-Oncology* **2021**, *23*, 304–313. [CrossRef]
21. Hrapșa, I.; Florian, I.A.; Șușman, S.; Farcaș, M.; Beni, L.; Florian, I.S. External Validation of a Convolutional Neural Network for IDH Mutation Prediction. *Medicina* **2022**, *58*, 526. [CrossRef] [PubMed]
22. Calabrese, E.; Rudie, J.D.; Rauschecker, A.M.; E Villanueva-Meyer, J.; Clarke, J.L.; A Solomon, D.; Cha, S. Combining radiomics and deep convolutional neural network features from preoperative MRI for predicting clinically relevant genetic biomarkers in glioblastoma. *Neuro-Oncol. Adv.* **2022**, *4*, vdac060. [CrossRef] [PubMed]

23. Kim, B.-H.; Lee, H.; Choi, K.S.; Nam, J.G.; Park, C.-K.; Park, S.-H.; Chung, J.W.; Choi, S.H. Validation of MRI-Based Models to Predict MGMT Promoter Methylation in Gliomas: BraTS 2021 Radiogenomics Challenge. *Cancers* **2022**, *14*, 4827. [CrossRef] [PubMed]
24. Sakly, H.; Said, M.; Seekins, J.; Guetari, R.; Kraiem, N.; Marzougui, M. Brain Tumor Radiogenomic Classification of $O^6$-Methylguanine-DNA Methyltransferase Promoter Methylation in Malignant Gliomas-Based Transfer Learning. *Cancer Control* **2023**, *30*, 10732748231169149. [CrossRef] [PubMed]
25. Saeed, N.; Ridzuan, M.; Alasmawi, H.; Sobirov, I.; Yaqub, M. MGMT promoter methylation status prediction using MRI scans? An extensive experimental evaluation of deep learning models. *Med. Image Anal.* **2023**, *90*, 102989. [CrossRef]

MDPI AG
Grosspeteranlage 5
4052 Basel
Switzerland
Tel.: +41 61 683 77 34

*Biomedicines* Editorial Office
E-mail: biomedicines@mdpi.com
www.mdpi.com/journal/biomedicines

www.ingramcontent.com/pod-product-compliance
Lightning Source LLC
Chambersburg PA
CBHW040803250926
51850CB00007BA/1067
*9783725861415*